Energy Efficiency of Manufacturing Processes and Systems

Energy Efficiency of Manufacturing Processes and Systems

Special Issue Editor

Konstantinos Salonitis

MDPI • Basel • Beijing • Wuhan • Barcelona • Belgrade • Manchester • Tokyo • Cluj • Tianjin

Special Issue Editor
Konstantinos Salonitis
Cranfield University
UK

Editorial Office
MDPI
St. Alban-Anlage 66
4052 Basel, Switzerland

This is a reprint of articles from the Special Issue published online in the open access journal *Energies* (ISSN 1996-1073) (available at: https://www.mdpi.com/journal/energies/special_issues/ energy_efficiency_manufacturing_processes_systems).

For citation purposes, cite each article independently as indicated on the article page online and as indicated below:

LastName, A.A.; LastName, B.B.; LastName, C.C. Article Title. *Journal Name* **Year**, *Article Number*, Page Range.

ISBN 978-3-03936-509-8 (Hbk)
ISBN 978-3-03936-510-4 (PDF)

Contents

About the Special Issue Editor

Konstantinos Salonitis, Ph.D., is a Mechanical Engineer and an expert in the modelling and simulation of manufacturing processes and systems. He has been leading research in the areas of the energy efficiency and the environmental impact of manufacturing for more than a decade. Konstantinos is a Professor of Manufacturing Systems at Cranfield University. He is also a visiting professor for Nanjing University of Aeronautics and Astronautics, as well as for Jiangsu University. He is the head of Sustainable Manufacturing Systems Centre. He has been working on a large number of projects funded by industries and research councils. He has published more than 250 papers and has an H factor of 34 (Google Scholar). He is a Chartered Engineer and a Fellow of the Institute of Mechanical Engineering and the Higher Education Academy.

Preface to "Energy Efficiency of Manufacturing Processes and Systems"

The availability and affordability of energy affect the whole life cycle of a product. The production phase of any product requires considerable amounts of energy. Manufacturing activities are responsible for one-third of the global total energy consumption and CO2 emissions. Thus, increasing the energy efficiency of production has been the focus of research in recent years and is nowadays considered one of the key decision-making attributes for manufacturing.

This book considers the energy efficiency of both manufacturing processes and systems. The scope of the book includes the following areas:

- Methods for the measurement of energy efficiency;

- Tools and techniques for the analysis and development of improvements with regard to energy consumption;

- Tools and techniques for the modelling and simulation of energy efficiency for both manufacturing processes and systems;

- Case studies on the management of such systems and the necessary practices to maintain;

- Green and lean manufacturing.

This book presents a breadth of relevant information, material, and knowledge to support research, policy-making, practices, and experience transferability to address the issues of energy efficiency.

Konstantinos Salonitis
Special Issue Editor

Editorial

Energy Efficiency of Manufacturing Processes and Systems—An Introduction

Konstantinos Salonitis

Sustainable Manufacturing Systems Centre, Manufacturing Department, Cranfield University, Cranfield, Bedfordshire MK43 0AL, UK; k.salonitis@cranfield.ac.uk; Tel.: +44-1234-758347

Received: 22 May 2020; Accepted: 3 June 2020; Published: 5 June 2020

Abstract: This Special Issue of Energies was devoted to the topic of "Energy Efficiency of Manufacturing Processes and Systems". It attracted significant attention of scholars, practitioners, and policy-makers from all over the world. Eighteen papers on this topic were submitted between 2018 and 2020, and a total of 10 papers were published. Main topics included the energy efficiency improvement in both the manufacturing process and system levels. Furthermore, new methodologies and analysis approaches in developing energy efficiency were presented.

Keywords: manufacturing energy efficiency; clean manufacturing; sustainable manufacturing; digital manufacturing

---

1. Introduction

For maintaining the quality of life that has been achieved in the developed countries, manufacturing is expected to further intensify activities, and scale up production. This will be probably required even more, as demand is expected to further rise due to living quality in developing countries catching up with that of the developed ones. This obviously means that more energy, and in general more resources, will be required for the production of higher volumes of products. However, it is evident that resources are finite, and we will need to manage to produce more with less. It is clear that producing with higher energy efficiency is an absolute requirement for the years to come.

Producing with higher energy efficiency has been the focus of research in recent years and is nowadays considered one of the key decision-making attributes for manufacturing. Higher energy efficiency is one of the key drivers in delivering a low-carbon economy. This has been highlighted at both international and national levels. Energy efficiency of manufacturing is aligned to a number of United Nations' sustainable goals, such as "goal 9" that is focused on promoting sustainable industrialization, "goal 13" on taking action to combat climate change and to a degree "goal 7" on affordable and clean energy. On this basis governments have set ambitious strategic plans for decarbonization of the whole economy, impacting obviously the manufacturing sector as well. As an example, UK was the first major economy in the world to pass a net zero emissions by 2050 law. For achieving such an ambitious goal, the manufacturing sector needs to adopt more energy efficient practices.

Energy efficiency is probably among the most cost-effective measures companies can take. Energy is a variable cost, and as such, contributes to the product's cost. Reducing the energy efficiency during production can happen in a number of ways, either at the process level, through the optimization of the process parameters for example, or at the systems' level. The present Special Issue has collected papers that deal with the techniques that can be used for reducing the energy consumption in both the process and the system level.

The response to the call for papers led to 18 submitted papers, of which 10 (55%) were accepted and eight (45%) rejected. The geographical distribution of the (first) author covers six countries, and is built as follows: China (three), Finland (one), Italy (one), Korea (one), Pakistan (one), United Kingdom (three).

2. Background and the Special Issue

Manufacturing energy efficiency can be approached in a number of different ways and at a number of different levels. Such levels can be more or less specific. In a number of studies five levels are considered: the device/process level, the line/cell/multi-machine system level, the facility level, the multi-factory system level and the enterprise/global supply chain level. In other studies, with a more high-level approach, two generic levels of analysis are considered, namely the manufacturing process or machine tool level and the manufacturing system level. In the analysis of the papers submitted in this Special Issue, the high-level classification is adopted.

The 10 papers collected in this Special Issue can broadly be divided into the following two categories: (a) manufacturing process energy efficiency studies, and (b) manufacturing systems energy efficiency studies. In both categories new methods and techniques for improving energy efficiency are presented. They will be described in the following subsections.

2.1. Manufacturing Processes Energy Efficiency Studies

A number of papers that focused on how to improve the energy efficiency of specific processes were published in this Special Issue. Papers focusing on processes such electro-discharge machining, laser drilling, laser welding and milling are included. The list of papers presented includes the following:

- Niamat, M.; Sarfraz, S.; Ahmad, W.; Shehab, E.; Salonitis, K. Parametric Modelling and Multi-Objective Optimization of Electro Discharge Machining Process Parameters for Sustainable Production. *Energies* 2020, 13, 38.
- Sarfraz, S.; Shehab, E.; Salonitis, K.; Suder, W. Experimental Investigation of Productivity, Specific Energy Consumption, and Hole Quality in Single-Pulse, Percussion, and Trepanning Drilling of IN 718 Superalloy. *Energies* 2019, 12, 4610.
- Um, J.; Stroud, I.A.; Park, Y.-K. Deep Learning Approach of Energy Estimation Model of Remote Laser Welding. *Energies* 2019, 12, 1799.
- Khan, A.M.; Jamil, M.; Salonitis, K.; Sarfraz, S.; Zhao, W.; He, N.; Mia, M.; Zhao, G. Multi-Objective Optimization of Energy Consumption and Surface Quality in Nanofluid SQCL Assisted Face Milling. *Energies* 2019, 12, 710.

In the following paragraphs, a brief review of these papers is provided. Niamat et al. [1] investigated the use of electro-discharge machining process for sustainable manufacturing. Their focus was on optimizing the process parameters, such as pulse on time, current and pulse off time and finding a tradeoff between quality of final produced part, productivity and cost. The work presented is mostly experimental, that led to setting up empirical models using response surface methodology for the control of the process.

Sarfraz et al. [2] focused on the laser drilling process optimization. Through a statistical design of experiments and analysis of variance (ANOVA) they presented empirical models for predicting the material removal rate, the specific energy consumption and the hole taper for the case of single pulse drilling, percussion and trepanning. A multi objective optimization algorithm was also used for the selection of the process parameters as well as the most appropriate drilling strategy.

Um et al. [3] investigated the use of deep learning for controlling the energy consumption due to remote laser welding process. Such a process is widely used in the automotive sector due to its flexibility and versatility. However, one of the key challenges when using remote laser welding is the high requirements in energy. Um et al. presented a neural network they developed using a deep learning approach for the prediction of the energy profile of the process.

Finally, Khan et al. [4] focused on the optimization of the process parameters for the energy consumption and surface quality of milling process. Two optimization methods were used, namely Grey Rational Analysis and the Non-Dominated Sorting Genetic Algorithm. The analysis followed allowed the optimization of the process parameters for reducing the energy consumption during the process.

All the aforementioned investigations attempt to improve the energy efficiency of the respective processes while at the same time maintain or even improve the surface quality, the productivity and the cost of processing. It is evident that energy efficiency is not a goal on its own but needs to be considered in a holistic way.

2.2. Manufacturing Systems Energy Efficiency Studies

As mentioned, another big group of papers were focused on the manufacturing system and not on a specific process. The papers published approached this from a number of different perspectives. The list of papers presented includes the following:

- Wang, J.; Fei, Z.; Chang, Q.; Li, S. Energy Saving Operation of Manufacturing System Based on Dynamic Adaptive Fuzzy Reasoning Petri Net. *Energies* 2019, 12, 2216.
- Salonitis, K.; Jolly, M.; Pagone, E.; Papanikolaou, M. Life-Cycle and Energy Assessment of Automotive Component Manufacturing: The Dilemma Between Aluminum and Cast Iron. *Energies* 2019, 12, 2557.
- Benedetti, M.; Bonfà, F.; Introna, V.; Santolamazza, A.; Ubertini, S. Real Time Energy Performance Control for Industrial Compressed Air Systems: Methodology and Applications. *Energies* 2019, 12, 3935.
- Kähkönen, S.; Vakkilainen, E.; Laukkanen, T. Impact of Structural Changes on Energy Efficiency of Finnish Pulp and Paper Industry. *Energies* 2019, 12, 3689.
- Peng, C.; Peng, T.; Zhang, Y.; Tang, R.; Hu, L. Minimising Non-Processing Energy Consumption and Tardiness Fines in a Mixed-Flow Shop. *Energies* 2018, 11, 3382.
- Saxeena, P; Stavropoulos, P.; Kechagias, J.; Salonitis, K. Sustainability assessment for manufacturing operations. *Energies* 2020, 13, 2730.

In the following paragraphs, a brief review of these papers is provided. Wang et al. [5] proposed a control method for manufacturing systems based on dynamic adaptive fuzzy reasoning Petri nets. The developed method allows the characterization of the state of machines in the manufacturing system for reducing idle times. They validated their method in a manufacturing line used for the serial production of automotive powertrains.

Salonitis et al. [6] also conducted research in the automotive sector. They developed a method for assessing the importance of the energy consumed during the manufacturing phase of components, such as the engine head of a car, in the overall environmental impact of a product. Their method relies on a thorough energy audit throughout the life cycle of a product, starting from the extraction of raw materials from earth. Through this analysis they compared the energy consumption associated with production of cast iron and aluminum engine heads and assessed the impact of lightening of automotive parts.

Benedetti et al. [7] focused on industrial compressed air systems. Air compressors are among the higher energy consumers in industry. Leaks in the delivery of pressurized air results in air compressors operating for longer times as to sustain the air pressure. Furthermore, an idle compressor can still use 40% to 70% of its full load. Benedetti et al. reported that in Europe, 10% of the electrical energy consumed in industry is due to compressed air systems. In their paper they developed a method for the real time energy performance monitoring and control of air compressors. They validated their method in a real industrial environment within a pharmaceutical plant, demonstrating how adopting such methods can reduce energy consumption and associated costs.

Kähkönen et al. [8] focused on analyzing the Finish pulp and paper industry. The restructuring of the sector is presented and the impact that this had on the energy efficiency is discussed. However, they also highlighted that the restructuring accounted for 20% of the energy efficiency whereas 80% of the improvement was due to other factors. Their analysis concluded suggesting that improving the existing mills can result in higher energy savings compared to replacing them with newer ones.

Pent et al. [9] approached the issue of energy efficiency of manufacturing systems through scheduling. For the case of a mixed-flow shop, they developed a scheduling approach for minimizing the energy consumption and tardiness fine of production. Their analysis investigated in detail the non-processing energy reduction. They validated the proposed method to a real case, and they were able to show improvements in the range of 70% for the non-processing energy.

Finally, Saxeena et al. [10] presented a holistic approach based on multi-criteria decision-making methods for assessing alternative process routes. The assessment method proposed allows the investigation of all three pillars of sustainability, namely the environmental, financial and social. For this reason, key performance indicators are proposed for each pillar and then measured or assessed. The method was demonstrated for the case of manufacturing of an automotive component that requires machining and heat treatment.

3. Concluding Remarks and Outlook

The Special Issue "Energy Efficiency of Manufacturing Processes and Systems" presents a collection of research articles covering relevant topics in the field. A number of different techniques and approaches were presented focusing on different levels; however, all approaches had the improvement of energy efficiency at their core.

The success of this Special Issue has motivated the editor to propose a new Special Issue that will complement the present one—Manufacturing Energy Efficiency and Industry 4.0. We invite the research community to submit novel contributions covering how Industry 4.0 and IIoT can help in improving the energy efficiency of manufacturing processes and systems.

Author Contributions: K.S. organized the Special Issue and wrote this editorial. All author have read and agreed to the published version of the manuscript.

Funding: This research did not receive any specific grant from funding agencies in the public, commercial or not-for-profit sectors.

Acknowledgments: The guest editor would like to thank the authors for submitting their excellent contributions to this Special Issue. Furthermore, the present Special Issue would not have been possible without the expert reviewers that carefully evaluated the manuscripts and provided helpful comments and suggestions for improvements. A special thank you is in order for the editors and the MDPI team for their outstanding management of this Special Issue.

Conflicts of Interest: The author declares no conflict of interest.

References

1. Niamat, M.; Sarfraz, S.; Ahmad, W.; Shehab, E.; Salonitis, K. Parametric Modelling and Multi-Objective Optimization of Electro Discharge Machining Process Parameters for Sustainable Production. *Energies* **2020**, *13*, 38. [CrossRef]
2. Sarfraz, S.; Shehab, E.; Salonitis, K.; Suder, W. Experimental Investigation of Productivity, Specific Energy Consumption, and Hole Quality in Single-Pulse, Percussion, and Trepanning Drilling of IN 718 Superalloy. *Energies* **2019**, *12*, 4610. [CrossRef]
3. Um, J.; Stroud, I.A.; Park, Y.-K. Deep Learning Approach of Energy Estimation Model of Remote Laser Welding. *Energies* **2019**, *12*, 1799. [CrossRef]
4. Khan, A.M.; Jamil, M.; Salonitis, K.; Sarfraz, S.; Zhao, W.; He, N.; Mia, M.; Zhao, G. Multi-Objective Optimization of Energy Consumption and Surface Quality in Nanofluid SQCL Assisted Face Milling. *Energies* **2019**, *12*, 710. [CrossRef]
5. Wang, J.; Fei, Z.; Chang, Q.; Li, S. Energy Saving Operation of Manufacturing System Based on Dynamic Adaptive Fuzzy Reasoning Petri Net. *Energies* **2019**, *12*, 2216. [CrossRef]
6. Salonitis, K.; Jolly, M.; Pagone, E.; Papanikolaou, M. Life-Cycle and Energy Assessment of Automotive Component Manufacturing: The Dilemma between Aluminum and Cast Iron. *Energies* **2019**, *12*, 2557. [CrossRef]
7. Benedetti, M.; Bonfà, F.; Introna, V.; Santolamazza, A.; Ubertini, S. Real Time Energy Performance Control for Industrial Compressed Air Systems: Methodology and Applications. *Energies* **2019**, *12*, 3935. [CrossRef]

8. Kähkönen, S.; Vakkilainen, E.; Laukkanen, T. Impact of Structural Changes on Energy Efficiency of Finnish Pulp and Paper Industry. *Energies* **2019**, *12*, 3689. [CrossRef]
9. Peng, C.; Peng, T.; Zhang, Y.; Tang, R.; Hu, L. Minimising Non-Processing Energy Consumption and Tardiness Fines in a Mixed-Flow Shop. *Energies* **2018**, *11*, 3382. [CrossRef]
10. Saxena, P.; Stavropoulos, P.; Kechagias, J.; Salonitis, K. Sustainability assessment for manufacturing operations. *Energies* **2020**, *13*, 2730. [CrossRef]

Article

Minimising Non-Processing Energy Consumption and Tardiness Fines in a Mixed-Flow Shop

Chen Peng [1], Tao Peng [1,*], Yi Zhang [2], Renzhong Tang [1] and Luoke Hu [1]

[1] Institute of Industrial Engineering, School of Mechanical Engineering, Zhejiang University, Hangzhou 310027, China; pcme@zju.edu.cn (C.P.); tangrz@zju.edu.cn (R.T.); 11125069@zju.edu.cn (L.H.)

[2] College of Information Science & Electronic Engineering, Zhejiang University, Hangzhou 310027, China; 21731108@zju.edu.cn

* Correspondence: tao_peng@zju.edu.cn; Tel.: +86-(0)571-8795-1145

Received: 2 November 2018; Accepted: 28 November 2018; Published: 3 December 2018

Abstract: To meet the increasingly diversified demand of customers, more mixed-flow shops are employed. The flexibility of mixed-flow shops increases the difficulty of scheduling. In this paper, a mixed-flow shop scheduling approach (MFSS) is proposed to minimise the energy consumption and tardiness fine (TF) of production with a special focus on non-processing energy (NPE) reduction. The proposed approach consists of two parts: firstly, a mathematic model is developed to describe how NPE and TF can be determined with a specific schedule; then, a multi-objective evolutionary algorithm with multi-chromosomes (MCEAs) is developed to obtain the optimal solutions considering the NPE-TF trade-offs. A deterministic search method with boundary (DSB) and a non-dominated sorting genetic algorithm (NSGA) are employed to validate the developed MCEA. Finally, a case study on an extrusion die mixed-flow shop is performed to demonstrate the proposed approach in industrial practice. Compared with three traditional scheduling approaches, the better performance of the MFSS in terms of computational time and solution quality could be demonstrated.

Keywords: energy consumption; scheduling approach; mixed-flow shop; multi-objective optimisation; tardiness fine

1. Introduction

Manufacturing accounts for about 25% of global energy consumption [1]. Energy Information Administration (EIA) reports that 90% of the processing electricity in industries was consumed by manufacturing [2]. Energy-efficient manufacturing, which can be implemented through production facility improvements, has been encouraged [3] but this increases the financial burden to build or purchase new facilities. Scheduling has been approved to be an effective and economic tool to reduce the manufacturing energy consumption of production facilities (MEPF) [4].

MEPF can be divided into the non-processing energy (NPE) and processing energy (PE) consumption of production facilities. NPE represents the energy consumption of production facilities during the non-processing phase and it is the integral of idle power over the relevant idle time [5]. PE represents the energy consumption of production facilities during the processing phase, which is associated with the processing power and processing time of the facilities. Research indicates that the non-processing power reaches up to 30% of the processing power [6]. However, the processing time of production facilities is normally shorter than their non-processing time. For example, Wiendahl investigated six industries and found that the processing time only accounted for 15% of the total manufacturing time [7]. Therefore, the NPE can be large.

Scheduling has been proven to be crucial in manufacturing and it plays an important role for companies to meet due dates committed to by customers [8]. Meanwhile, the existing literature suggests that the NPE can be effectively reduced by scheduling at the production plan stage in job shops [9] and in flow shop environments [10], while application scenarios in mixed-flow shop environment are limited.

The mixed-flow shop environment is critical to quickly respond to the diversified demand of customers in today's manufacturing, such as an extrusion die workshop. Compared with the traditional flow shop, the jobs in mixed-flow shops have standard as well as customised processes. A standard process is defined as a process with the same operations among multiple jobs in which all the jobs need to be executed on the machines with the same route during the standard process. A customised process is defined as the specific operations of jobs, operations that are applied for processing the personalised parts of the products. For the extrusion die making industry, the diversification is obvious, therefore, building a mixed-flow shop environment is necessary. Extrusion dies are customised for various Aluminium profiles. Different dies share some standard processes, such as cylindrical turning, end milling, and other basic processing, while the manufacturing of customised holes in dies is diverse. Furthermore, a real-operating extrusion die workshop needs to produce more than thirty thousand types of dies in one year, while the daily average yield is only seventy. The current productivity cannot meet the requirement and the tardiness fine (TF) is severe. The TF is the compensation (financial or other forms) to clients caused by the tardiness and the amount of compensation per unit time is based on the contracts. Besides, the idle machines are common in workshops. Scheduling has been proven to be an effective method to guarantee punctual deliveries [11] and reduce the NPE consumed by machines while waiting for jobs [12]. To minimise the NPE and TF in a mixed-flow shop, this paper formulates a multi-objective optimisation problem and proposes a mixed-flow shop scheduling approach (MFSS) to solve it. The MFSS is a scheduling approach aimed at minimising the NPE and TF, and it contains two parts: a mathematic model describing how the NPE and TF are influenced by a specific schedule, and a multi-objective evolutionary algorithm with multi-chromosomes (MCEAs) to obtain the optimal solutions considering the NPE-TF trade-offs.

When employing a scheduling approach to optimise the NPE and TF in mixed-flow shops, there are several difficulties: (1) the process routes can be different, thus, uncertainties for production facilities exist when selecting the next job to process with considerations toward the NPE and TF; (2) the job sequence on each facility can be different, thus, the computation scale is larger than that of a typical flow shop with the same amount of jobs and machines; (3) a trade-off between the NPE and TF should be considered because there exists a conflict between the NPE and TF. For example, when the tardiness fine is minimised, the NPE of the production facilities may increase due to the continuous running of the machines to ensure production readiness. The proposed approach to address the above difficulties is the main contribution of this paper.

NPE is first characterised in the mixed-flow shop environment and its relationship with TF is then analysed. The relationship between the scheduling scheme and the TF and NPE are reflected, and then a TF–NPE bi-objective optimisation model is developed. The bi-objective optimisation in this research is to achieve optimal trade-offs in completing a production task by properly scheduling the jobs. The MCEA is proposed to search for the Pareto optimum. An optimal solution represents a scheduling scheme that results in the optimal trade-offs between the two objectives. Through a case study, the developed models and optimisation approaches are demonstrated, compared, and discussed.

In the remainder of this paper, the literature review is presented in the following section. The description of the research problem and the bi-objective model are given in Section 3. In Section 4, the working procedure of the MCEA for solving this optimisation problem is described. A case study is conducted to demonstrate the applicability and effectiveness of the developed approach in Section 5. Finally, a brief summary and future works are given in Section 6.

2. Related Work

An increasing amount of energy-aware scheduling research has been conducted intensively in the recent twenty years, from which four typical scenarios can be found: the single machine, parallel machine, flow shop, and job shop scenarios. In single machine scheduling, Mehmet et al. provided a methodology for decision-makers to choose the most efficient scheduling with the appropriate energy consumption of a single machine and an efficient genetic algorithm was developed [13]. Shrouf et al. presented a scheduling method to minimise the energy consumption costs whilst considering the variable energy prices during the day [14]. Yin et al. reduced the total earliness/tardiness cost and energy consumption of a single machine by controlling the processing times and turn off/on [15]. In parallel machines scheduling, Zhantao Li considered the unrelated parallel scheduling problem in the background of big data, with the total tardiness and energy consumption as two objectives. Ten heuristic algorithms were devcloped to solve the mathematic model, and their performance was tested by designing computational experiments [16]. Moon et al. [17] developed a method to optimise both the makespan of production and time-dependent energy costs of the unrelated parallel machine. Ding et al. [18] studied the unrelated parallel machine scheduling problem in a time-of-use pricing scheme, aiming to minimise the total electricity cost. However, the energy-aware scheduling research on a single machine and parallel machines is not sufficient to solve the problem of mixed-flow shops because the NPE of a machine is affected by other machines in its involved processes. In the mixed-flow shop, jobs and machines should be considered systematically.

For flow shop and job shop scheduling, a large amount of multi-objective optimisation research can be referred to. Lu et al. formulated a mathematic model concerning the transportation and sequence-dependent setup stage and proposed a hybrid multi-objective backtracking search algorithm to solve the model [10]. Zhang et al. proposed a time-indexed integer programming formulation to optimise the electricity cost and the CO_2 emissions at the same time [19]. Liu et al. developed a model that minimises the total non-processing energy consumption and total weighted tardiness in a job shop, and the turn off/on strategy was employed as an energy saving approach [20]. Zhang et al. solved the bi-objective, total weighted tardiness, and energy consumption in the job shop scheduling problem with a multi-objective genetic algorithm including the local improvement strategies and compared its performance to that of NSGA-II [21]. Chen et al. studied the energy consumption reduction in Bernoulli serial lines with finite buffers and machines through the effective scheduling of machine startups and shutdowns [4].

The actual production is not limited to traditional job shops or flow shops. Energy-aware scheduling research is carried out in more complex production scenarios. Tang et al. and Dai et al. modified the particle swarm optimisation algorithm and genetic-simulated annealing algorithm to get the optimal schedule in flexible flow shops [22,23]. Li et al. proposed a scheduling approach based on Petri net models and a genetic algorithm to reduce the total energy consumption in flexible manufacturing systems [24]. Mouzon et al. investigated the impact of dispatching rules on reducing the energy consumption of manufacturing equipment [25]. Zhang et al. presented a novel approach of dynamic rescheduling in flexible manufacturing systems, concerned with energy consumption and schedule efficiency [26]. Liu et al. introduced a mixed-integer nonlinear programming model for the hybrid flow shop scheduling problem by minimising the energy consumption and setting up a constraint to require all the jobs to be delivered on time, though this constraint can sometimes not be satisfied in practical production [27]. Tong et al. and Li et al. analysed the production characteristics of forging shops and welding shops, respectively, and put forward corresponding energy-saving scheduling schemes [28,29]. As in the aforementioned study in complex production scenarios, energy consumption can be reduced observably by the schedule. However, the models or algorithms of one scheduling approach cannot be universally applied to other types of scenarios.

According to the literature reviewed, energy-aware scheduling research in various scenarios has been conducted. While the research on diversified customised production scenarios is rarely considered, Zhou et al. described the characteristics of multi-varieties and the small-batch production scheduling mode and proposed an improved genetic annealing algorithm to shorten the production cycle, maximise the resource utilization rate, etc. [30]. Huang et al. designed a polynomial-time dynamic programming algorithm to minimise the makespan in a flow shop for mass customization [31]. Wang et al. established a Petri net based real-time model for hybrid flow shops to satisfy the small-batch customised production [32]. Huang et al. discussed the re-scheduling problems of a mixed-line flow shop and a drum-buffer-rope technique is applied to minimise the longest tardiness for the customers' orders [33]. The research articles above studied the scheduling problem in customised production scenarios with some of them being the mixed-flow shop scenario. However, the makespan and resource allocation are the main objectives and energy consumption is rarely concerned.

Based on the above discussion, the energy-aware scheduling research in mixed-flow shops is limited and the existing scheduling approaches in the other types of workshops cannot be applied in this environment because each scenario has its own features, including mixed-flow shop scenarios. These features should be taken into consideration when proposing a scheduling approach. In a mixed-flow shop, both standard processes and customised processes exist among the jobs' routes. As a result, the scheduling models and algorithms in a mixed-flow shop are more complex than in a traditional flow shop. To bridge such a gap, a scheduling approach specifically for the mixed-flow shop to minimise TF and NPE is introduced in the following sections.

3. Problem Statement

The problem addressed in this paper is a mixed-flow shop scheduling problem with NPE and TF objectives. In this scenario, n jobs $\left\{J_p\right\}_{p=1}^{n}$ are to be processed on m machines $\left\{M_k\right\}_{k=1}^{m}$ in sequence and a job contains a batch of the same workpieces. Each job J_p has an established process route $\left\{O_p^j\right\}$. O_p^j is equal to k in number and all the jobs flow according to the ascending machine numbers. t_p^k denotes the processing time of J_p on M_k. Each job has a release time r_p and a due date d_p. In manufacturing, a feasible job sequence $S_k = \left\{J_{k_i}\right\}_{i=1}^{r_k}$ will be arranged on M_k, where J_{k_i} is the i-th job processed on M_k and r_k is the total number of jobs using M_k. In addition, the jobs in $\left\{J_{k_i}\right\}$ have a one-to-one relationship with those jobs to be processed on M_k. An example of the directed graph of a mixed-flow shop is illustrated in Figure 1. The idle time and NPE vary along with the scheduling approaches for the jobs' processing time. Therefore, the key issue is to find the job permutation strategy for each machine that minimises the NPE as well as TF.

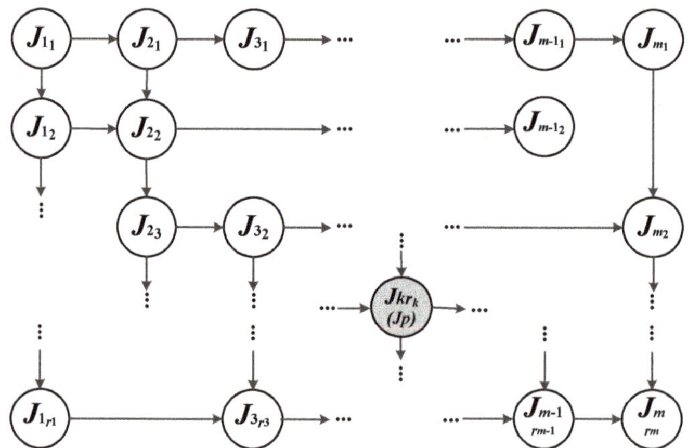

Figure 1. An example of the directed graph of a mixed-flow shop.

NPE is the integration of the machines' idle power in their idle time, respectively. To reduce the NPE, an optimal job sequence can be adopted, and this planned schedule should ensure the due dates of all jobs. However, the relationship between minimising NPE and ensuring the due date is not positive. In other words, a conflict exists between them. The following scheduling situation is taken as an example (see Figure 2). When J_5 on M_3 and M_4 is arranged to B, the other jobs' arrangements are not affected. The TF of other jobs and the NPE of M_1, M_2, and M_5 are unchanged while the replaced job's TF decreases and the NPE values of M_3 and M_4 increase. As a result, the total TF decreases and the NPE rises.

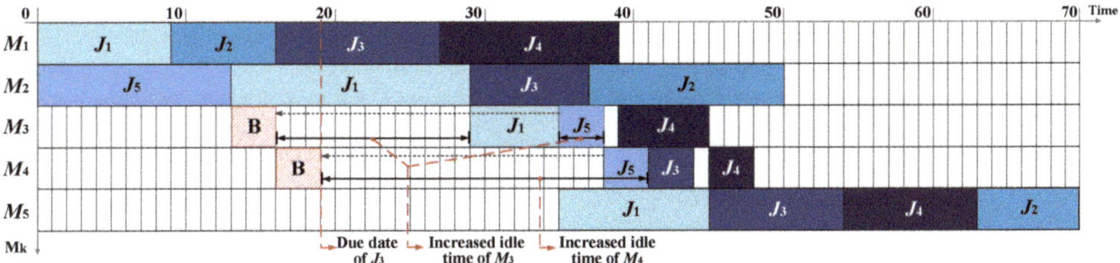

Figure 2. An exemplifying situation for the conflict between NPE and TF.

Hence, a multi-object mathematic model to get an optimal schedule strategy is developed. The assumptions are as follows:

(1) All the machines run from the first workpiece and come without the priority among the jobs;
(2) The release time of jobs is the same;
(3) One machine is able to process one workpiece simultaneously;
(4) All processes are non-preemptive and no machine needs to be used twice for one job;
(5) The processing time of each job is generated as a statistical average value;

The process route, processing time, release time, and due date of each job and the idle power of the machines are given. Only when a job is completed before the due date is the order considered to be completed punctually without paying a tardiness fine. The value of the tardiness fine per unit time has been determined by the contracts.

The multi-objective model aims to find the most energy-saving scheduling scheme of jobs with the minimum overdue fine. The multi-objective model *Multi* aims to *minimise* the NPE and TF at the same time:

$$Multi = Minimise\{\text{TF}(s), \text{NPE}(s)\} \tag{1}$$

The model has two parts. One part is the NPE model. NPE can be minimised by shortening the non-processing time of the machines caused by queuing. The non-processing time is the idle time when the previous job on a machine has finished and the next job has yet arrived. The other part is the TF model. The total TF is a sum of fines in all considered jobs, which is affected by the job makespan, due date, and penalty mechanism.

For n jobs $\{J_p\}_{p=1}^{n}$ to be processed on m machines, $\{M_k\}_{k=1}^{m}$ in a mixed-flow shop, the job sequence S_k on each M_k cannot be the same. Given a feasible scheduling scheme s, the total TF of all jobs in this scheduling can be calculated as

$$\text{TF}(s) = \sum_{p=1}^{n} T_p \times F_p \tag{2}$$

where T_p and F_p are the tardiness time and tardiness fine per unit time of J_p, respectively. The total NPE of all machines can be calculated as

$$\text{NPE}(s) = \sum_{k=1}^{m} P_k^{idle} \times t_k^{idle} \tag{3}$$

where P_k^{idle} is the idle power of machine M_k, and t_k^{idle} is the total idle time on it.

The detailed mathematic models are shown in the appendix. To achieve the optimal scheduling schemes of minimising the TF and NPE with the model, a feasible algorithm needs to be adopted.

4. Multi-Objective Optimisation

To obtain the optimal job sequence, heuristic or deterministic algorithms can be employed to solve the model. The global optima can be acquired by the deterministic algorithm. However, its computation burden is heavy. Although heuristic algorithms can find a feasible sequence within a short time, it normally traps the sequence into the local optimum. To obtain a high-quality solution within a tolerant computation time, a multi-objective evolutionary algorithm with multi-chromosomes (MCEsA) is proposed. Besides, a deterministic search method with boundaries is used to validate the accuracy of the MCEA.

MCEA is designed based on a genetic algorithm, which is a robust meta-heuristic that imitates the process of natural selection. As stated in the aforementioned problem, the job sequence and the job quantity to be processed on each machine are different. To code the scheduling scheme feasibly, a two-dimensional array is proposed as an individual. Each row vector, regarded as a chromosome, represents a job processing sequence on a machine. One individual represents a feasible scheduling scheme s and can be expressed as follows:

$$
\text{Individual}: \left\{
\begin{array}{l}
\left[J_{1_1},\ J_{1_2},\ J_{1_3},\ \cdots\ J_{1_v},\ \cdots\ J_{1_{r(1)}} \right] \\
\left[J_{2_1},\ J_{2_2},\ J_{2_3},\ \cdots\ J_{2_v},\ \cdots\ J_{2_{r(2)}} \right] \\
\cdots \\
\left[J_{k_1},\ J_{k_2},\ J_{k_3},\ \cdots\ J_{k_v},\ \cdots\ J_{k_{r(k)}} \right]
\end{array}
\right\}
\tag{4}
$$

where J_{k_v}, regarded as a gene, denotes the v-th job processed on M_k. $r(k)$ is the total amount of jobs on the machine. The flow chart of MCEA is shown in Figure 3.

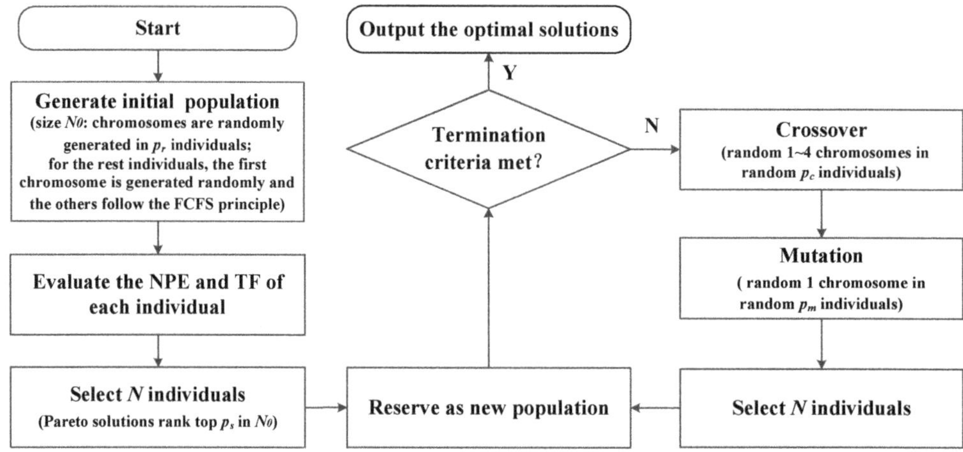

Figure 3. The flowchart of the MCEA.

First, the chromosome's coding method and fitness function should be given. In the MCEA, a gene, a chromosome, and an individual represent a certain job, a job sequence, and a scheduling scheme s, respectively. Second, the initial population with N_0 individuals is generated. Two principals are applied in generating the initial population: one is that all the chromosomes in individuals are generated randomly; the other one is that the first chromosome in the individual is randomly generated and the other chromosomes follows; first come first served. The p_r of the N_0 individuals are generated by the first principal and the rest are generated by the second one. The initial population is selected by the fitness function, where fitness $1 = NPE(s)$ and fitness $2 = TF(s)$. Then N_0 individuals whose Pareto solutions rank the top p_s in N_0 as a new population are reserved. Third, the children population is produced by crossovers and mutations: (1) Crossover: exchanging a random 1–4 chromosomes between two individuals. The crossover rate is p_c. (2) Mutation: modifying 1 random chromosome in the individuals. The mutation rate is p_m. Fourth, the next generation is selected from the combination of the individual parent and children by evaluating the fitness function. Once the iteration achieves G, the process ceases and outputs the non-dominant optimal solutions. If the termination criteria are not met, step four and step five repeat once more. In addition, all of the random selections are created by the roulette wheel approach.

The Deterministic Search Method with Boundaries (DSB) is an improved enumeration algorithm which can validate the accuracy of the MCEA. DSB traverses all the paths based on a depth-first search with an initial bound which is a set of non-dominant Pareto solutions output by MCEA. In the searching process, new schemes with better Pareto solutions replace the inferior ones within the boundaries and the else-scheduling schemes are pruned. When the traversal is completed, the Pareto solutions within the latest updated boundaries are reported as the global optimum.

5. Case Studies and Discussion

Aluminium profiles are customised products. To meet the diversified aluminium profiles, the matched dies need to be manufactured. The production of extrusion dies is carried out through multi-specifications and small batch manufacturing. Extrusion dies are shown in Figure 4. Both standard processes and customised processes exist among the different dies' processing routes and the extrusion die workshop is a typical mixed-flow shop. Three cases in the extrusion die workshop are studied in this paper to verify the performance of the MCEA and demonstrate the MFSS, respectively. Algorithms are developed on the Visual Studio 2013 software with the C++ programming language. The computing platform is an Intel (R) Core (TM) i5-4210U CPU with a 1.70 GHz frequency; 8.00 GB of RAM; Windows 10 (64 bit) operating system.

(**a**)　　　　　　　　　　(**b**)

Figure 4. Extrusion dies with customised holes for different aluminium profiles (**a**) and (**b**).

5.1. Comparison of the MCEA with the DSB

To evaluate the performance of the MCEA, two cases including 4 jobs × 5 machines and 5 jobs × 5 machines are tested, respectively. The production data are generated based on the production tasks in an extrusion die workshop; see Tables 1 and 2.

Table 1. The basic data in case 1 (4 jobs × 5 machines).

Job	Due Date	Tardiness Fine per Unit Time	M_1 $P_1^{idle}=3.2$	M_2 $P_2^{idle}=3.5$	M_3 $P_3^{idle}=2.8$	M_4 $P_4^{idle}=2.1$	M_5 $P_5^{idle}=3.7$
J_1	50	2	9	16	6	-	10
J_2	39	3	11	8	-	3	9
J_3	32	1	8	13	3	3	-
J_4	46	5	12	-	6	3	9

Table 2. The basic data in case 2 (5 jobs × 5 machines).

Job	Due Date	Tardiness Fine	M_1 $P_1^{idle}=3.2$	M_2 $P_2^{idle}=3.5$	M_3 $P_3^{idle}=2.8$	M_4 $P_4^{idle}=2.1$	M_5 $P_5^{idle}=3.7$
J_1	41	2	8	6	8	-	5
J_2	39	1	7	3	-	6	10
J_3	46	5	11	8	-	3	9
J_4	35	3	9	7	3	5	-
J_5	52	4	12	5	6	3	9

The results of DSB and MECA in the two cases are shown in Figure 5. To assess the solution quality of MCEA, the hypervolume indicator is applied [34,35]. The maximum value of the NPE and TF in the results is taken as the reference point's horizontal and vertical coordinates, respectively [36]. The results of one test in 4 jobs × 5 machines is reflected in Figure 6 and the point R is the reference point. The area enclosed in red straight lines represents the dominated subspaces of solutions in DSB. The smaller area enclosed in blue dotted lines represents the dominated subspaces of the solutions in MCEA.

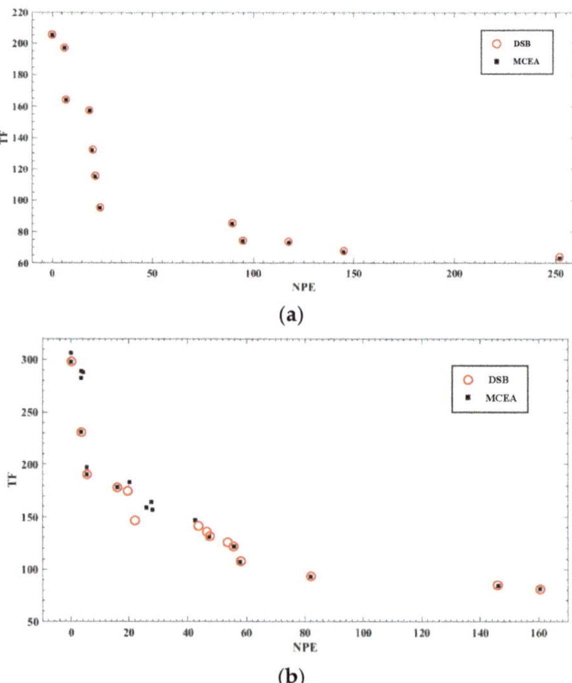

Figure 5. The results of DSB and MCEA, (**a**) in 4 jobs × 5 machines, (**b**) in 5 jobs × 5 machines.

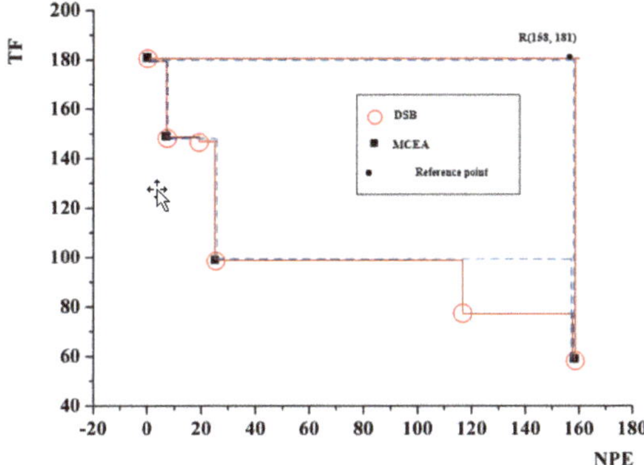

Figure 6. The calculation method schematic diagram of the hypervolume indicator.

The hypervolume indicator can be calculated as the dominated subspaces ratio of MCEA to DSB. The average hypervolume indicators of MCEA in case 2 and 3 are 99.791% and 97.950%, respectively, while the hypervolume indicator of the MCEA varies with the cases because of the randomness of this algorithm. The larger the case, the more indeterminate the results of the algorithm. For MCEA, the global optimal solutions cannot be guaranteed as the DSB. The feasibility of MCEA in a larger case will be discussed in Section 5.3. Meanwhile, in actual practice, the computation time of the algorithms is also considered. In MCEA, the average computation time of the tests under 5 jobs × 5 machines is 5.976 s. While in DSB, the average computation time of the tests under 5 jobs × 5 machines is 553.869 s. Once the number of jobs or machines increased, the computing burden is heavier and the runtime of DSB will be over one hour, which cannot meet the real-time scheduling demand. Thus, when the scheduled task is larger than 5 jobs × 5 machines, the MCEA could be employed to find the optimal scheduling schemes. Otherwise, both MCEA and DSB are capable.

5.2. Application of the MFSS

To demonstrate the MFSS, a case concerning 12 jobs and 5 machines in an extrusion die mixed-flow shop is selected. MFSS consists of the mathematic models and MCEA. The due date and tardiness fine data are randomly set. Each job contains a batch of the same parts and all the jobs are released at the same time. The effectiveness of the MFSS is demonstrated based on the case, as follows. The basic data of the case are listed in Table 3.

Table 3. The basic data in case 3 (12 jobs × 5 machines).

Job	Due Date	Tardiness Fine per Unit Time	M_1 $P_1^{idle}=5.2$	M_2 $P_2^{idle}=3.5$	M_3 $P_3^{idle}=2.8$	M_4 $P_4^{idle}=2.1$	M_5 $P_5^{idle}=3.7$
J_1	106	5	9	12	6	9	6
J_2	113	7	6	10	3	6	-
J_3	120	4	7	8	5	8	5
J_4	147	5	9	9	7	11	6
J_5	132	9	8	6	8	-	5
J_6	122	6	7	9	-	10	4
J_7	134	8	9	8	6	12	7
J_8	102	9	8	10	-	7	-
J_9	168	8	5	9	4	10	6
J_{10}	153	7	9	7	3	8	4
J_{11}	174	9	8	12	4	-	5
J_{12}	158	5	6	10	-	9	3

After fine turning, the parameters in MCEA are as follows: the initial population size $N_0 = 1000$, population size $N = 1000$, crossover rate $p_c = 0.8$, mutation rate, $p_m = 0.15$, and iteration = 1000. A set of Pareto solutions is generated at the end of each run. One solution contains four parts: the scheduling schemes, TF, NPE, and a Gantt chart. A solution among the Pareto sets is chosen to verify the correctness of the model and algorithm. The details of the solution are shown in Table 4 and Figure 7. The start and completion time of each job and the idle time of each machine are visible in the Gantt chart.

Table 4. The job sequences, NPE and TF, of the example solution.

Machine	Job Sequence
M_1	J_5-J_4-J_8-J_7-J_3-J_2-J_9-J_{12}-J_6-J_1-J_{11}-J_{10}
M_2	J_4-J_5-J_8-J_3-J_9-J_{12}-J_2-J_7-J_1-J_6-J_{11}-J_{10}
M_3	J_9-J_4-J_3-J_2-J_7-J_1-J_5-J_{11}-J_{10}
M_4	J_9-J_8-J_2-J_4-J_7-J_6-J_3-J_{12}-J_{10}-J_1
M_5	J_5-J_6-J_{11}-J_3-J_9-J_4-J_{12}-J_7-J_1-J_{10}
	NPE = 92.3 TF = 715

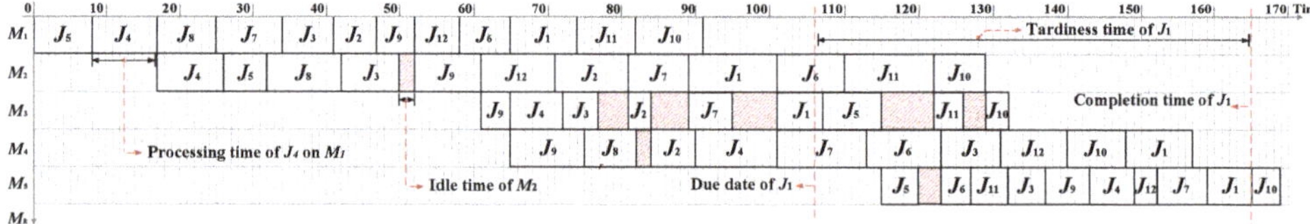

Figure 7. The Gantt chart of the example solution.

Based on the basic data of this case and the job sequences of the aforementioned solution, the calculation of NPE and TF are demonstrated in Tables 5 and 6, respectively.

Table 5. The NPE on the machines in the given schedule.

Machine	M_1	M_2	M_3	M_4	M_5	Total
Idle time	0	2	25	2	3	32
NPE	0	7	70	4.2	11.1	92.3

Table 6. The TF of the jobs in the given schedule.

Job	J_1	J_2	J_3	J_4	J_5	J_6	J_7	J_8	J_9	J_{10}	J_{11}	J_{12}	Total
TF	295	0	68	10	0	30	200	0	0	112	0	0	715

5.3. Results Comparison and Discussion

The non-dominated Sorting Genetic Algorithm (NSGA) and its related improved algorithms are widely applied in the multi-objective optimisation of workshop scheduling problems [37,38]. To assess the solutions' quality of MCEA, the NSGA is compared in this study with the input of case 3. The results of the MCEA and NSGA after 10 independent runs are shown in Figure 8 and they are mainly distributed in the two zones enclosed within the red dotted lines and yellow dotted lines, respectively. The intersection zone contains the results of both algorithms, as enclosed with the blue dotted lines. The computation times of the MCEA and NSGA are approximates, ranging from 150 s to 200 s. The length of the computation time is mainly influenced by the parameters including the scale of the case, the iteration times, and the size of the population. As demonstrated in Figure 8, the results of MCEA dominate those of NSGA.

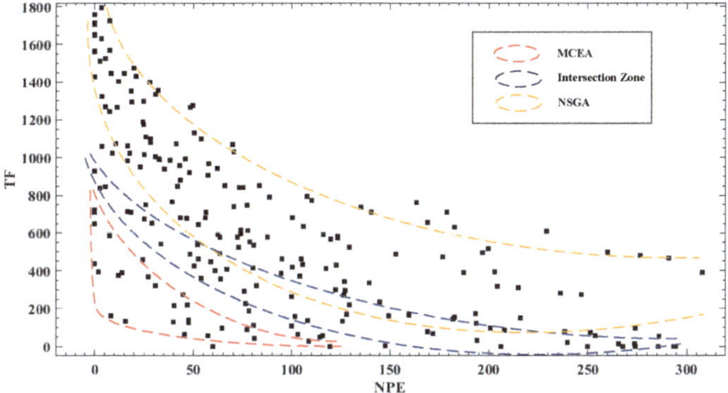

Figure 8. The results obtained by MECA and NSGA in case 3.

To verify the effectiveness of the MFSS, three traditional scheduling approaches are used in this case [39]. FCFS indicates first come first served and the jobs are processed in the sequence of appearance. SPT is the shortest process time first principle and it aims to shorten the processing time of the jobs. EDD represents the earliest due date first rule, which considers the earliest due date as a priority. In this study, all jobs are released at the same time and they are processed in the sequence of being received. The description and comparison of each approach are shown in Table 7.

Table 7. The comparison with three scheduling approaches.

Approach	Description	Result
MFSS	Using the proposed mathematic models and MCEA algorithm to schedule	NPE: 0, TF: 438; NPE: 2.1, TF: 392; NPE: 8.4, TF: 163; NPE: 15.8, TF: 133; NPE: 39.9, TF: 130; NPE: 45.4, TF: 63; NPE: 57.4, TF: 55; NPE: 59.8, TF: 0
FCFS	First come first served in the queue	NPE: 314.7, TF: 0
SPT	Shortest process time first	NPE: 433, TF: 165
EDD	Earliest due date first	NPE: 324.1, TF: 0

According to the table, the MFSS performs best in saving energy and decreasing TF. FCFS is relatively simple and accessible, while the randomness of the jobs' release sequence has a negative effect on the scheduling efficiency. Energy saving and time optimisation scheduling are difficult to achieve. SPT minimises the cycle time, while the due date and tardiness fine of the jobs are different. Scheduling without considering the variant urgency and importance of jobs is not efficient to reduce the tardiness and the total TF. EDD takes the due date into account, which contributes to the on-time delivery. While the energy consumption is not saved using this approach, based on the data in the chart and table, the MFSS solutions dominate those of FCFS, SPT, and EDD. The MFSS is an effective approach for obtaining optimal scheduling schemes that minimise the TF and NPE. TF can be controlled at a low value and about 69.97% of NPE can be reduced using the MFSS solutions compared to FCFS, SPT, and EDD.

6. Conclusions and Future Work

A mixed-flow shop scheduling problem derived from a customised manufacturing environment is described in this paper. In the mixed-flow shop scenario, jobs with both standard and customised processes are processed in a workshop. Three main difficulties of scheduling in this scenario are solved: (1) more uncertainties are brought to the machines in selecting the next job to process because of the difference of the jobs' process routes with the purpose of minimising the NPE and TF; (2) compared with a typical flow shop, the job sequence on each machine can be different, which causes a higher computational burden; (3) a conflict exist between the TF and NPE, thus the trade-off of the two objectives should be considered. We propose a scheduling approach to provide feasible schemes for decision-makers to minimise the NPE and TF in such scenarios. We developed a mathematic model to describe the effect of scheduling schemes on NPE and TF. Then, we proposed the heuristic algorithm MCEA to obtain the Pareto solutions by the trade-offs between NPE and TF. To validate its feasibility, we use the deterministic algorithm, DSB. A case study with 12 processing jobs on 5 machines has been conducted using the MFSS. The MCEA is compared with the NSGA and the dominant solutions of the MCEA are demonstrated in this scenario. The results of the MFSS have been compared with those of three traditional schedule approaches. By comparison, 69.97% of NPE can be reduced by using the MFSS and there is a 10% possibility to reduced TF to zero. This demonstrates the effectiveness of the proposed approach in reducing TF and NPE. This research has a range of limitations: for example, we made some assumptions of the same release and constant processing time that limit the versatility of our method. Additionally, the robustness and precision of the algorithm can be improved through testing with additional cases.

For future works, the stochastic arrival and the real processing time of the jobs should be taken into consideration so that the approach can react quickly to the dynamic and changing workshop. Besides, in practical production, the jobs and machines can be even larger. For the extrusion die workshop mentioned in the introduction, the daily yield required is as high as ninety. Thus, the algorithm needs to be improved to guarantee both the computing speed and effectiveness in practical large-scale scheduling cases. Finally, parallel machines and more factors including transportation and setup should be incorporated into the mathematical model to make the model more applicable to actual production.

Author Contributions: Conceptualization, C.P. and L.H.; Methodology, C.P.; Software, C.P. and Y.Z.; Formal Analysis, T.P.; Investigation, T.P.; Resources, R.T., T.P. and L.H.; Writing-Original Draft Preparation, C.P.; Writing-Review & Editing, C.P. and T.P.; Supervision, R.T. and T.P.; Funding Acquisition, R.T. and T.P.

Funding: This research was funded by National Natural Science Foundation of China grant number No. 51805479 and No. U151248.

Acknowledgments: The authors would like to thank the researchers in the Institute of Industrial Engineering, Zhejiang University, for their inputs.

Conflicts of Interest: The authors declare no conflict of interest.

Abbreviations and Nomenclature

Abbreviations	Description
MFSS	the proposed Mixed-Flow Shop Scheduling approach
NPE	Non-Processing Energy
PE	Processing Energy
TF	Tardiness Fine
MEPF	Manufacturing Energy consumption of Production Facilities
MCEA	multi-objective Evolutionary Algorithm with Multi-Chromosome
NSGA-II	Non-dominated Sorting Genetic Algorithm II
DSB	Deterministic Search method with Boundary
NSGA	Non-dominated Sorting Genetic Algorithm
FCFS	First Come First Served in the queue
SPT	Shortest Process Time first
EDD	Earlies Due Date first

Nomenclature	Description
p	indices for jobs
k	indices for machines
n	number of jobs in a schedule
m	number of machines in a schedule
i	indices for the jobs on one machine
j	indices for the operations of a job
$\{J_p\}_{p=1}^n$	a finite set of n jobs
$\{M_k\}_{k=1}^m$	a finite set of m machines
J_{k_i}	the i-th job processed on M_k
r_k	total amount of jobs processed on M_k
$\{O_p^j\}$	process route of J_p
S_k	a feasible job sequence on M_k
t_p^k	processing time of J_p on M_k
r_p	release time of J_p
d_p	due date of J_p
C_p	completion time of J_p
T_p	tardiness time of J_p
F_p	tardiness fine of J_p
P_k^{idle}	idle power of the machine M_k
t_k^{idle}	total idle time of the machine M_k
$C(J_p, k)$	completion time of job J_p on the machine M_k
$S(J_p, k)$	start time of job J_p on the machine M_k
s	a feasible scheduling scheme

Appendix A

Number	Model and Constrains	Description
1	$Multi = Minimise\{\mathrm{TF}(s), \mathrm{NPE}(s)\}$	Multi-objective function *Multi* is to minimise the NPE and TF at the same time.
2	$\mathrm{TF}(s) = \sum_{p=1}^n T_p \times F_p$	Given a feasible scheduling scheme s, the total TF TF(s) is the sum of each job's TF.
3	$\mathrm{NPE}(s) = \sum_{k=1}^m P_k^{idle} \times t_k^{idle}$	Given a feasible scheduling scheme s, the total NPE NPE(s) is the sum of each machine's NPE.
4	$T_p = \max\{0, C_p - d_p\}$	Tardiness time of J_p is measured by the time over the due date.
5	$C_p = C\left(J_p, \max\{O_p^j\}\right)$	Completion time of J_p is the moment that its last operation finishes.
6	$t_k^{idle} = C\left(J_{k_{r_i}}, k\right) - S(J_{k_1}, k) - \sum_{i=1}^{r_k} t_{k_i}^k$	Total idle time of the machine M_k equals the completion moment of its last job minus the start moment of its first job minus the total processing time of all the jobs processed on it.
7	$\mathrm{TF}(s) = \sum_{p=1}^n max\left\{0, C\left(J_p, max\{O_p^k\}\right) - d_p\right\} \times F_p$	Given a feasible scheduling scheme s, the total TF TF(s) equals the sum of each job's tardiness time multiply by its unit tardiness fine.
8	$\mathrm{NPE}(s) =$ $\sum_{k=1}^m \left[max\{C(J_p, k)\} - min\{S(J_p, k)\} - \sum_{p=1}^n t_p^k X_{pk}\right] \times P_k^{idle}$	Given a feasible scheduling scheme s, the total NPE NPE(s) equals the sum of each machine's idle time multiply by its idle power.
9	$\sum_{p=1,k=1}^{p=n,k=m} X_{pk} = \begin{cases} 1, & \exists O_p^j \in \{O_p^j\} = k \\ 0, & else \end{cases}$	Restrict the jobs to process according to the process routes. If one procedure of J_p is on M_k, then $X_{pk} = 1$. No job need use one machine twice.
10	$C(J_{k_i}, k) \geq C(J_{k_{i-1}}, k) + t_{k_i}^k, \ i \in (2, r_k); \ k_i \in (2, r_k)$	Define that only one job can be processed on a machine at the same time and all processes are non-preemptive.
11	$C\left(J_p, O_p^j\right) \geq C\left(J_p, O_p^{j-1}\right) + t_p^{O_p^j}, \ p \in (1, n); \ j \in (2, k); \ O_p^j \in (1, k)$	Guarantee that the job cannot be ceased machining unless the whole job has been finished
12	$C\left(J_p, O_p^1\right) \geq t_p^{O_p^1} + r_p, \quad p \in (1, n)$	Allows all of the jobs to be processed at as soon as released

References

1. Hu, L.; Peng, C.; Evans, S.; Peng, T.; Liu, Y.; Tang, R.; Tiwari, A. Minimising the machining energy consumption of a machine tool by sequencing the features of a part. *Energy* **2017**, *121*, 292–305. [CrossRef]
2. Administration UEI. *Annual Energy Review*; Government Printing Office: Washington, DC, USA, 2011.
3. Jia, S.; Yuan, Q.; Lv, J.; Liu, Y.; Ren, D.; Zhang, Z. Therblig-embedded value stream mapping method for lean energy machining. *Energy* **2017**, *138*, 1081–1098. [CrossRef]
4. Chen, G.; Zhang, L.; Arinez, J.; Biller, S. Energy-efficient production systems through schedule-based operations. *IEEE Trans. Autom. Sci. Eng.* **2013**, *10*, 27–37. [CrossRef]
5. He, Y.; Li, Y.; Wu, T.; Sutherland, J.W. An energy-responsive optimization method for machine tool selection and operation sequence in flexible machining job shops. *J. Clean. Prod.* **2015**, *87*, 245–254. [CrossRef]
6. Kordonowy, D.N. *A Power Assessment of Machining Tools*; Massachusetts Institute of Technology: Cambridge, MA, USA, 2002.
7. Wiendahl, H.-P. *Load-Oriented Manufacturing Control*; Springer: Berlin/Heidelberg, Germany, 1995.
8. Pinedo, M.L. *Scheduling: Theory, algorithms, and Systems*; Springer: New York, NY, USA, 2016.
9. Liu, Y.; Dong, H.; Lohse, N.; Petrovic, S.; Gindy, N. An investigation into minimising total energy consumption and total weighted tardiness in job shops. *J. Clean. Prod.* **2014**, *65*, 87–96. [CrossRef]
10. Lu, C.; Gao, L.; Li, X.; Pan, Q.; Wang, Q. Energy-efficient permutation flow shop scheduling problem using a hybrid multi-objective backtracking search algorithm. *J. Clean. Prod.* **2017**, *144*, 228–238. [CrossRef]
11. Bierwirth, C.; Kuhpfahl, J. Extended grasp for the job shop scheduling problem with total weighted tardiness objective. *Eur. J. Oper. Res.* **2017**, *261*, 835–848. [CrossRef]
12. May, G.K.; Stahl, B.; Taisch, M.; Prabhu, V. Multi-objective genetic algorithm for energy-efficient job shop scheduling. *Int. J. Prod. Res.* **2015**, *53*, 7071–7089. [CrossRef]
13. Yildirim, M.B.; Mouzon, G. Single-machine sustainable production planning to minimize total energy consumption and total completion time using a multiple objective genetic algorithm. *IEEE Trans. Eng. Manag.* **2012**, *59*, 585–597. [CrossRef]
14. Shrouf, F.; Ordieres-Meré, J.; García-Sánchez, A.; Ortega-Mier, M. Optimizing the production scheduling of a single machine to minimize total energy consumption costs. *J. Clean. Prod.* **2014**, *67*, 197–207. [CrossRef]
15. Yin, L.; Li, X.; Lu, C.; Gao, L. Energy-efficient scheduling problem using an effective hybrid multi-objective evolutionary algorithm. *Sustainability* **2016**, *8*, 1268. [CrossRef]
16. Li, Z.; Yang, H.; Zhang, S.; Liu, G. Unrelated parallel machine scheduling problem with energy and tardiness cost. *Int. J. Adv. Manuf. Technol.* **2015**, *84*, 213–226. [CrossRef]
17. Moon, J.-Y.; Park, J.; Shin, K. Optimization of production scheduling with time-dependent and;machine-dependent electricity cost for industrial energy efficiency. *Int. J. Adv. Manuf. Technol.* **2013**, *68*, 523–535. [CrossRef]
18. Ding, J.Y.; Song, S.; Zhang, R.; Chiong, R.; Wu, C. Parallel machine scheduling under time-of-use electricity prices: New models and optimization approaches. *IEEE Trans. Autom. Sci. Eng.* **2016**, *13*, 1138–1154. [CrossRef]
19. Zhang, H.; Zhao, F.; Fang, K.; Sutherland, J.W. Energy-conscious flow shop scheduling under time-of-use electricity tariffs. *CIRP Ann.-Manuf. Technol.* **2014**, *63*, 37–40. [CrossRef]
20. Liu, Y.; Dong, H.; Lohse, N.; Petrovic, S. A multi-objective genetic algorithm for optimisation of energy consumption and shop floor production performance. *Int. J. Prod. Econ.* **2016**, *179*, 259–272. [CrossRef]
21. Zhang, R.; Chiong, R. Solving the energy-efficient job shop scheduling problem: A multi-objective genetic algorithm with enhanced local search for minimizing the total weighted tardiness and total energy consumption. *J. Clean. Prod.* **2015**, *112*, 3361–3375. [CrossRef]
22. Tang, D.; Dai, M.; Salido, M.A.; Giret, A. Energy-efficient dynamic scheduling for a flexible flow shop using an improved particle swarm optimization. *Comput. Ind.* **2016**, *81*, 82–95. [CrossRef]
23. Dai, M.; Tang, D.; Giret, A.; Salido, M.A.; Li, W.D. Energy-efficient scheduling for a flexible flow shop using an improved genetic-simulated annealing algorithm. *Robot. Comput.-Integr. Manuf.* **2013**, *29*, 418–429. [CrossRef]
24. Li, X.; Xing, K.; Wu, Y.; Wang, X.; Luo, J. Total energy consumption optimization via genetic algorithm in flexible manufacturing systems. *Comput. Ind. Eng.* **2017**, *104*, 188–200. [CrossRef]

25. Mouzon, G.; Yildirim, M.B.; Twomey, J. Operational methods for minimization of energy consumption of manufacturing equipment. *Int. J. Prod. Res.* **2007**, *45*, 4247–4271. [CrossRef]
26. Zhang, L.; Li, X.; Gao, L.; Zhang, G. Dynamic rescheduling in fms that is simultaneously considering energy consumption and schedule efficiency. *Int. J. Adv. Manuf. Technol.* **2013**, *87*, 1–13. [CrossRef]
27. Liu, X.; Zou, F.; Zhang, X. Mathematical model and genetic optimization for hybrid flow shop scheduling problem based on energy consumption. In Proceedings of the Control and Decision Conference CCDC2008, Yantai, China, 2–4 July 2008; pp. 1002–1007.
28. Tong, Y.; Li, J.; Li, S.; Li, D. Research on energy-saving production scheduling based on a clustering algorithm for a forging enterprise. *Sustainability* **2016**, *8*, 136. [CrossRef]
29. Li, X.; Lu, C.; Gao, L.; Xiao, S.; Wen, L. An Effective Multi-Objective Algorithm For Energy Efficient Scheduling In A Real-Life Welding Shop. In *IEEE Transactions on Industrial Informatics*; IEEE: Piscataway, NJ, USA, 2018.
30. Zhou, D.C.; Zeng, L. Intelligent scheduling method oriented to multi-varieties and small-batch production mode. In *Applied Mechanics and Materials*; Trans Tech Publications: Zurich, Switzerland, 2013; pp. 1269–1274.
31. Huang, T.-C.; Lin, B.M. Batch scheduling in differentiation flow shops for makespan minimisation. *Int. J. Prod. Res.* **2013**, *51*, 5073–5082. [CrossRef]
32. Wang, M.; Zhong, R.Y.; Dai, Q.; Huang, G.Q. A mpn-based scheduling model for iot-enabled hybrid flow shop manufacturing. *Adv. Eng. Inform.* **2016**, *30*, 728–736. [CrossRef]
33. Huang, H.-H.; Pei, W.; Wu, H.-H.; May, M.-D. A research on problems of mixed-line production and the re-scheduling. *Robot. Comput.-Integr. Manuf.* **2013**, *29*, 64–72. [CrossRef]
34. Zitzler, E.; Knowles, J.; Thiele, L. Quality assessment of pareto set approximations. In *Multiobjective Optimization*; Springer: Berlin, Germany, 2008; pp. 373–404.
35. Auger, A.; Bader, J.; Brockhoff, D.; Zitzler, E. Hypervolume-based multiobjective optimization: Theoretical foundations and practical implications. *Theor. Comput. Sci.* **2012**, *425*, 75–103. [CrossRef]
36. Hu, L.; Tang, R.; Liu, Y.; Cao, Y.; Tiwari, A. Optimising the machining time, deviation and energy consumption through a multi-objective feature sequencing approach. *Energy Convers. Manag.* **2018**, *160*, 126–140. [CrossRef]
37. Mousavi, S.; Mahdavi, I.; Rezaeian, J.; Zandieh, M. An efficient bi-objective algorithm to solve re-entrant hybrid flow shop scheduling with learning effect and setup times. *Oper. Res.* **2018**, *18*, 123–158. [CrossRef]
38. Khan, B.; Hanoun, S.; Johnstone, M.; Lim, C.P.; Creighton, D.; Nahavandi, S. Multi-objective job shop scheduling using i-nsga-iii. In Proceedings of the 2018 Annual IEEE International Systems Conference (SysCon), Vancouver, BC, Canada, 24–26 April 2018; pp. 1–5.
39. Faccio, M.; Nedaei, M.; Pilati, F. A comparative analysis of job scheduling for optimum performance of parallel machines by considering the energy consumption. *Eur. J. Eng. Res. Sci.* **2018**, *3*, 6–11. [CrossRef]

Article

Multi-Objective Optimization of Energy Consumption and Surface Quality in Nanofluid SQCL Assisted Face Milling

Aqib Mashood Khan [1], Muhammad Jamil [1], Konstantinos Salonitis [2,*], Shoaib Sarfraz [2], Wei Zhao [1], Ning He [1], Mozammel Mia [3] and GuoLong Zhao [1]

[1]   College of Mechanical and Electrical Engineering, Nanjing University of Aeronautics and Astronautics, Nanjing 210016, China; dr.aqib@nuaa.edu.cn (A.M.K.); engr.jamil@nuaa.edu.cn (M.J.); nuaazw@nuaa.edu.cn (W.Z.); drhe515@hotmail.com (N.H.); zhaogl@nuaa.edu.cn (G.Z.)

[2]   Manufacturing Department, School of Aerospace, Transport and Manufacturing, Cranfield University, Cranfield, Bedfordshire, MK43 0AL, UK; Shoaib.Sarfraz@cranfield.ac.uk

[3]   Mechanical and Production Engineering, Ahsanullah University of Science and Technology, Dhaka 1208, Bangladesh; mozammelmiaipe@gmail.com

*   Correspondence: k.salonitis@cranfield.ac.uk; Tel.: +44-(0)-1234-758347

Received: 16 December 2018; Accepted: 17 February 2019; Published: 21 February 2019

Abstract: Considering the significance of improving the energy efficiency, surface quality and material removal quantity of machining processes, the present study is conducted in the form of an experimental investigation and a multi-objective optimization. The experiments were conducted by face milling AISI 1045 steel on a Computer Numerical Controlled (CNC) milling machine using a carbide cutting tool. The Cu-nano-fluid, dispersed in distilled water, was impinged in small quantity cooling lubrication (SQCL) spray applied to the cutting zone. The data of surface roughness and active cutting energy were measured while the material removal rate was calculated. A multi-objective optimization was performed by the integration of the Taguchi method, Grey Relational Analysis (GRA), and the Non-Dominated Sorting Genetic Algorithm (NSGA-II). The optimum results calculated were a cutting speed of 1200 rev/min, a feed rate of 320 mm/min, a depth of cut of 0.5 mm, and a width of cut of 15 mm. It was also endowed with a 20.7% reduction in energy consumption. Furthermore, the use of SQCL promoted sustainable manufacturing. The novelty of the work is in reducing energy consumption under nano fluid assisted machining while paying adequate attention to material removal quantity and the product's surface quality.

Keywords: energy consumption; energy efficiency; sustainable machining; multi-objective optimization; multi-criteria decision making method; small quantity cooling lubrication SQCL; cu nanofluid

1. Introduction

Recently, immoderate energy consumption has become one of the most severe problems in the manufacturing industry, and conserving energy has become a necessity for the industry. In manufacturing processes, machine tools are the primary energy consuming devices [1]. Gutowski et al. [2] have concluded that 14.8% of total energy was utilized during the material removal process. According to Newman, 6–40% of energy can be saved by choosing optimum cutting parameters, cutting tools and cutting paths [3]. Additionally, the energy consumption of machine tools is also associated with an increased environmental cost. Therefore, there is a dire need for the optimization of cutting parameters based on minimum energy criteria. Material removal rate, cutting force, tool life, surface roughness (SR), power and energy consumption are the responses measured during the milling process. However, the optimization of a single objective has many limitations.

The conclusion of the above discussion is that this problem necessitates the need for a framework to optimize conflicting objectives simultaneously.

The manufacturing industry is facing a new challenge to reduce energy consumption due to the higher demand forenergy and increasing concerns regarding fossil fuel depletion [4]. However, more difficulties arise when reducing energy consumption without compromising on the surface quality of the product. In the machining process, the typical interactions between the cutting tool-work surface and the cutting tool-chip cause severe friction and intensive material deformation. Even, the harder cutting tool becomes worn due to an excessive mechanical load on the tool edges. Moreover, the thermal adversities caused by friction at the adjacent surfaces instigate unfavourability in machining. This unfavourability demeans the machining productivity and consumes higher amounts of energy. A significant portion of consumed power is lost due to friction during the cutting process [5].

Thermal upsetting at machining interfaces can be pacified by using cooling agents. Likewise, the effects of friction can be limited by employing lubricating agents. Owing to this fact, the use of cooling-lubricating agents in machining is widespread to date. However, the growing concerns for the environment and human health-related problems have forced manufacturers to look for an alternative as well as better cooling-lubrication modes [6–8].

Considering those aspects, for decades, the alternative mode of cooling-lubrication in machining has been the use of small droplets of agents under high air pressure so that the impinged sprays penetrate in between the interfaces. These droplets serve two functions: (i) they reduce friction by creating a thin film in between the material bodies [9], and (ii) they enhance heat dissipation by absorbing and removing heat from that zone [10]. This approach is reported in the literature under a number of different terms such as: small quantity cooling lubrication (SQCL), minimum quantity cooling lubrication (MQCL), minimum quantity lubrication (MQL), and near dry machining (NDM) etc., although the different terms exhibit slight differences in the way the cooling is delivered. Many research works show the positive sides of SQCL [11,12]. However, more advantages are gained when SQCL was applied using nano-fluids in suspension. The improved tribological properties of the nano-fluids themselves result in several other benefits in machining, such as a reduced coefficient of friction, lower cutting forces, improved surface finish, reduced cutting temperature, and lower tool wear associated with prolonged tool life.

The use of nanoparticles such as Cu, CuO, Al_2O_3, MoS_2, SiO_2, and diamond/graphite nanoparticles in the base fluids in the machining process under MQL conditions has become a new trend. However, in depth analysis of the application of water-based nano-fluids in the machining process is still a common research area [13]. Mintsa et al. [14] measured the thermal conductivity of alumina and copper oxide mixed in water (water-based nano-fluids). They showed that the effective thermal conductivity increased when increasing the particle volume fraction and when reducing the size of nanoparticles. In another study, Zhang et al. [15] also measured the effective thermal conductivity of distilled water based nanofluids. The results revealed that effective thermal conductivity increases with the increase of nanoparticle concentration. Lvet al. [16] utilized graphene oxide/silicon dioxide hybrid nanoparticle in water-based nano-fluids in evaluating the machining characteristics of MQL technology. Results revealed a significant reduction in the worn scar diameter (WSD) and the coefficient of friction (COF) using hybrid GO/SiO_2 water-based MQL technology in the milling process.

Modern manufacturing aims at reducing energy consumption as well as achieving high quality within small timeframes [17]. The International Organization for Standardization is currently working on an ISO/DIS 14955 and ISO/DIS 50001 series which would assess the energy efficiency of machine tools and their environmental effects. Kara and Li [18] developed an empirical model to characterize the relationship between energy consumption and machining parameters and reached the apparent conclusion that specific energy consumption decreases when increasing the material removal rate. Abhang and Hameedullah [19] investigated the productivity and power consumption of machine tools in the turning of EN-31 steel. Results showed that when the tool nose radius, cutting speed, feed rate and depth of cut were at lower levels, less power was consumed by the machine tool. Bhattacharya et al. [20] performed experiments to study surface quality and power consumption during the high-speed machining of AISI 1045 steel. The cutting speed was found to be the most significant cutting parameter for both surface roughness and power consumption. However, all experiments were performed under dry cutting conditions. Camposeco-Negrete [21] optimized the cutting parameters for minimum energy consumption in the turning of AISI 6061. It was found that feed rate has the most significant effect on the surface roughness and energy consumption; lower values of feed rate resulted in better surface finish, but with higher energy consumption. Pusavec et al. [22] conducted many experiments in the dry, MQL, cryogenic and hybrid cooling assisted machining of nickel-based Inconel alloy 718. Cutting forces, power, tool life, material removal rate and cooling and lubrication performance were studied. The cooling and lubrication techniques were found useful for reducing cutting forces and improving tool life.

Velchevet al. [23] developed an approach for the optimization of cutting parameters for minimum energy in the turning process. A mathematical model was developed that considered the cutting parameters to express the direct energy consumption of machine tool. Kumar et al. [24] performed a multi-objective optimization by assigning different weights to energy consumption, surface quality, and productivity. The authors have concluded that the depth of cut and the use of radius were vital factors for analytical hierarchy process (AHP) and entropy weight methods respectively. Maruda et al. [25] studied the influence of cooling conditions on-chip formation and tool wear. The authors have concluded that the application of MQCL reduced cutting tool wear.

To conserve resources while maintaining improved machinability, there is a need for multiple objectives optimization using different algorithms. Grey relation algorithm (GRA) is one of the most efficient algorithms used for multi-objective optimization [26]. The GRA is based on a simple calculation, and it can provide an accurate analysis of small sample size. Moreover, it can assign variable weights to each response. It reduces complex multi-objective objective problems into a single object effectively and efficiently. Mia et al. [27] successfully employed a GRA method to optimize the cutting force and surface roughness in end milling under MQL conditions. On the other side, Deb et al. [28] proposed a fast and elitist heuristic method know as Non-dominated Sorting Genetic Algorithm NSGA-II; researchers have been successfully using the NSGA II to optimize machining parameters. Rao et al. [29] employed NSGA II for a multi-objective optimization for multiple machining processes. Then, Li et al. [30] reported the use of NSGA-II for the machining of Ti-6Al-4V alloy machining.

It is evident that the chip-tool interface endures friction, which requires lubrication and cooling. At the same time, energy reduction is a prime motto of the sustainable production system. For that reason, the whole machining system needs to be optimized considering multiple objectives. On the one hand, GRA cannot provide us with a Pareto front solution to find trade-off relationships. However, on the other hand, NSGA-II cannot provide information on a complex problem as a single characteristic. So, to overcome the drawbacks of each algorithm and to present a solution to a complex problem in detail, the multiple objectives are simultaneously optimized in this study using both a conventional method (GRA) as well as using a heuristic one (NSGA-II). Besides these aspects, other novelties include the use of Cu nano-fluid in distilled water and its use in the form of SQCL, reduction of energy consumption, and improvement of material removal rate.

2. Experimental Procedure

The face milling experiments were performed on a CNC machine tool (Carver 400M_RT) with a spindle power of 5.6 KW and a maximum rotation speed of 6000 rpm. AISI-1045 steel was used as work material. A three fluted carbide cutting tool with a diameter of 24 mm was used for machining. Figure 1 shows the schematic, experimental setup, and equipment used in measurements.

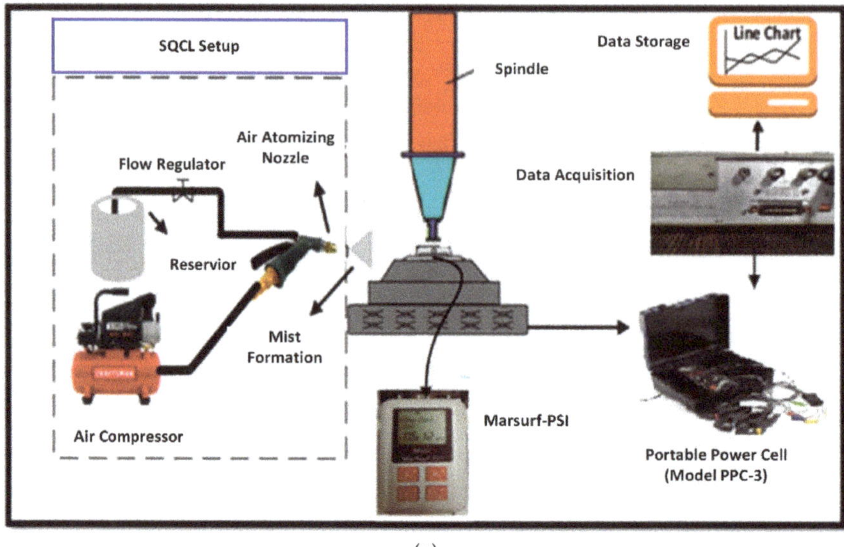

(**a**)

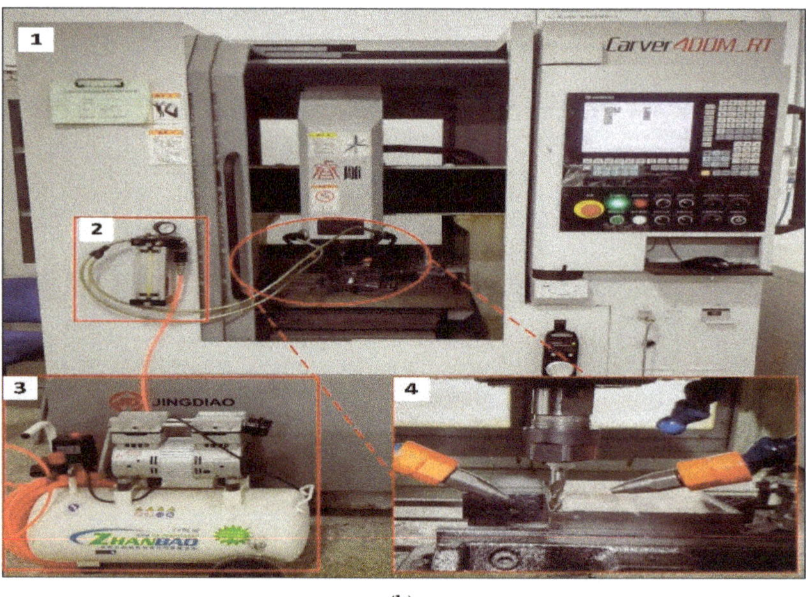

(**b**)

Figure 1. Experimental setup for a small quantity cooling lubrication (SQCL) assisted milling process (**a**) schematic diagram (**b**) Experimental setup.

For nano-fluid based SQCL, a solution of copper nanoparticles suspended in distilled water was employed. Diionized water (DI water) was used as a coolant due to the fact that it has no harmful effects on workers' health, and it is of low cost compared to other lubricants. Moreover, the addition of copper nanoparticles in distilled water increases its thermal stability, as reported by Mintsa et al. [14]. Copper nanoparticles with a 60 nm diameter were mixed with distilled water to prepare the solution (encouraged by reference [31]). The dispersion stability of nanoparticles is essential to investigate their performance. Thereby, the sodium dodecyl sulfate (SDS) was used as the surfactant in the mixture to achieve good stability. The stability mechanism of nano-fluid was measured in terms of zeta potential Z. Zeta potential values of more than 30 mV showed the high stability of the nano-fluid and its value was found to be positive 45 mV, depicting that the nano-fluid was highly stable. Figure 2 shows the preparation procedure of the Cu based nanofluid.

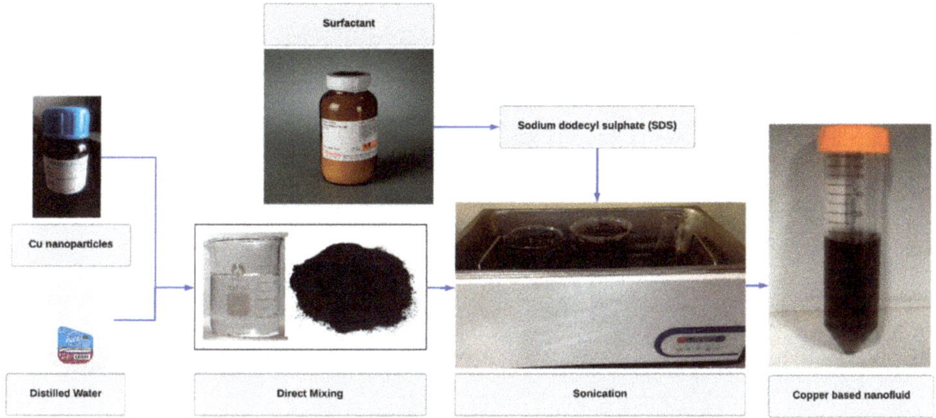

Figure 2. The preparation of copper based nanofluid.

The focus of the study is to investigate the effect of cutting parameters on the responses above. Therefore, the technological parameters related to the SQCL system were kept constant by doing preliminary experiments and following the existing literature [11,12]. The nano-fluid SQCL was sprayed using a single nozzle at an air pressure of 4 bars, a lubricant flow rate of 300 mL/h; and a nozzle angle of 45°. The maximum material removal volume (MRV) was set to 2400 mm^3. Experiments were designed according to the Taguchi L27 orthogonal array. Cutting parameters and their levels were selected by making preliminary trial runs and also from the published literature [32].

Figure 3 shows the cause-effect diagram for the efficiency of the machining process. In the current study, machining parameters (spindle speed, feed rate, depth of cut, and width of cut) were selected. Table 1 contains the levels of cutting parameters.

Table 1. Cutting parameters and their levels.

Factors	Cutting Parameters	Level 1	Level 2	Level 3
Cutting Speed	N (rev/min)	1200	1700	2200
Feed Rate	f (mm/min)	220	270	320
Depth of Cut	a_p (mm)	0.3	0.4	0.5
Width of Cut	a_e (mm)	5	10	15

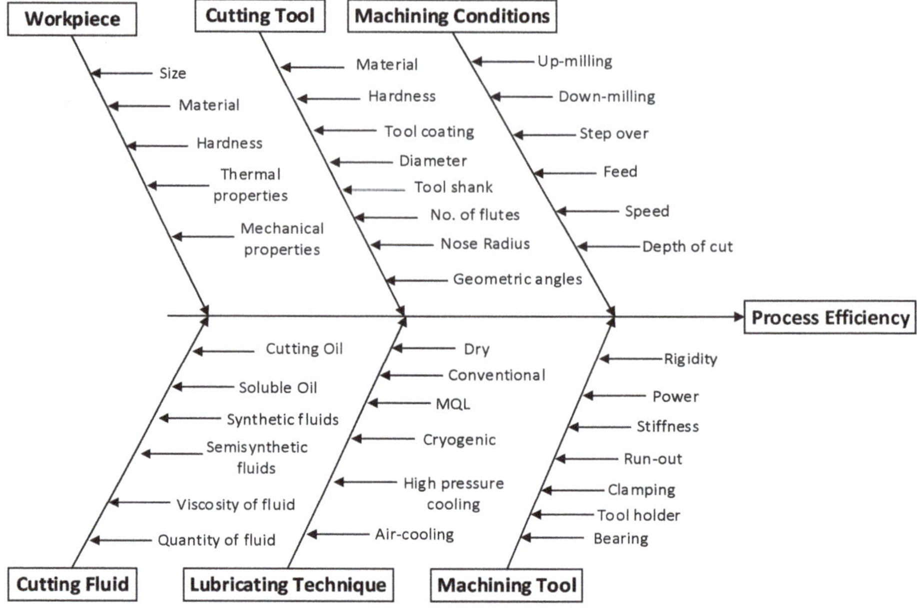

Figure 3. The effect of process parameters on process efficiency.

3. Response Measurements

The responses were measured for three successive runs, and finally, the average of the measured data was adopted as the mean value for further computation.

3.1. Production Rate

The production rate was determined by the material removal rate (mm^3/min) which was computed using Equation (1) [32].

$$MRR = f_z \times z \times N \times a_p \times a_e \tag{1}$$

where z, a_p, a_e are the number of tool flutes, depth of cut and width of cut, respectively.

3.2. Surface Quality

The quality of the produced machined surface was defined by the average of the roughness profile that the machining process was generating. For that purpose, the average surface roughness parameter (SR) was selected as the representative index of surface quality. It was measured using a portable stylus profilometer Mitutoyo Surftest SJ-410 at three equally separated points, and the average values were taken to mitigate errors [33].

3.3. Active Cutting Energy

The Portable Power Cell model PPC-3 (Figure 4) was used to measure power demand during the cutting process. To measure the active energy consumption, a power sensor was attached to the main bus of the electrical cabinet of the CNC machine at a sampling frequency of 10Hz. Equation (2) can be used to estimate active energy consumption [32].

$$ACE = \sum_{i=1}^{k} \frac{P_i \times t_i}{1000} \tag{2}$$

where ACE; active cutting energy (kJ), P_i; active power measured in watts (W), t_i; time, and k; data samples during the process. In the experimental analysis, the lubricant delivery system used a separate power resource that was assumed to be constant. The measurement system for surface roughness and power consumption is shown in Figure 4.

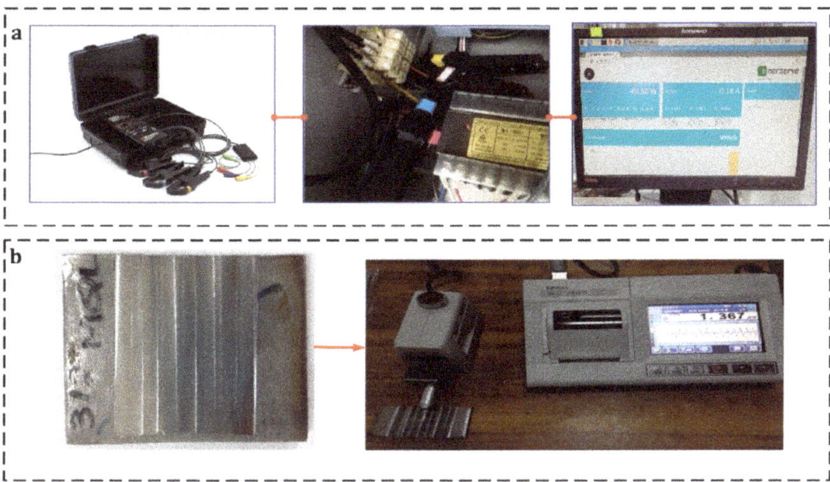

Figure 4. Experimental equipment used, showing: (**a**) power measurement using a Portable Power Cell PPC-3 (**b**) surface roughness measurement.

4. Results and Discussion

In this section, the multi-objective optimization is performed using Grey relational analysis (GRA) and NSGA-II methods to deal with difficult decision-making situations. Furthermore, the influence of the milling parameter on the surface quality, material removal rate and the cutting energy consumption is discussed. Finally, the mechanism of nano fluid assisted machining favorability is presented. The framework of the multi-objective optimization problem is shown in Figure 5. The framework explains the importance of energy consumption as a response. A comparison has been performed between the conventional method when only two responses (surface quality and productivity) are considered and a proposed method where energy is a sustainable parameter has been added. In the conventional method, equal weights are given to both responses to give them equal importance. However, in the proposed method an entropy method has been employed to the weights of each response according to the sensitivity of the response. In the proposed method, optimization has been performed using two algorithms: GRA and NSGA-II. The reason for employing both optimization algorisms is to get benefits from both methods to understand the complex multi-objective problem fully. The application of GRA is straight forward and quick. It converts the multi-objective optimization problem into a single objective and provides Grey relational grades (GRG).The GRG is a performance index of the GRA and provides information as a single characteristic in optimization problems. For example, a GRG value of 0.8678 highlights the best experiment in the L27 design. However, the heuristics are mostly recommended for the optimization instead of classic approaches, due to their capacity for a more precise global search. It is nearly impossible to get the optimal values of all responses simultaneously. So, NSGA-IIhas been used to obtain Pareto fronts to find the tradeoff relationships between a pair of two objects. GRA provides us the local optimal results. However, NSGA-II provides global optimal results.

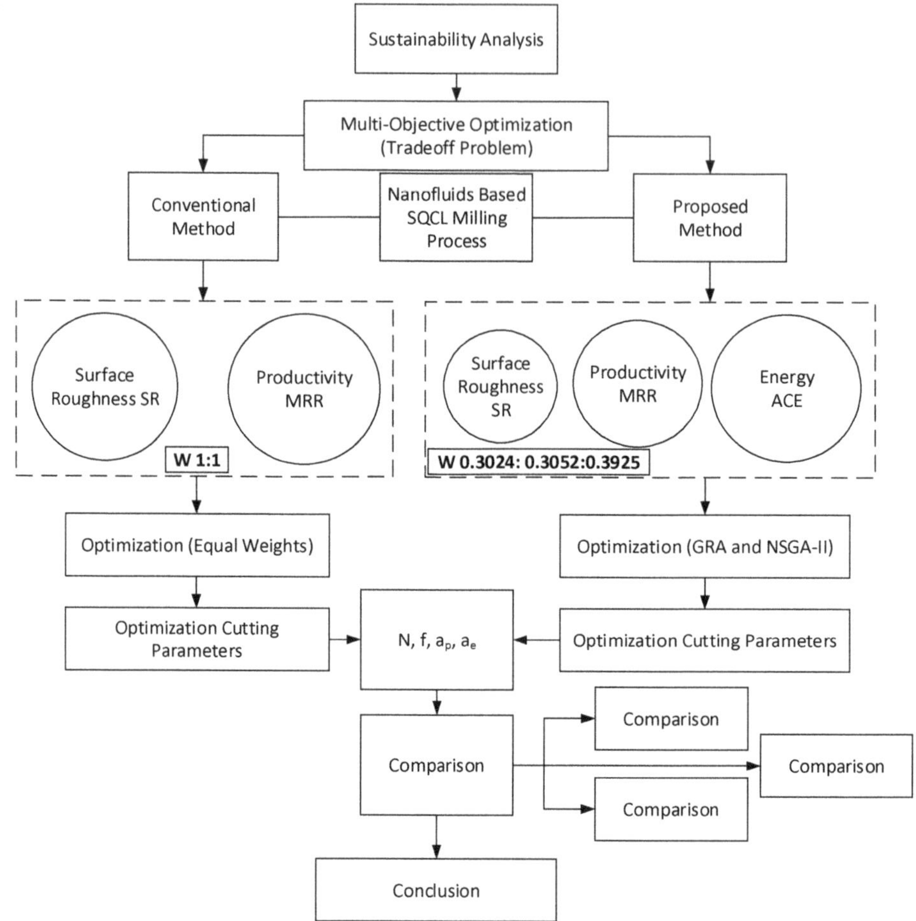

Figure 5. A framework for the energy reduction problem in the milling process.

Experimental design L27 along with process parameters and obtained results are presented in Table 2.

Table 2. Experimental design with measured values of responses.

Serial No.	N (rev/min)	f (mm/min)	a_p (mm)	a_e (mm)	MRR (mm³/min)	R_a (μm)	ACE (kJ)
1	1200	220	0.3	5	330	3.30	535.802
2	1200	220	0.4	10	880	2.95	184.929
3	1200	220	0.5	15	1650	1.41	88.519
4	1200	270	0.3	5	405	3.83	426.109
5	1200	270	0.4	10	1080	3.87	146.050
6	1200	270	0.5	15	2025	1.68	69.823
7	1200	320	0.3	5	480	3.97	361.832
8	1200	320	0.4	10	1280	3.53	122.976
9	1200	320	0.5	15	2400	2.29	53.988
10	1700	220	0.3	10	660	1.81	337.042
11	1700	220	0.4	15	1320	1.13	142.727
12	1700	220	0.5	5	550	3.47	299.031
13	1700	270	0.3	10	810	2.85	269.604
14	1700	270	0.4	15	1620	1.41	113.648

Table 2. Cont.

Serial No.	N (rev/min)	f (mm/min)	a_p (mm)	a_e (mm)	MRR (mm³/min)	R_a (μm)	ACE (kJ)
15	1700	270	0.5	5	675	3.91	238.476
16	1700	320	0.3	10	960	2.55	213.559
17	1700	320	0.4	15	1920	1.39	92.551
18	1700	320	0.5	5	800	4.12	193.109
19	2200	220	0.3	15	990	1.76	244.303
20	2200	220	0.4	5	440	3.33	425.797
21	2200	220	0.5	10	1100	2.36	165.620
22	2200	270	0.3	15	1215	1.17	193.939
23	2200	270	0.4	5	540	3.72	338.579
24	2200	270	0.5	10	1350	2.58	131.343
25	2200	320	0.3	15	1440	1.41	160.886
26	2200	320	0.4	5	640	3.86	286.850
27	2200	320	0.5	10	1600	2.76	108.147

4.1. Grey Relational Analysis

Grey relational analysis was used to solve the complicated multi-objective optimization problem. There are three simple steps for this algorithm. In the first step, to avoid variability and different units, each response value was normalized. If the required objective is 'the *larger the better*', then the results were normalized using Equation (3). If the objective is 'the *smaller the better*', then the results were normalized using Equation (4) [26],

$$x_i^*(k) = \frac{x_i^0(k) - \min\left(x_i^0(k)\right)}{\max(x_i^0(k) - \min\left(x_i^0(k)\right)} \tag{3}$$

$$x_i^*(k) = \frac{\max\left(x_i^0\right) - x_i^0(k)}{\max(x_i^0(k) - \min\left(x_i^0(k)\right)} \tag{4}$$

The grey relation coefficients (GRC) for each factor were calculated by Equation (5). Finally, Grey relational grades (GRG) were computed using Equation (6) [26].

$$\xi(k) = \frac{\Delta_{min} + \xi\Delta_{max}}{\Delta 0_i(k) + \xi\Delta_{max}} \tag{5}$$

$$\gamma_i = \sum_{k=1}^{n} w_k \xi_i(k) \tag{6}$$

$\Delta 0_i(k)$ is known as the deviation sequence and it is the absolute value. Minimum and maximum values of $\Delta 0_i(k)$ are denoted by Δ_{min} and Δ_{max}, respectively γ_i represents the GRG of each experiment provided in Appendix A. The values of ξ lies between 0 and 1 and it is called a distinguishing coefficient, normally it is taken as 0.5. The w_k represents the normalized weightage of factor k.

Traditionally, equal weights are assigned to objectives in simple problems. However, in complex problems, it is not easy to assume equal weights for each criterion. In complex multi-attribute decision making (MADM) problems, the weight of each criterion is calculatedby prioritization, expert opinion, and evaluation [34]. Finding an appropriate and effective way to assign the weights is a dire need of the MADM. Subjective and objective are the two well-known weight assignment methods. The examples of subjective weight assignment methods are the AHP, Delphi method, and weight least square method; in these methods, weights are assigned according to the preference of the decision makers.

Moreover, subjective weight methods are less accurate and have some drawbacks [35]. However, in the objective based weight assignment method, weights are assigned on the inherent information of indexes to determine weights of indexes. Entropy-based weight assignment is an objective weight assignment method which not only avoids human-made disturbances and assumptions but also provides results according to the sensitivity of the objectives. In that way, more realistic and accurate weights are calculated [35]. For these reasons, in this paper, the weight of each response was calculated in a systematic manner using entropy weighting.

In common practice, the concept of entropy is originated from thermo-science. This method finds the weight factor based on the influence of various input parameters on responses [24]. According to which, a higher weight should be assigned to the objective with higher variation in its values. In the first step, the mean values of each coefficient for each response were calculated based on Table 3. For example, to calculate the mean values of the Grey relation coefficients of the width of cut at cutting level 1 (5 mm), all experiments were selected with a width of cut at 5 mm and then the mean was taken.

Table 3. Weight calculation for each response using Grey entropy weight.

Parameter	MRR				SR				ACE			
	Level 1	Level 2	Level 3	R	Level 1	Level 2	Level 3	R	Level 1	Level 2	Level 3	R
Cutting speed	0.518	0.450	0.442	0.075	0.492	0.450	0.442	0.075	0.680	0.631	0.611	0.069
feed rate	0.416	0.463	0.530	0.114	0.416	0.463	0.530	0.114	0.575	0.644	0.703	0.128
Cutting depth	0.402	0.419	0.549	0.147	0.402	0.419	0.549	0.147	0.515	0.655	0.751	0.236
Width of cut	0.358	0.404	0.607	0.248	0.358	0.404	0.607	0.248	0.469	0.575	0.787	0.318
$\sum R$				0.585				0.580				0.753
Weight				0.305				0.302				0.392

Level = Cutting parameters level; R = Range.

Similarly, the mean was taken at level 2 (10 mm) and level 3 (15 mm), respectively. The range was calculated from the difference between the maximum and minimum values of the mean of GRC. The Ranges (R) for each factor were computed using Equation (7).

$$R_{i,j} = \max\{Q_{i,j,1}, Q_{i,j,2}, \dots Q_{i,j,s}\} - \text{Min}\{Q_{i,j,1}, Q_{i,j,2} \dots Q_{i,j,s}\} \tag{7}$$

$$wi = \sum_{i=1}^{c} R_{i,j} / \sum_{i=1}^{c} \sum_{j=1}^{p} R_{i,j} \tag{8}$$

where, $I = 1, 2, \dots c$, c: number of response. $j = 1, 2, \dots p$, p: cutting parameters. $s = 1, 2, \dots l$, l: cutting parameter's level. Similarly, R: GRC range, Q: mean Grey relational coefficient of each cutting parameter at three different levels (Appendix B).

Weights for each response were computed using Equation (8). Finally, Equation (9) was used to calculate grey relation grades and is given in Table 4.

$$\text{GRG} = 0.305 \times \text{GRC}_{(MRR)} + 0.302 \times \text{GRC}_{(SR)} + 0.392 \times \text{GRC}_{(ACE)} \tag{9}$$

$\text{GRC}_{(MRR)}$, $\text{GRC}_{(SR)}$, $\text{GRC}_{(ACE)}$ in Table 4 are the grey relation coefficients of material removal rate, surface roughness, and active cutting energy. GRG is an overall all Grey relation grade.

Table 4. Grey relation coefficients (GRC) and Grey relation grade (GRG) values for each response.

No.	$\text{GRC}_{(MRR)}$	$\text{GRC}_{(SR)}$	$\text{GRC}_{(ACE)}$	GRG
1	0.333	0.408	0.333	0.355
2	0.405	0.451	0.648	0.514
3	0.580	0.842	0.875	0.774
4	0.342	0.356	0.393	0.366
5	0.439	0.353	0.724	0.524
6	0.734	0.731	0.938	0.813
7	0.350	0.345	0.439	0.383
8	0.480	0.384	0.777	0.567
9	1.000	0.563	1.000	0.867
10	0.373	0.687	0.460	0.502
11	0.489	1.000	0.731	0.738

Table 4. *Cont.*

No.	GRC$_{(MRR)}$	GRC$_{(SR)}$	GRC$_{(ACE)}$	GRG
12	0.359	0.390	0.496	0.421
13	0.394	0.465	0.528	0.468
14	0.570	0.842	0.802	0.743
15	0.375	0.350	0.566	0.442
16	0.418	0.513	0.602	0.518
17	0.683	0.852	0.862	0.804
18	0.393	0.333	0.634	0.469
19	0.423	0.704	0.559	0.561
20	0.346	0.405	0.393	0.382
21	0.443	0.549	0.683	0.569
22	0.466	0.974	0.633	0.685
23	0.358	0.366	0.458	0.399
24	0.496	0.508	0.757	0.602
25	0.519	0.842	0.693	0.684
26	0.370	0.354	0.508	0.419
27	0.564	0.478	0.816	0.637

4.2. Development of RSM Models and Analysis of Variance

The second order Response Surface Methodology (RSM) models for the three objective functions: material removal rate, surface roughness, and active cutting energy were formulated using the measured data presented in Table 2. Process parameters in coded form affected the response more evenly, thereby accelerating the convergence speed of the algorithm and improving the precision of the algorithm. Coded variables were sometimes plotted on the *x*-axis and can easily be calculated by Equations (10)–(13).

$$A = (N - 1700)/500 \text{ Coded Variables} = 1,\ 0,\ -1 \tag{10}$$

$$B = (f - 270)/50 \text{Coded Variables} = 1,\ 0,\ -1 \tag{11}$$

$$C = (a_p - 0.4)/0.1 \text{Coded Variables} = 1,\ 0,\ -1 \tag{12}$$

$$D = (a_e - 10)/5 \text{Coded Variables} = 1,\ 0,\ -1 \tag{13}$$

It is necessary to understand the relationship between cutting parameters and Grey relational grades forthe optimization of cutting parameters. Response surface methodology was employed to generate the second-order RSM model for MRR, SR, and ACE. Mia et al. [36] worked on the investigation of machinability characteristics and concluded that well-fitting mathematical models could be obtained using RSM. Equations (14)–(16) describe a second-order empirical model for MRR, SR, ACE respectively. Moreover, Equation (17) shows the mathematical relation of Grey relational grade regarding input parameters.

$$\begin{aligned} MRR(N,f,a_p,a_e) = {}& 3048 - 1.45 \times (N) - 3.15 \times (f) - 10665 \times (a_p) + 51.3 \times (a_e) + \\ & 0.0002 \times (N^2) - 2.9 \times 10^{-17} \times (f^2) + 6750 \times (a_p{}^2) + 2.7 \times (a_e{}^2) - 5.0 \times 10^{-4} \times (N \times f) + \\ & 2.7 \times (N \times a_p) - 0.054 \times (N \times a_e) + 10.0 \times (f \times a_p) + 0.4 \times (f \times a_e) \end{aligned} \tag{14}$$

$$\begin{aligned} SR(N,f,a_p,a_e) = {}& -8.9 - 5.56 \times 10^{-4} \times (N) + 0.043 \times (f) + 37.67 \times (a_p) - 0.06 \times (a_e) + \\ & 1.0 \times 10^{-6} \times (N^2) - 5.8 \times 10^{-5} \times (f^2) - 32.44 \times (a_p{}^2) + 0.015 \times (a_e{}^2) - 5.16 \times 10^{-6} \times \\ & (N \times f) - 0.007 \times (N \times a_p) + 0.0001 \times (N \times a_e) + 0.014 \times (f \times a_p) - 3.53 \times 10^{-4} \times (f \times a_e) \end{aligned} \tag{15}$$

$$\begin{aligned} ACE(N,f,a_p,a_e) = {}& 2985.65 - 0.29 \times (N) - 4.87 \times (f) - 5911.40 \times (a_p) - 71.63 \times (a_e) \\ & + 5.4 \times 10^{-5} \times (N^2) + 0.0036 \times (f^2) - 3794.33 \times (a_p{}^2) + 2.67 \times (a_e{}^2) - 3.12 \times 10^{-5} (N \times f) \\ & + 0.6535 \times (N \times a_p) - 0.0133 \times (N \times a_e) + 3.04 \times (f \times a_p) + 0.083 \times (f \times a_e) \end{aligned} \tag{16}$$

$$GRG(N, f, a_p, a_e) = 0.895 - 0.00016 \times (N) - 0.0015 \times (f) + 1.91 \times (a_p) + 0.0057$$
$$\times (a_e) + 2.0 \times 10^{-6} \times (f^2) + 1.17 \times (a_p{}^2) + 0.0018 \times (a_e{}^2) - 0.00002 \times (f \times N) + 0.00067 \qquad (17)$$
$$\times (f \times a_p) + 0.00057 \times (f \times a_e)$$

Figure 6 presents the relationship between the measured and predicted values of GRG. Measured values were close to predicted values, and the average percentage deviation was found to be 4.69, which indicated that experimental and predicted results were in good agreement.

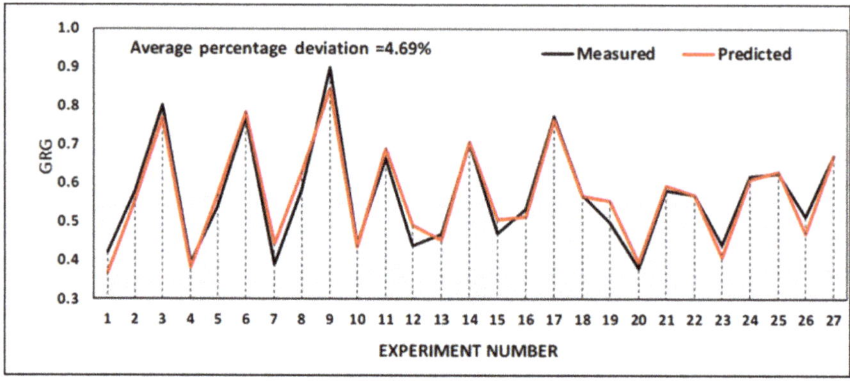

Figure 6. Comparison of measured GRG and predicted GRG values.

Table 5 presents the analysis of variance (ANOVA) results for the regression model. ANOVA results illustrated that the coefficient of determination R-square value was 99.02%, and R-(adj) was 98.03%, which indicated the model's excellent compatibility with experimental results. Experiment No. 9 (in bold in Table 4) was identified as the optimal trade-off having combination ($N = 1200 \frac{rev}{min}, f = 320 \frac{mm}{min}, a_p = 0.5$ mm, $a_e = 15$ mm) and can be verified from Figure 6 as having the highest *GRG* value.

Table 5. ANOVA table for the Grey relational grade (GRG).

Factor	DOF	Seq-SS	Adj-MS	FValue	P-Value	$F_{0.01}$
Regression model	10	0.608	0.0467	100.75	0.000	3.705
Error	16	0.006	0.0004	–	–	–
Total	26	0.6143	–	–	–	–

S = 0.021; R-Sq = 99.02%; R-Sq (adj) = 98.03%; R-Sq (pred) = 95.40%.

4.3. Non-Dominated Sorting Genetic Algorithm NSGA-II

The genetic algorithm is a heuristic algorithm to optimize NP-hard problems inspired by a natural selection process in bio-science systems. Srinivas and Deb [37] proposed NSGA, and later Deb et al. [38] proposed NSGA-II, which elevates all the short-comings of NSGA, NSGA-II can obtain true Pareto optimal solutions based on the rank of non-domination. In NSGA-II, minimization of fitness is assumed, and each solution is given a rank which is equal to its non-domination level, e.g., one is considered the best level. Therefore, it obtains solutions not just within the experimental runs, but global space is used for searching for the optimal solution. A flowchart of GRA coupled with NSGA-II has been shown in Figure 7.

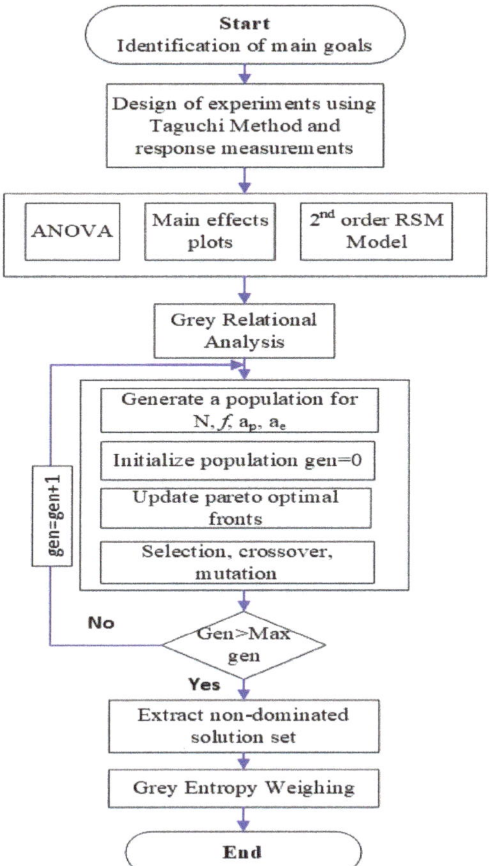

Figure 7. Flow chart of the milling process optimization algorithm with GRA Coupled with theNon-Dominated Sorting Genetic Algorithm (NSGA-II).

To convert multi-objective optimization into a single objective optimization problem, each objective function is multiplied with its respective weight. Improvement in the response highly depends on the weight assigned. In weight-based GA (WBGA), weights are random; this random assignment of weights does not consider the priority of responses. Therefore, the optimal solution is not guaranteed. In random weight assignment methods, weights are assigned to each objective function and then combined them to make a single objective. However, although these methods are simple to use, weight selection is still a challenging problem. In this paper, the weight of each response was calculated in a systematic manner using entropy weighting of GRA as discussed in Section 4.1.

4.3.1. Optimization Variables and Objective Function

Equations (14)–(16) serve as the objective function of NSGA-II. In this method, NSGA-II was the main technique, and GRA entropy was utilized to generate the weight of each response. NSGA-II was used to find four optimized milling parameters (spindle speed, feed rate, depth of cut, and width of cut) so that the best performance measures (MRR, SR, and ACE) were achieved without violating any of the constraints considered for this problem. For the simultaneous optimization of goals, the three objective functions were given as follows: Objective 1 = material removal rate; Objective 2 = surface roughness; Objective 3 = active cutting energy.

4.3.2. Constraints and Parameters for the NSGA-II Algorithm

The measured response must meet the constraints in the CNC milling process. Following the practical situation, the following constraints were considered.

Machine tool: CNC milling conditions must be within an allowable range of machine tool parameter values; Equation (18) explains this constraint for processing.

$$x_{min} < x_i < x_{max}(i = 1, 2, 3, 4) \tag{18}$$

$$P_c = \frac{F_c Dn}{318} < \eta P_{max} \tag{19}$$

x_i denotes milling parameters, F_c represents the maximum cutting force, P_{max} is nominal motor power, and η themachine efficiency. In the optimization procedure using NSGA-II, the following parameters were listed based on the study to get optimal solutions with low computational effort and have been set by trial and error through many repeats of simulations and also determined from the literature [39]: crossover probability: 0.95; distribution index for the crossover: 10; mutation probability: 0.01; distribution index formulation: 10; population size: 100.

4.3.3. Pareto Optimal Front

Pareto-optimal fronts were obtained from NSGA-II and are displayed in Figure 8. The data points in the Pareto-fronts do not dominate each other, and they represent the optimal combinations of milling parameters. The selection of any data points from Pareto fronts depended upon desired SQCL milling criterion. It is clear from Figure 8 the active energy consumption values decreased as the surface roughness increased.

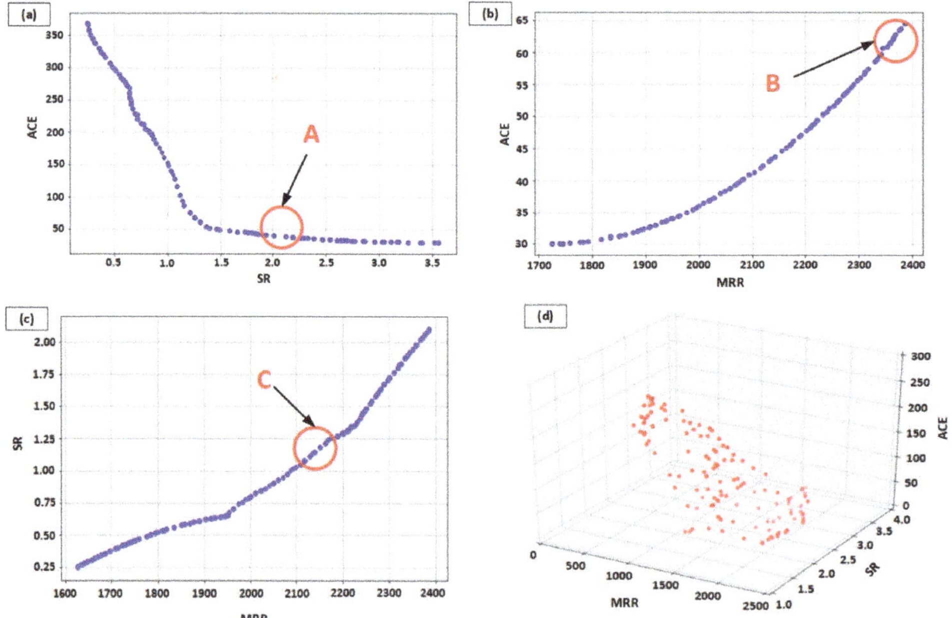

Figure 8. Pareto optimal-front for optimization objectives: (**a**) Active cutting energy (ACE) vs. Surface roughness (SR), (**b**) Active cutting energy (ACE) vs. Material removal rate (MRR) and (**c**) Surface roughness (SR) vs. Material removal rate (MRR) (**d**) 3D Pareto front.

However, ACE increased as the MRR increased. On the other hand, surface roughness values also increased as the MRR increased. As seen in Figure 8a, region A consisted of a Pareto-front solution, which was good for reducing ACE and at the same time not causing an increase in R_a values. The optimized values of cutting parameters and response values were: (N:1622.38; f:300.394; ap:0.4999; ae:13.7012; SR:2.022; ACE:40.2928) at point A. In addition, the Pareto optimal solutions in region B were good for getting maximum productivity. However, the ACE was higher in this region. The optimized milling parameters at point B were: (N:1200; f:320; ap:0.5; ae:15) and the corresponding response values were: (MRR:2387.5; ACE:64.60). Similarly, point C was a tradeoff between SR and MRR with optimized milling parameters at point: (N:1217; f:320.001; ap:0.3; ae:15) and the response values were: (MRR:2135.73; SR:0.66). Besides, other solutions can be selected after considering the desired criteria. A wide range of Pareto-optimum solutions demonstrated the applicability of this method in machining optimization that considers conflicting objectives.

4.4. Influence of Milling Parameters on MRR, ACE, and SR

It is clear from Figure 9 that for active cutting energy (ACE) and MRR, the width of the cut had the most significant effect, while for spindle speed, this effect was the least significant.

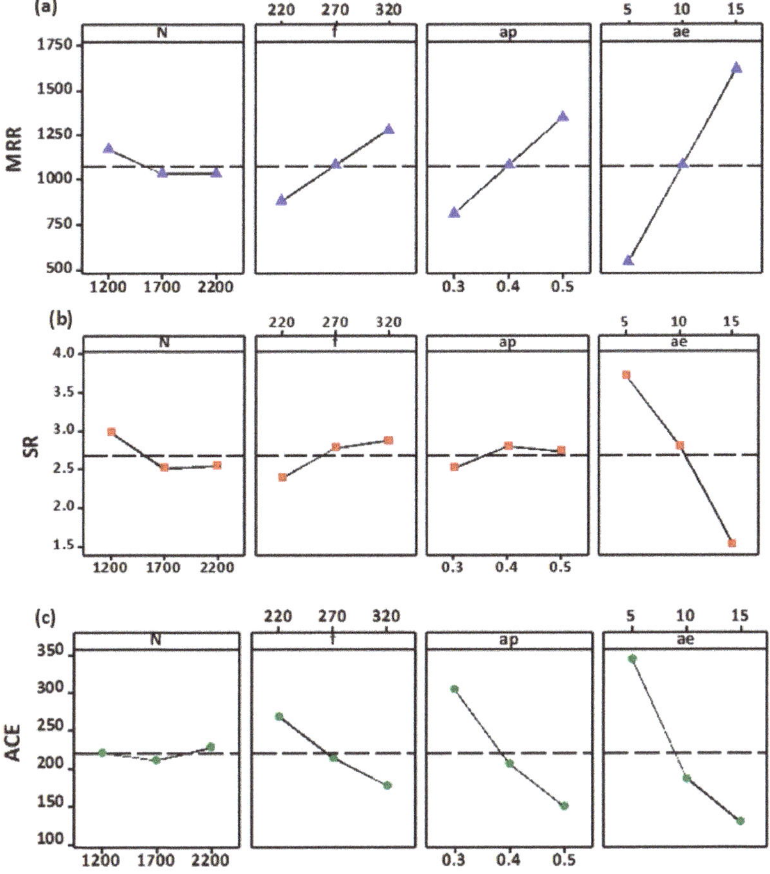

Figure 9. Main effect plots for, (a) MRR, (b) SR, (c) ACE.

Furthermore, GRC of ACE and MRR were in a linear relationship with the width of cut, so the highest value of the width of cut produced the lowest values of ACE and highest MRR. Similarly, for SR, the width of the cut also showed the most significant effect. From Table 2, the maximum SR values were found at a 5 mm width of cut, and the lowest SR value was found at 15 mm.

Extensive dispersion of Cu nanoparticles in the distilled water facilitated by cool high-pressure mist in workpiece and tool interface showed an excellent performance in reducing surface roughness. The mist from nano-fluid assisted SQCL systems exhibited more efficient feeding into the machining zone than the conventional cooling method [40]. This was because, in conventional wet cooling, lubricant splashes off by the rotating tool, but, mist sticks to the tool-workpiece interface and thus assists in improving cutting performance. In addition to this, the Cu nanoparticle in the SQCL provided a polishing effect, since an interacting force was then induced between the tool-particle interface, which is shown in Figure 9a [41]. This induced interacting force travelled upon the cutting tool interacting surface and produced power on the surface. Hence, the surface energy of the workpiece was significantly reduced due to a high interacting force between the particle workpiece interface [42].

On the one hand, the breaking of nanoparticles required a certain energy for the removal of asperities from the workpiece surface, which generated a new surface with better finish. On the other hand, the process of making Cu nanoparticles transferred potential energy from a cutting tool, which later became kinetic energy of workpiece surface atoms. Therefore, a 300 ml/h flow rate provided a number of Cu nanoparticles, which transferred more kinetic energy to surface atoms and enhanced heat dissipation from the cutting zone. Figure 9b explains how nanoparticles behave as a spacer and effectively minimize the frictional effects between interacting surfaces. In the high speed-milling process, heat generated in the cutting zone altered elasto-hydrodynamic lubrication into boundary lubrication, and thus, this may create a possibility of rolling effect (Figure 9b) which could result in a reduction in coefficient of friction [43].

Moreover, the increase in the concentration of Cu particles in distilled water strengthened the thin protective film on the workpiece surface. Thiswas due to the strong chemical interaction between the newly generated surface and nanoparticles. This phenomenon increased the surface quality of the workpiece [42]. The decrease in surface roughness was mainly due to the nanoparticles between the workpiece-tool interface. A good surface finish was mainly due to the excellent lubrication that slides the chips away from tool surface; better chip quality resulted in less surface roughness. Furthermore, a mist formation that entered the machining zone reduced the friction due to a ball bearing effect due to billions of rolling nanoparticles, and high-pressure spray removed the chips, ultimately leading to a better surface finish [44]. A higher concentration of solid fine Cu nanoparticles of a large size provided effective lubrication under aggravated sliding conditions. Because these particles easily penetrated the surface between rubbing face, the original sliding friction divided into sliding as well as rolling friction, and that reduced the coefficient of friction [45].

4.5. Comparative Analysis and Validation

Traditionally, specialized handbooks and worker experience are used to determine the desired cutting parameters, but these sources provide starting values of cutting parameters, and these values are not necessarily suitable for optimal conditions. Initially, the levels of milling parameters were set at ($N2$, $f2$, a_p1, a_e1), and this combination is shown in Table 2 at experiment number 13. In this study, Figure 2 describes conventional multi-objective optimization, different from our purposed optimization method, only surface roughness, and material removal rate were selected as the optimization objectives in conventional optimization. Milling parameters ($N = 2200$ rev/min, $f = 320$ mm/min, $a_p = 0.5$ mm and $a_e = 15$ mm) were identified as the optimal setting for conventional multi-objective optimization. In order to perform the comparison between conventional and proposed multi-objective optimization method, we have also obtained optimal milling parameters from conventional optimization with two objectives. Table 6 shows the comparison of the experimental results of two optimization methods.

Table 6. Optimal responses using various optimization techniques.

Optimization Method	MRV	Optimal Parameter Combination				Optimal Response		
		N	f	a_p	a_e	MRR	SR	ACE
Initial parameter [a]	2400	1700	270	0.3	10	810	2.85	269.6
Conventional optimization [b]	2400	2200	320	0.5	15	2400	2.23	68.15
Grey relational analysis [c]	2400	1200	320	0.5	15	2400	2.29	53.98
GRA coupled with NSGA-II [d]	2400	1200.01	320	0.5	15	2387.5	2.09	64.4

[a] Experiment 13; [b] MRR & SR (with equal weights); [c] Experiment 9, MRR + SR + ACE with weights (0.3052:0.3024:0.3925); [d] MRR + SR + ACE with weights (0.3052:0.3024:0.3925).

Initially, the GRA based optimization results compared with the initial settings of milling parameters. For the comparison of two methods, material removal volume (MRV) was fixed to the maximum level, i.e., 2400 mm³. As a result, SR and ACE decreased from 2.85 μm to 2.29 μm and 269.6 kJ to 53.98 kJ respectively. Parameters settings based on the proposed optimization method resulted in a 79.9% reduction in energy consumption as compared to the initial setting. It is clear that the proposed optimization method can improve multi-response in the cutting process. The feasibility of the proposed method is indicated by the promising results obtained.

As shown in Table 6, the GRA based optimization resulted in a 20.7% reduction in ACE as compared with conventional objectives optimization. Moreover, along with traditional objectives, the proposed method also added a sustainable objective, i.e., active energy consumption, which promotes environmental benefits and saves energy.

5. Conclusions

With an objective to simultaneously reduce the energy consumption and improve the surface quality, this study presented an experimental and optimization approach to reduce energy consumption and improve surface quality without compromising the material removal quantity under the use of nano-fluid assisted SQCL. The optimization was performed as a 'multi-objective' using GRA and NSGA-II methods. The following concluding remarks are found:

- The GRA based optimization method has revealed that a cutting speed of 1200 rev/min, a feed rate of 320 mm/min, a depth of cut of 0.5 mm, and a width of cut of 15 mm were the optimal cutting levels. It also found that the width of the cut hold had the most significant effect on response measurements.
- Comparison of experimental results revealed that a lower spindle speed and higher width of cut were suitable for energy efficiency in the practiced milling process. The reduction of energy consumption was 20.7% as compared to the conventional optimization method.
- In NSGA-II, three conflicting objective functions, i.e., MRR, SR, and ACE, were considered for the optimization problem. The optimal solution for milling process was found from Pareto-fronts. The GRA converted the complex multi-objective problem into a single objective, and a maximum value of GRG represented the initial optimal cutting condition.
- The results showed that NSGA-II gave an efficient result and similarly, GRA predicted results accurately with less than 5% error.
- Comparison of experimental results revealed that lower spindle speed and higher levels of the width of cut were suitable for energy efficiency of machine tools. Hence, the proposed algorithm in this paper provides broader solutions for engineers to choose cutting parameters according to optimal results; therefore, the proposed method demonstrated applicability for manufacturing engineers working in the metal processing industries.

In conclusion, the implementation of the current study in the industry can reduce a significant amount of energy consumption.

Author Contributions: Conceptualization, A.M.K. and N.H.; Data curation, G.L.Z., A.M.K. and M.J.; Formal analysis, A.M.K.; Funding acquisition, K.S.; Investigation, A.M.K. and M.J.; Methodology, A.M.K.; Project administration, S.S., W.Z. and N.H.; Resources, W.Z. and N.H.; Software, A.M.K., M.J., S.S. and M.M.; Supervision, N.H., G.L.Z.; Validation, K.S., M.J. and M.M.; Visualization, M.J. and M.M; Writing—original draft, A.M.K.; Writing—review, editing, layout formatting K.S., S.S., and M.M., G.L.Z.

Funding: The work is supported by the National Natural Science Foundation of China (No. 51875285), the Natural Science Foundation of Jiangsu Province (No. BK20160792), the Fundamental Research Funds for the Central Universities (NO. NP2018302).

Acknowledgments: The authors would like to acknowledge the efforts made by Prof. Kornel F. Ehmann of Northwestern University (USA) who reviewed the manuscript and gave some valuable suggestions.

Conflicts of Interest: The authors declare no conflict of interest.

Appendix A

Table A1. The normalization, grey relational coefficient and grey relational grade generation.

Sr.	Normalization			Grey Relational Coffiecient			GRG
	MRR (mm^3/min)	R_a (µm)	ACE (kJ)	MRR	R_a	ACE	
1	0.000	0.274	0.000	0.333	0.408	0.333	0.3559
2	0.266	0.391	0.728	0.405	0.451	0.648	0.5142
3	0.638	0.906	0.928	0.580	0.842	0.875	0.7748
4	0.036	0.097	0.228	0.342	0.356	0.393	0.3662
5	0.362	0.084	0.809	0.439	0.353	0.724	0.5248
6	0.819	0.816	0.967	0.734	0.731	0.938	0.8132
7	0.072	0.050	0.361	0.350	0.345	0.439	0.3834
8	0.459	0.197	0.857	0.480	0.384	0.777	0.5677
9	1.000	0.612	1.000	1.000	0.563	1.000	0.8678
10	0.159	0.773	0.413	0.373	0.687	0.460	0.5021
11	0.478	1.000	0.816	0.489	1.000	0.731	0.7385
12	0.106	0.217	0.491	0.359	0.390	0.496	0.4219
13	0.232	0.425	0.552	0.394	0.465	0.528	0.4680
14	0.623	0.906	0.876	0.570	0.842	0.802	0.7432
15	0.167	0.070	0.617	0.375	0.350	0.566	0.4424
16	0.304	0.525	0.669	0.418	0.513	0.602	0.5187
17	0.768	0.913	0.920	0.683	0.852	0.862	0.8043
18	0.227	0.000	0.711	0.393	0.333	0.634	0.4694
19	0.319	0.789	0.605	0.423	0.704	0.559	0.5611
20	0.053	0.264	0.228	0.346	0.405	0.393	0.3821
21	0.372	0.589	0.768	0.443	0.549	0.683	0.5693
22	0.428	0.987	0.710	0.466	0.974	0.633	0.6850
23	0.101	0.134	0.409	0.358	0.366	0.458	0.3996
24	0.493	0.515	0.839	0.496	0.508	0.757	0.6020
25	0.536	0.906	0.778	0.519	0.842	0.693	0.6848
26	0.150	0.087	0.517	0.370	0.354	0.508	0.4195
27	0.614	0.455	0.888	0.564	0.478	0.816	0.6372

Explanation for Experiment No. 1

Step 1: Normalized Data using Equations (A1) and (A2)

The response MRR is 'the larger the better', so Equation (A1) was used for normalization of MRR, and SR and ACE are 'the minimum the better', so Equation (A2) was used for the normalization of SR and ACE.

$$x_i^*(k) = \frac{x_i^0(k) - \min\left(x_i^0(k)\right)}{\max\left(x_i^0(k) - \min\left(x_i^0(k)\right)\right)} \tag{A1}$$

$$x_i^*(k) = \frac{\max\left(x_i^0\right) - x_i^0(k)}{\max\left(x_i^0(k) - \min\left(x_i^0(k)\right)\right)} \tag{A2}$$

Normalization of MRR Experiment 1: $x_i^*(k) = \left(\frac{330-330}{2070}\right) = 0$

Normalization of SR Experiment 1: $x_i^*(k) = \left(\frac{4.12-3.3}{2.99}\right) = 0.274$

Normalization of SR Experiment 1: $x_i^*(k) = \left(\frac{535.82 - 535.82}{481}\right) = 0$

"Reversing operator" was applied on the normalized data to get final values

Step 2: Calculation of Grey relational coffiecient (GRC)

GRC was calculated using Equation (A3)

$$\xi(k) = \frac{\Delta_{min} + \xi\Delta_{max}}{\Delta 0_i(k) + \xi\Delta_{max}} \tag{A3}$$

GRC of Experiment. 1:

$$\text{GRC}_{\text{MRR}} = \frac{(0+0.5*1)}{(1+0.5*1)} = 0.333$$
$$\text{GRCSR} = \frac{(0+0.5*1)}{(0.726+0.5*1)} = 0.408$$
$$\text{GRC}_{\text{ACE}} = \frac{(0+0.5*1)}{(1+0.5*1)} = 0.333$$

Step 3: Calculation of Grey relational grade (GRG)

GRG was calculated using Equation (A4).

$$\gamma_i = \sum_{k=1}^{n} w_k \xi_i(k) \tag{A4}$$

GRG of Experiment 1:

$$\text{GRG} = \frac{(0+0.5*1)}{(1+0.5*1)} = (0.305*0.333 + 0.302*0.408 + 0.393*0.333) = 0.355$$

Appendix B

Calculation of Weights

In this example weights of MRR were calculated, the other weights were calculated in the same fashion.

1. Each cutting parameter had 3 levels and each level possessed its own Grey rational coefficient (GRC) value. For example, Spindle Speed (N) had three levels: 1200, 1700 and 2200. So, GRC values for each levels were added and then its averagewas taken. Effect of N on MRR; GRC values at level 1 of N: = (0.3333 + 0.4051 + 0.5798 + 0.3416 + 0.4395 + 0.7340 + 0.3503 + 0.4803 + 1.00)/9
2. Similarly, these average values were taken for 1700 and 2200 levels of spindle speed.

Level 1(1200) = 0.5182	Level 2 (1700) = 0.4505	Level 3 (2200) = 0.4428	See Table 3

3. Step 1 and 2 was repeated for each level of feed, depth of cut and width of cut.
4. Ranges were found by the difference of maximum and minimum values using Equation (A5).

$$\text{Ri,j} = \max\{\text{Qi,j,1, Qi,j,2, ... Qi,j,s}\} - \text{Min}\{\text{Qi,j,1, Qi,j,2 ... Qi,j,s}\} \tag{A5}$$

For example, Range R for spindle speed in MRR = 0.5182 − 0.4405 = 0.0777

5. Step 4 was repeated for feed, depth of cut and width of cut.
6. All ranges values were summed up. For example, in the case of MRR0.07 + 0.114 + 0.1472 + 0.2489 = 0.5801
7. Similarly, sum ranges for Surface roughness (SR) and Active cutting energy (ACE) were also determined.

8. In the end Equation (A6) was used to find the weights of each factor.

$$wi = \sum_{i=1}^{c} R_{i,j} / \sum_{i=1}^{c} \sum_{j=1}^{p} R_{i,j} \tag{A6}$$

where, I = 1, 2, … c, c is the number of response. j = 1, 2, … p, p is the number of cutting parameters. s = 1, 2, … . l. l is the number of cutting parameter's level. Similarly, R represents the Grey relational coefficient range. Q denotes the mean Grey relational coefficient of each cutting parameter at three different levels.

References

1. Liu, F. Content Architecture and Future Trends of Energy Efficiency Research on Machining Systems. *J. Mech. Eng.* **2013**. [CrossRef]
2. Gutowski, T.; Dahmus, J.; Thiriez, A. Electrical Energy Requirements for Manufacturing Processes. In Proceedings of the 13th CIRP International Conference on Life Cycle Engineering, Leuven, Belgium, 31 May–2 June 2006. [CrossRef]
3. Newman, S.T.; Nassehi, A.; Imani-Asrai, R.; Dhokia, V. Energy efficient process planning for CNC machining. *CIRP J. Manuf. Sci. Technol.* **2012**, *5*, 127–136. [CrossRef]
4. Salonitis, K.; Jolly, M.R.; Zeng, B.; Mehrabi, H. Improvements in energy consumption and environmental impact by novel single shot melting process for casting. *J. Clean. Prod.* **2016**, *137*, 1532–1542. [CrossRef]
5. Hegab, H.; Abdelfattah, W.; Rahnamayan, S.; Mohany, A.; Kishawy, H. Multi-Objective Optimization During Machining Ti-6Al-4V Using Nano-Fluids. *CSME Conf. Proc.* **2018**. [CrossRef]
6. Cetin, M.H.; Ozcelik, B.; Kuram, E.; Demirbas, E. Evaluation of vegetable based cutting fluids with extreme pressure and cutting parameters in turning of AISI 304L by Taguchi method. *J. Clean. Prod.* **2011**, *19*, 2049–2056. [CrossRef]
7. Kuram, E.; Ozcelik, B.; Demirbas, E.; Şık, E. Effects of the Cutting Fluid Types and Cutting Parameters on Surface Roughness and Thrust Force. *World Congr. Eng.* **2010**, *2*, 978–988.
8. Khan, A.M.; Jamil, M.; Mia, M.; Pimenov, D.Y.; Gasiyarov, V.R.; Gupta, M.K.; He, N. Multi-objective optimization for grinding of AISI D2 steel with Al_2O_3 wheel under MQL. *Materials* **2018**, *11*, 2269. [CrossRef]
9. Maruda, R.W. Influence of cooling conditions on the machining process under MQCL and MQL conditions. *Tehnički Vjesnik* **2015**, *22*, 965–970. [CrossRef]
10. Maruda, R.W.; Krolczyk, G.M.; Nieslony, P.; Wojciechowski, S.; Michalski, M.; Legutko, S. The influence of the cooling conditions on the cutting tool wear and the chip formation mechanism. *J. Manuf. Process.* **2016**, *24*, 107–115. [CrossRef]
11. Manojkumar, K.; Ghosh, A. Assessment of cooling-lubrication and wettability characteristics of nano-engineered sunflower oil as cutting fluid and its impact on SQCL grinding performance. *J. Mater. Process.Technol.* **2016**, *23*, 55–64. [CrossRef]
12. Verma, N.; ManojKumar, K.; Ghosh, A. Characteristics of aerosol produced by an internal-mix nozzle and its influence on force, residual stress and surface finish in SQCL grinding. *J. Mater. Process. Technol.* **2017**, *240*, 223–232. [CrossRef]
13. Gu, Y.; Zhao, X.; Liu, Y.; Lv, Y. Preparation and tribological properties of dual-coated TiO2nanoparticles as water-based lubricant additives. *J. Nanomater.* **2014**, *2*. [CrossRef]
14. Mintsa, H.A.; Roy, G.; Nguyen, C.T.; Doucet, D. New temperature dependent thermal conductivity data for water-based nanofluids. *Int. J. Therm. Sci.* **2009**, *48*, 363–371. [CrossRef]
15. Xue, Q.-Z. Model for effective thermal conductivity of nanofluids. *Phys. Lett. A* **2003**, *307*, 313–317. [CrossRef]
16. Lv, J.; Tang, R.; Tang, W.; Liu, Y.; Zhang, Y.; Jia, S. An investigation into reducing the spindle acceleration energy consumption of machine tools. *J. Clean. Prod.* **2017**, *143*, 794–803. [CrossRef]
17. Salonitis, K.; Ball, P. Energy efficient manufacturing from machine tools to manufacturing systems. *Procedia CIRP* **2013**, *7*, 634–639. [CrossRef]
18. Kara, S.; Li, W. Unit process energy consumption models for material removal processes. *CIRP Ann. Manuf. Technol.* **2011**, *60*, 37–40. [CrossRef]

19. Abhang, L.B.; Hameedullah, M. Power prediction model for turning EN-31 steel using response surface methodology. *J. Eng. Sci. Technol. Rev.* **2010**, *3*, 116–122. [CrossRef]

20. Bhattacharya, A.; Das, S.; Majumder, P.; Batish, A. Estimating the effect of cutting parameters on surface finish and power consumption during high speed machining of AISI 1045 steel using Taguchi design and ANOVA. *Prod. Eng.* **2009**, *3*, 31–40. [CrossRef]

21. Camposeco-Negrete, C. Optimization of cutting parameters for minimizing energy consumption in turning of AISI 6061 T6 using Taguchi methodology and ANOVA. *J. Clean. Prod.* **2013**, *53*, 195–203. [CrossRef]

22. Pusavec, F.; Deshpande, A.; Yang, S.; M'Saoubi, R.; Kopac, J.; Dillon, O.W.; Jawahir, I.S. Sustainable machining of high temperature Nickel alloy—Inconel 718: Part 1—Predictive performance models. *J. Clean. Prod.* **2014**, *81*, 255–269. [CrossRef]

23. Velchev, S.; Kolev, I.; Ivanov, K.; Gechevski, S. Empirical models for specific energy consumption and optimization of cutting parameters for minimizing energy consumption during turning. *J. Clean. Prod.* **2014**, *80*, 139–149. [CrossRef]

24. Kumar, R.; Bilga, P.S.; Singh, S. Multi objective optimization using different methods of assigning weights to energy consumption responses, surface roughness and material removal rate during rough turning operation. *J. Clean. Prod.* **2017**, *164*, 45–57. [CrossRef]

25. Maruda, R.W.; Krolczyk, G.M.; Feldshtein, E.; Pusavec, F.; Szydlowski, M.; Legutko, S.; Sobczak-Kupiec, A. A study on droplets sizes, their distribution and heat exchange for minimum quantity cooling lubrication (MQCL). *Int. J. Mach. Tools Manuf.* **2016**, *100*, 81–92. [CrossRef]

26. Jomaa, W.; Lévesque, J.; Bocher, P.; Divialle, A.; Gakwaya, A. Optimization study of dry peripheral milling process for improving aeronautical part integrity using Grey relational analysis. *Int. J. Adv. Manuf. Technol.* **2017**, *91*, 931–942. [CrossRef]

27. Mia, M.; Khan, M.A.; Rahman, S.S.; Dhar, N.R. Mono-objective and multi-objective optimization of performance parameters in high pressure coolant assisted turning of Ti-6Al-4V. *Int. J. Adv. Manuf. Technol.* **2017**, *90*, 109–118. [CrossRef]

28. Deb, K.; Pratap, A.; Agarwal, S.; Meyarivan, T. A fast and elitist multiobjective genetic algorithm: NSGA-II. *IEEE Trans. Evol. Comput.* **2002**, *6*, 182–197. [CrossRef]

29. Pawar, P.J.; Rao, R.V. Parameter optimization of machining processes using teaching-learning-based optimization algorithm. *Int. J. Adv. Manuf. Technol.* **2013**, *67*, 995–1006. [CrossRef]

30. Li, J.; Yang, X.; Ren, C.; Chen, G.; Wang, Y. Multiobjective optimization of cutting parameters in Ti-6Al-4V milling process using nondominated sorting genetic algorithm-II. *Int. J. Adv. Manuf. Technol.* **2014**, *76*, 941–953. [CrossRef]

31. Yu, W.; Xie, H. A review on nanofluids: Preparation, stability mechanisms, and applications. *J. Nanomater.* **2012**, 1. [CrossRef]

32. Li, C.; Xiao, Q.; Tang, Y.; Li, L. A method integrating Taguchi, RSM and MOPSO to CNC machining parameters optimization for energy saving. *J. Clean. Prod.* **2016**, *135*, 263–275. [CrossRef]

33. Khan, A.M.; Jamil, M.; Ul Haq, A.; Hussain, S.; Meng, L.; He, N. Sustainable machining. Modeling and optimization of temperature and surface roughness in the milling of AISI D2 steel. *Ind. Lubr. Tribol.* **2018**. [CrossRef]

34. Winkler, R.L.; Cochrane, J.L.; Zeleny, M. Multiple Criteria Decision Making. *J. Am. Stat. Assoc.* **1975**. [CrossRef]

35. Lotfi, F.H.; Fallahnejad, R. Imprecise shannon's entropy and multi attribute decision making. *Entropy* **2010**, *12*, 53–62. [CrossRef]

36. Mia, M.; Rifat, A.; Tanvir, M.F.; Gupta, M.K.; Hossain, M.J.; Goswami, A. Multi-objective optimization of chip-tool interaction parameters using Grey-Taguchi method in MQL-assisted turning. *Meas. J. Int. Meas. Confed.* **2018**, *129*, 156–166. [CrossRef]

37. Srinivas, N.; Deb, K. Muiltiobjective Optimization Using Nondominated Sorting in Genetic Algorithms. *Evol. Comput.* **1994**, *2*, 221–248. [CrossRef]

38. Qu, S.; Zhao, J.; Wang, T. Experimental study and machining parameter optimization in milling thin-walled plates based on NSGA-II. *Int. J. Adv. Manuf. Technol.* **2017**, *89*, 2399–2409. [CrossRef]

39. Yang, S.H.; Natarajan, U. Multi-objective optimization of cutting parameters in turning process using differential evolution and non-dominated sorting genetic algorithm-II approaches. *Int. J. Adv. Manuf. Technol.* **2010**, *49*, 773–784. [CrossRef]

40. Liew, W.Y.H. Low-speed milling of stainless steel with TiAlN single-layer and TiAlN/AlCrN nano-multilayer coated carbide tools under different lubrication conditions. *Wear* **2010**, *269*, 617–631. [CrossRef]
41. Yan, J.; Zhang, Z.; Kuriyagawa, T. Effect of nanoparticle lubrication in diamond turning of reaction-bonded SiC. *Int. J. Autom. Technol.* **2011**, *5*, 307–312. [CrossRef]
42. Lee, K.; Hwang, Y.; Cheong, S.; Choi, Y.; Kwon, L.; Lee, J.; Kim, S.H. Understanding the role of nanoparticles in nano-oil lubrication. *Tribol. Lett.* **2009**, *35*, 127–131. [CrossRef]
43. Sayuti, M.; Sarhan, A.A.D.; Hamdi, M. An investigation of optimum SiO2nanolubrication parameters in end milling of aerospace Al6061-T6 alloy. *Int. J. Adv. Manuf. Technol.* **2013**, *67*, 833–849. [CrossRef]
44. Tawakoli, T.; Hadad, M.J.; Sadeghi, M.H.; Daneshi, A.; Stöckert, S.; Rasifard, A. An experimental investigation of the effects of workpiece and grinding parameters on minimum quantity lubrication-MQL grinding. *Int. J. Mach. Tools Manuf.* **2009**, *49*, 924–932. [CrossRef]
45. Tao, X.; Jiazheng, Z.; Kang, X. The ball-bearing effect of diamond nanoparticles as an oil additive. *J. Phys. D Appl. Phys.* **1996**, *29*, 2932. [CrossRef]

Article

Deep Learning Approach of Energy Estimation Model of Remote Laser Welding

Jumyung Um [1,*], Ian Anthony Stroud [2] and Yong-keun Park [1]

[1] Department of Industrial & Management System Engineering, Kyung Hee University,
     1732 Deogyeong-daero, Yongin-si 17104, Korea; yongkp@khu.ac.kr
[2] SkAD Labs SA, Chemin de la Raye 13, 1024 Ecublens, Switzerland; ianstroud52@gmail.com
*   Correspondence: jayum@khu.ac.kr; Tel.: +82-31-201-3695; Fax: +82-31-203-4004

Received: 22 March 2019; Accepted: 1 May 2019; Published: 11 May 2019

Abstract: Due to concerns about energy use in production systems, energy-efficient processes have received much interest from the automotive industry recently. Remote laser welding is an innovative assembly process, but has a critical issue with the energy consumption. Robot companies provide only the average energy use in the technical specification, but process parameters such as robot movement, laser use, and welding path also affect the energy use. Existing literature focuses on measuring energy in standardized conditions in which the welding process is most frequently operated or on modularizing unified blocks in which energy can be estimated using simple calculations. In this paper, the authors propose an integrated approach considering both process variation and machine specification and multiple methods' comparison. A deep learning approach is used for building the neural network integrated with the effects of process parameters and machine specification. The training dataset used is experimental data measured from a remote laser welding robot producing a car back door assembly. The proposed estimation model is compared with a linear regression approach and shows higher accuracy than other methods.

Keywords: remote laser welding; energy-efficient process; machine learning; welding process; neural network

---

1. Introduction

Manufacturing activities account for one-third of the total energy consumption and CO_2 emissions worldwide. Thus, energy-efficient production has been the focus of research over the last decade and is considered one of the major decision-making attributes of manufacturing today. This paper takes into account both the energy efficiency of the manufacturing process and the system. In mass production industries energy-efficient processes have received much interest, for example from the automotive industry.

The energy use depends on the process technology of the machine conducting each process. This paper concerns energy efficiency in welding, specifically with RLW (Remote Laser Welding) as compared to spot welding. Figure 1 illustrates one aspect of movement differences between spot welding and remote welding. A spot welding robot visits all positions of stitches (Figure 1 (Left)) and consumes energy for movement to welding positions as well as for the welding itself whereas in remote welding the robot performs welding at a distance and hence can use simpler movement paths (Figure 1 (Right)).

Bearing this in mind, it can be seen that the process technology influences the energy consumption of other components in a machine. To reduce the total energy use, measurement and estimation of the whole process are required. As pointed out in [1], when new processes are introduced into factory production lines, energy consumption is a significant factor in machine selection in addition to productivity needs. In this respect, RLW is interesting because it dramatically reduces both lead time and energy use in the assembly phase of production. Because of these benefits, RLW is of interest in mass production, notably in automotive assembly lines. However, it is necessary to be able to evaluate accurately potential benefits in order to see when the high initial investment is defrayed with savings. The current approach to new machine selection is to compare the technical specifications of energy use, which is a simple static number. The specification usually seldom shows the variation of the energy profile occurring in the process of specific products.

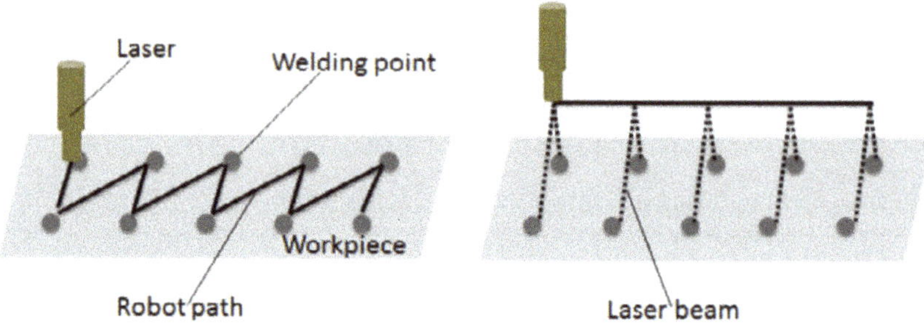

Figure 1. (**Left**) Short-distance welding path; (**Right**) remote welding path. The differences between short-distance and remote welding.

It is necessary for approaches to estimate the energy profile occurring when a welding robot attaches a specific set of stitches, as well as the energy efficiency of the overall processes. To do this, an energy estimation model is required before a new welding robot is introduced into the assembly line. The literature on laser welding focuses on measuring the energy profile of welding a specific assembly. Most research about welding energy consumption concerns estimation of the robot kinetic energy. Currently, there are many energy models of robot arms in the literature. The major energy consumers of RLW are the robot arm, the laser machine because laser generation energy is higher than conventional lasers, and cooling equipment. The ratio of robot motion among major consumers is relatively smaller than for a spot welding robot, which is measured with only kinetic energy use. For this reason, the energy use of each assembly process is different from others because the laser equipment is an additional variable causing fluctuations in the energy profile compared with the spot welding process.

The mechanism of the laser machine consists of the series of laser generation, concentration, delivery, and focusing, which are relatively more complicated than robot motion. The energy consumption model also reflects the mechanism and main consumption by using weighting factors. Using rule-based approaches, it is difficult to develop common models for different processes and machines. Thus, an energy estimation model for laser machines is proposed in this paper using a machine learning approach. In particular, a deep learning model is utilized for training and correlation with process parameters and extracted feature data.

This paper proposes a comprehensive energy estimation model for RLW and its energy efficiency model. In Section 2, the basic concept and existing models for energy estimation are introduced. Section 3 describes the deep learning approach to simulate estimated energy use, followed by Section 4 explaining the results of experiments with the welding process for an automotive rear door assembly. Then, in Section 5, the difference between the proposed approach and conventional models is described. Conclusions are given in Section 6.

2. Literature Survey

In this section, previous models of energy efficiency of manufacturing resources, which are specifically for machine tools, are discussed. Then, some methods and strategies for estimating energy use, which is a key input for calculation of energy efficiency, are introduced from aspects of energy efficiency and energy consumption of functional analysis and process analysis. Finally, the use of machine learning to estimate energy profiles is evaluated in terms of pros and cons.

2.1. Energy Efficiency of Machine Tools

Fundamentally, energy efficiency started from a power-based energy efficiency model proposed by Sebastian [2]. Zhou et al. [3] also discussed other approaches to distinguish instantaneous energy efficiency and process energy efficiency to categorize changes in mechanical design and process optimization. Relatively speaking, instantaneous energy efficiency can be estimated using the machine specification. However, process energy efficiency is calculated using process parameters, as well as machine configurations. As evaluated by Zhou et al. [3], the energy efficiency method has improved from optimization of machine design and process planning. Both approaches led to integrated design and planning to incorporate real-time data, a kinematic-based model, the consideration of the NC code, etc.

Pagone et al. [4] also pointed out the limitation of total energy consumption in the energy efficiency of ferrous material processing. In particular, the material characteristics, such as the aspect of thermal dynamics, are comprised of the internal energy storage, which are seldom measured by external sensors. Comparison between two different processes is also needed to decompose a whole process into the components and unit process.

2.2. Energy Consumption Model of Machine Tools

Improving the accuracy of efficiency models requires an accurate energy decomposition model even though process parameters and machine configurations are changed. Gustawski et al. [5–7] illustrated the spectrum of energy consumption of a three-axis CNC milling machine and the different contributions of machine components to energy use. Gontarz et al. [8] described the need for measurement at the component level with an electric analogue structure of a milling machine. In terms of laser processes, Um et al. [1] compared the machine components of RLW with electric analogue structures as well.

2.2.1. Early Estimation of Energy Consumption

The dominant approach to early energy estimation is direct measurement of specific conditions. The purpose of this approach is to find optimal process parameters. For process optimization, Winter [9] measured energy use of a grinding machine and extended the concept to eco-efficiency [10]. In terms of laser processes, the laser device of an additive process has been investigated. Ma et al. [11] built the formulation of energy consumption for the laser device of a Selective Laser Sintering (SLS) machine. The energy consumption correlated with material characteristics was discussed by Zhu et al. [12].

2.2.2. Process-Driven Approaches

Practically, simple measurement provides only an intuitive view of the energy consumption of a new process. It is necessary to decompose it into process aspects. Gregory [13] investigated robot trajectory planning and optimized energy use with an estimation model. Avram [14] proposed an energy evaluation model of a 2.5D milling machine by decomposing the process into machining working steps. Duflow [15] emphasized the process approach to find more accurate energy consumption estimates because of the need to compare a variety of machine tools. Newman et al. [16] proposed energy efficiency process planning with consideration of the cutting parameters, which are the feed rate, depth of cut, and spindle speed.

2.3. Integrated Approaches and Neural Networks

As with Duflow [15], the integration of process-driven approaches and measurement-driven approaches is proposed by discretizing the energy profile into unit blocks. Recent research on eco-efficiency focuses on estimating consumed energy depending on the process plan and structure of the machine tool. Um et al. [1] proposed the integration of process and machine specification, but it is still necessary to have an advanced method for determining the energy profile in order to improve the accuracy of the energy consumption model, as well as for energy efficiency.

In order to fill the accuracy gap, machine learning is used in integrated approaches. The major application of deep learning concerns homes. The field of energy estimation models was introduced in the building industry to estimate the energy profile of electricity use. This is intended to draw a continuous and smooth curve relating time and energy use. This energy estimation model is typically applied to buildings and houses where a large amount of uncontrollable parameters influences the profile with repeated daily or monthly patterns. The characteristics of these applications lead to time series data analysis, such as recurrence neural networks applied to home electricity use [17]. This application focuses on observation of the uncontrollable patterns of energy use [18]. With energy estimation of building energy use, the following features can be discerned: (1) multiple variables are considered; (2) the distribution of events influencing energy consumption is high.

In contrast, manufacturing resources are more controllable than building systems because of repeatable, planned operations. However, the energy profiles of production facilities are determined by changes of the product model, machine configuration, line balance, scheduling, etc. The quantity of measured data is relatively smaller than for buildings because optimized machine programs are achieved prior to reaching the factory, so they do not lead to loss by cost and failure of quality. Being controllable and predictable, the factory facility is reflected in the energy estimation model. Since around 2010, an increasing amount of research has been conducted in deep learning applications over a diverse range of areas from home to factory. This paper introduces the current state of deep learning applications for energy estimation and in particular assesses contributions made in the field of manufacturing systems.

3. Methodology of Estimating Energy Use of RLW

3.1. Energy Efficiency of RLW

The intention of this section is two-fold. Firstly, we introduce the formulation of energy efficiency for RLW, which is a new area of research on energy estimation models in manufacturing facilities. Secondly, the energy decomposition derived from the analysis of the e-scheme is introduced to give the details of the proposed energy efficiency formulation. We will then discuss how to apply both models to RLW.

In this section, we introduce some definitions that support easier understanding of the laser welding process based on power demand (Figure 2). Sebastian [2] proposed the formulation of energy efficiency of machine tools that remove volumes of material. The opposite direction to metal cutting is to produce products by combining material, in the welding process by combining two or more parts. For this reason, there is a series of questions about how to calculate production rates.

- How do we count the amount of work? Typically, the work of metal cutting is measured by the removal volume. This factor is directly correlated to the feed rate, depth of cut, and cutting speed of milling and turning.
- How do we measure the productivity of each shape of a stitch? Each stitch has a shape, i.e., a line, S-line, C-curve, etc., with different depths and widths. The energy use depends on all aspects.
- The velocity of the robot arm is another factor, but with the same effect on energy consumption as for metal cutting.
- Even though the same process parameters of laser welding are applied, the absorption rate of different materials is a factor that should be considered.

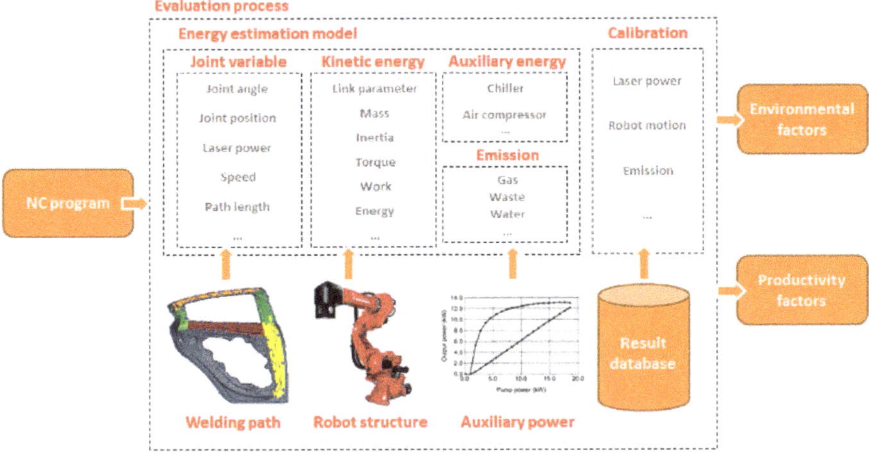

Figure 2. Overall process of the eco-efficiency of the Remote Laser Welding (RLW) process.

3.2. Energy Consumption Estimation of RLW

The energy decomposition of the e-scheme shows the major consumer of electricity. However, to estimate an accurate profile, the mechanism of energy consumption based on a theoretical model, such as kinematics or thermal dynamics, is required. Modeling all aspects of a specific machine is hard to achieve for machine users. Compensation with experimental data and process parameters is necessary. Additionally, experimental data of unit processes support adjustment of specific factors, which are difficult to represent as a theoretical model. Both approaches are breakthroughs in the limitations of theoretical models of energy consumption.

In this paper, the authors propose an integrated approach connecting function and process approaches by utilizing theoretical models and experimental compensation. Figure 3 shows the proposed evaluation model, which is a procedure to merge function and process data and derive a compensated estimated energy profile. The functional approach consists of deriving the main functions, decomposing machine components, and establishing the theoretical model of major components. The process approach starts by analyzing the NC program and continues by measuring the energy profiles of each component and classifying the profiles into unit processes. Compensation links the output of the theoretical model with the energy profile of unit processes by a regression approach.

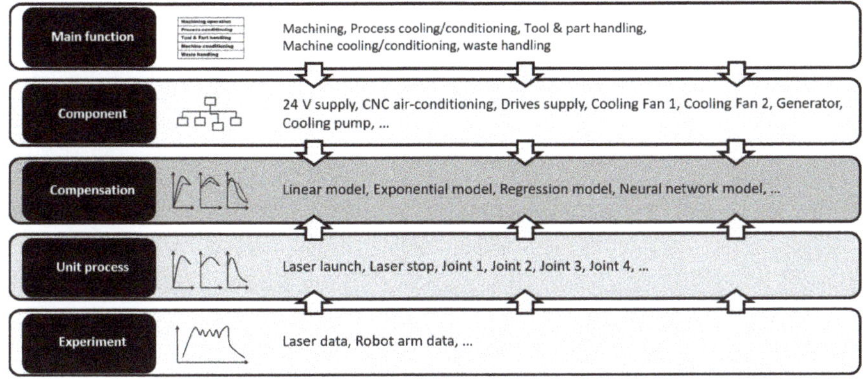

Figure 3. Compensation between the functional approach and the process approach.

3.2.1. Functional Approach

In this section, the procedure proposed by Um et al. [1] is described. It consists of functional analysis and component decomposition. The purpose of this approach is to find major energy consumers among components and to draw the approximate energy profile. However, component specifications given by machine builders provide insufficient information to estimate the energy profile of the specific product.

- Function level: This approach classifies functions into predefined general parts that consist of machine operation, workpiece manipulation, process conditioning, tool handling and die change, machine cooling and heating, and recyclables and waste handling. In the case of RLW, the machine operation function is laser generation.
- Component level: RLW consists of a 24-V power supply, CNC air-conditioning, drive power supply, two cooling fans, a generator, and a cooling pump, which are described in the e-scheme of RLW.

Each component was assigned to predefined functions according to ISO 14955-1 as shown in Figures 4 and 5. Representing the connected load of the component, the e-scheme was used for selecting components from the whole machine. The estimation results showed which components were significant energy consumers. The proposed functional analysis was that machining and cooling conditioning were major consumers. Physically, machining functions were conducted by a robot arm and laser generator. For cooling conditioning, the chiller was the major part. The sub-components of major consumers were joint motors, laser diodes, laser disks, and the water compressor. The laser scanner of the robot arm was a relatively minor consumer compared with joint motors.

CD Laser C5G (COMAU)	Machine operation (machining process, motion and control)	Process conditioning and cooling	Workpiece handling	Tool handling, or die change	Recyclables and waste handling	Machine cooling/heating
24 V supply	100%					
CNC air-conditioning						100%
Drives supply	100%					
Cooling Fan 1					100%	
Cooling Fan 2					100%	
Generator	100%					100%
Cooling pump						100%

Figure 4. Cluster matrix for functional analysis according to ISO 14955 [19].

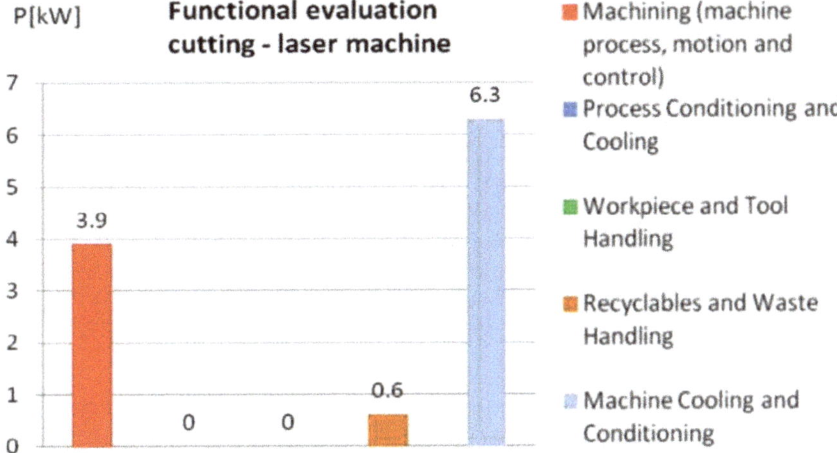

Figure 5. Functional analysis according to ISO 14955 with RLW [19].

The structure of the laser source influences the energy produced by laser generation. For the case study, a Trumpf TruDisk 4002 was used. This is a high-power laser source concentrating 176 low-power diode laser sources through four disks [20]. A single stack consisted of 11 diode bars generating 100 W of laser power. Sixteen stacks transferred the laser into 4fourserial disks chilled by the coolants. The overall structure is illustrated in Figure 6. Each disk was connected with the optical structure to collect small diode laser sources. Collected laser power reached 4 KW and a 1-Mfocal length. This performance enabled welding robots to cover a wide area with only laser scanner movement [21].

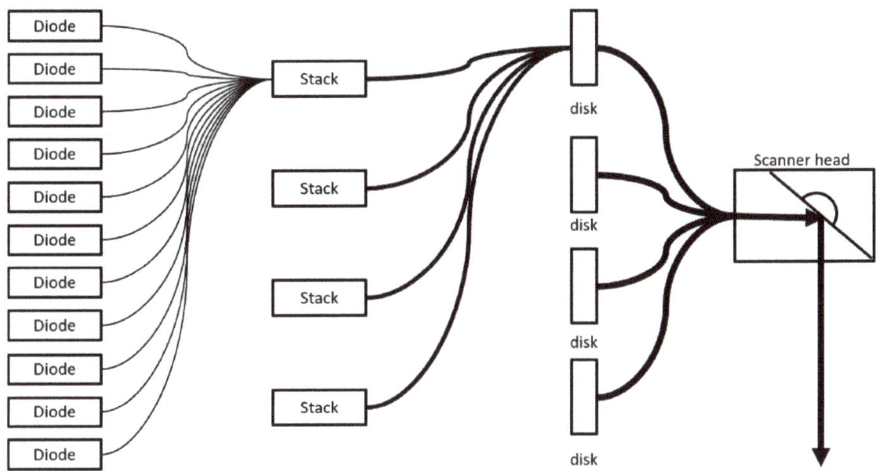

Figure 6. Structure of the laser source.

With this performance specification, the proposed energy estimation model was designed with the following assumptions.

- Each laser diode was activated when a laser machine was ready to launch. Only optical control, such as mirror blocking, was conducted without deactivating the didoes if the laser was not used.

- Otherwise, a laser machine was equipped with a chiller to cool disks by coolant. The chiller consumed more energy if laser disks concentrated more laser power than before. This means that the chiller energy increased if laser output power rose.

3.2.2. Process Approach

The functional approach started from the static characteristics of the welding machine. However, there are various factors that affect energy use, such as the number of stitches, product shape, and type of stitches, which are dynamic characteristics in specific products. The process parameters of the welding machine need to be adjusted according to these factors. The unit process model found the correlation between the energy profile and the process parameters. The objective of the process approach was to make the link between process parameters and variations of unit processes. The process approach consisted of interpreting the controller program as unit processes and of analyzing the impact of process parameters on energy profiles of unit processes.

- Parameters of unit processes: As discussed previously, process-oriented approaches are improved by using unit block models and their experimental data. The behavior of the laser machine was composed of ready state, first laser launch, and continuous laser launch and cooling. Laser output power was the parameter to be considered in energy estimation.
- Experimental data of each process: The part program was converted into a sequence of unit processes. To estimate the energy consumption, each unit process was mapped to the experimental data of the energy profile.

For this reason, compensation was necessary for filling the gap between theory and reality. For the experiments, the LabView data acquisition module from the energy meter was implemented. Preliminary experimental planning involved determining a set of sample geometries and welding conditions that could be measured to provide an initial database. Figure 7 shows the setup of the measurement device.

Collection of compensation data involved integrating an energy measurement system with the welding equipment. The system included a power cell, a data acquisition device, and an energy monitoring program. The power cell was connected with the power supply of the robot controller and the laser machine to measure the current and voltage used. The calculated power was transferred to the data acquisition device, which converted the power signal into digital values sampled at a rate of 5000 Hz. The energy monitoring program stored the resulting energy profile with a time series dataset.

3.2.3. Compensation by Machine Learning Model

Theoretical estimation is not enough to represent sensitive changes in energy profiles. A compensation step used experimental data from unit processes stored in a database. Each set of experimental data was sorted by energy profile and process parameters. The database scheme represented the relationship with component specifications and input parameters. This section investigates the various methods to draw the estimated profile. Linear regression and deep neural networks were used for developing the estimation model with the feature engineering reflecting the machine configuration of RLW.

Profile Generation Algorithm

First of all, feature generation and model development were carried out in this layer, as shown in Figure 8. Process data and measured data were combined into single tables along a time-series. Measured data were counted at 5000 Hz and converted into second units. This was necessary for calculating the duration time of each command execution.

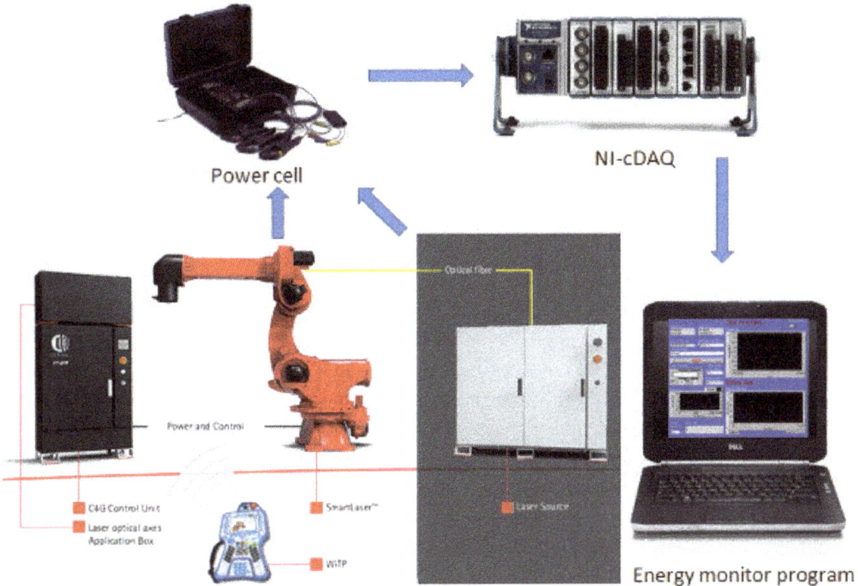

Figure 7. Setup of measurement equipment.

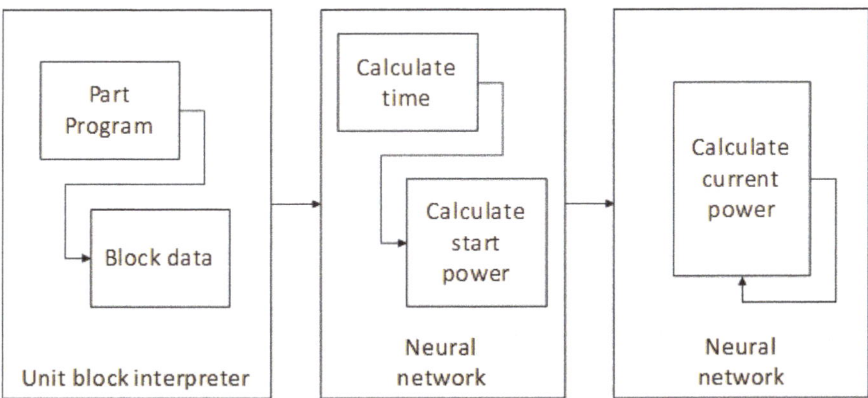

Figure 8. Profile generation algorithm.

Feature Engineering

Feature engineering is a means for utilizing the process parameters so as to estimate the detail of the energy profile by overcoming the limitation of localization. In the case of RLW, the given parameters were smaller than for the other cases of big data problems. It is necessary to focus on the specific data and find the important relationships. As mentioned in the analysis of the electric connection, cooling and generation of the laser were the major components. Laser generation showed rapid jumps in energy consumption. Cooling followed the motor behaviors to reduce the temperature of disks and diodes. The key features of laser energy consumption were the current status of the laser machine and laser control data. In order to build a common model of laser cooling, the details of the cooling system were included in the formulation. For this reason, the model followed the assumption that the electrical energy of the motor had a positive correlation with the current status of the laser machine. The given data had the current power as the current status.

As discussed in the paper by Zhou et al. [3], the localization of the deep learning model is a limitation even though its accuracy is high. The derivation of key features is needed to overcome the localized learning. The authors of this paper proposed that the following features be used as the means to train the neural network with the characteristics of laser welding: (1) mode is the feature to separate each operation of the laser controller or status of the laser machine; (2) starting time and (3) starting power: starting time and power are the factors influencing the current power by each initial status of laser launch.

Generally, a time-series dataset is applied to neural networks because of the correlation between current status and previous status. The energy profile of laser welding has fluctuations at the microscopic level from 2500 Hz–5000 Hz. This feature easily causes an over-fitting problem of the neural network. The collected data included the laser controller program. By using the controller program, the fluctuation was reduced. The procedure of feature engineering consisted of smoothing, peak detection, and synchronizing with the controller program.

The Savitzky–Golay filter was used as the smoothing filter. The filter windows were estimated by the welding time derived from the controller program. For example, the window was set as 500 Hz if each welding process time was 0.1 s. This number was the half of each peak profile. The welding energy profile followed the fourth polynomial. This was determined by the analysis of experimental data. Figure 9 shows raw data and the smoothing result.

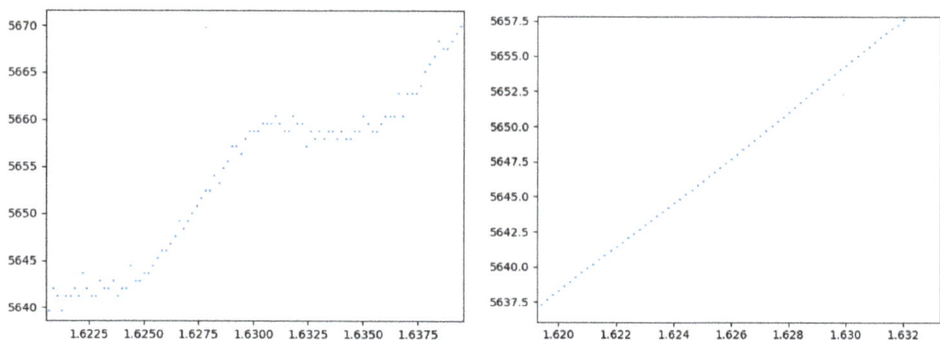

Figure 9. Raw data (**left**) and the result (**right**) after the smoothing filter.

The filtered data were applied to the peak detection algorithm, which utilized continuous wavelet transform-based pattern matching to find the highest point in each welding process. The algorithm was described by Pan et al. [22] with detail. Low peaks were collected in $\{t_{li}|i = 1...n\}$, while high peaks were stored in $\{t_{hi}|i = 1...n\}$. The measured power data were represented in $\{p(t)|t = 1...n\}$. Mode 1 used process time $\Delta t (= t - t_i)$ and starting power $P(t_{li})$ of each welding profile as the input of neural network. Mode 2 divided each welding profile into welding status and cooling status. Welding mode and cooling mode are defined in the following equation.

$$m_i(t) = \left\{ \begin{array}{c|c} 0 & \text{if } t \leq t_{hi} \\ 1 & \text{if } t > t_{hi} \end{array} \right\} \tag{1}$$

The next section describes the proposed estimation methods of energy consumption and the calculation of the energy efficiency of RLW.

Linear Regression

The first estimation model was linear regression using the modes of laser launching and cooling defined in the feature engineering section. The energy profile consisted of a series of curves. Therefore, the regression model was applied to find coefficients in order to generalize the experimental data. After trying a linear model, exponential model, and power model, the highest accuracy model was utilized as the estimation model. Each piece of the estimated profile was merged into a single profile to show the overall energy consumption.

Neural Network Model

Secondly, the proposed neural network models were trained with two mode types in order to show how much the distinction between cooling and laser launch affected the accuracy. The variations of the network model generated up to 10 nodes and three layers, as shown in Figure 10. The random seed was changed automatically. Each variation was tested with different layer configurations so as to find the best configuration.

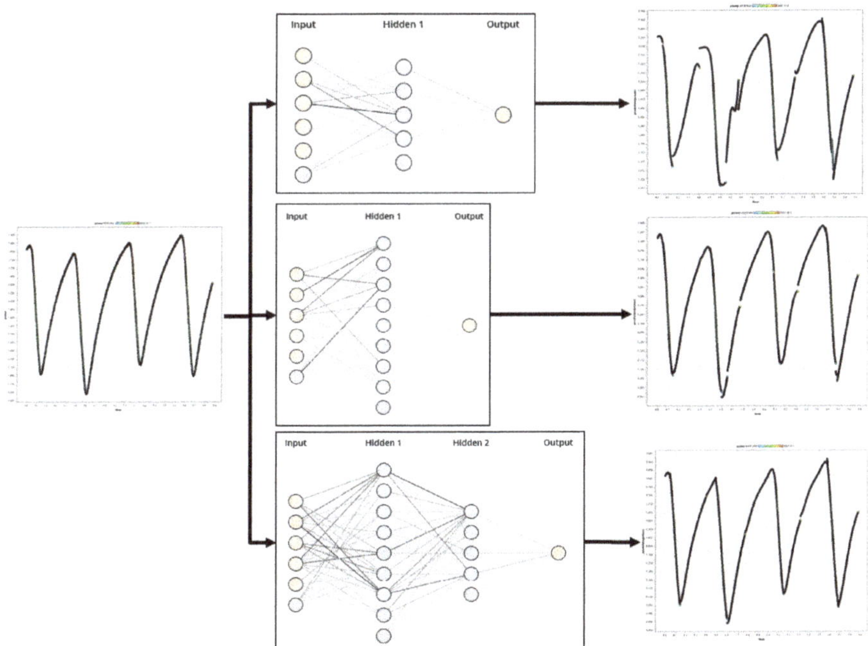

Figure 10. Variations of the neural network models.

The learning rate is one of the significant factors in neural network optimization. In a pre-test, 1000 cycles, a 0.1 learning rate, and 0.2 momentum were effective for the convergence of the network learning. Each model was compared with linear regression and compared with two different modes. The root mean squared and R squared were calculated on each model and were the criteria for selection of the optimal model to be used for energy efficiency.

In the efficiency calculation, comparison with alternative processes was needed. The welding speed of each stitch was already fixed because of the effect on welding quality. The rapid movement between stitches was the means to vary the process characteristics. The case study included variation of rapid movement.

4. Case Study

The proposed methodology described in the previous section was applied to the RLW of a car rear door assembly. This case study introduced the method for application of the integrated approach to a real product and showed how the proposed approach can improve the accuracy. The estimation model was implemented using RapidMiner 9.2. The process approach was applied to robot programming written in COMAUPDL2. Assembly of the car rear door parts involved a combination of welding positions in a set of 10 continuous stitches and 19 dimples made by a single laser welding robot with a fixed work-in-process on a single fixture (Figure 11). The COMAU SmartLaser consisted of a robot arm and laser machine. The robot arm had two parts. The first part had four revolute joints and the second part another three laser manipulators. Figure 12 shows the experimental setting of the SmartLaser and the rear door assembly on the fixture. Further details of the robot arm can be found in Erdős et al. [23].

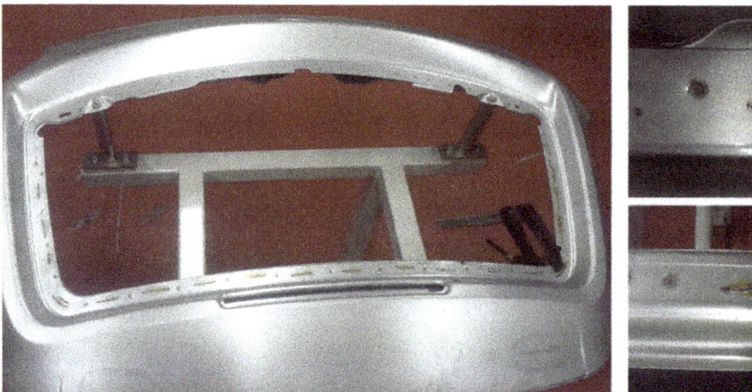
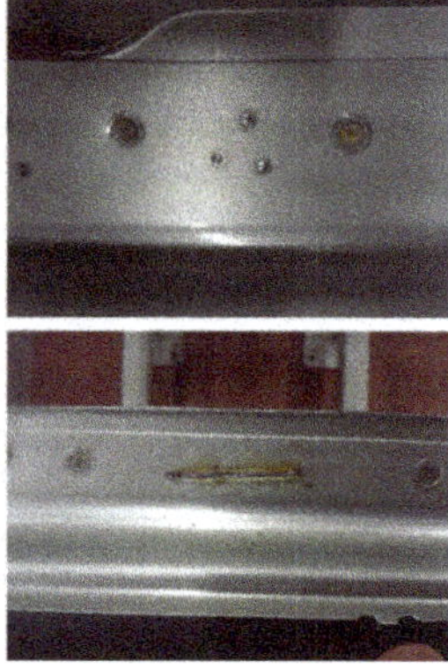

Figure 11. Target product of the case study.

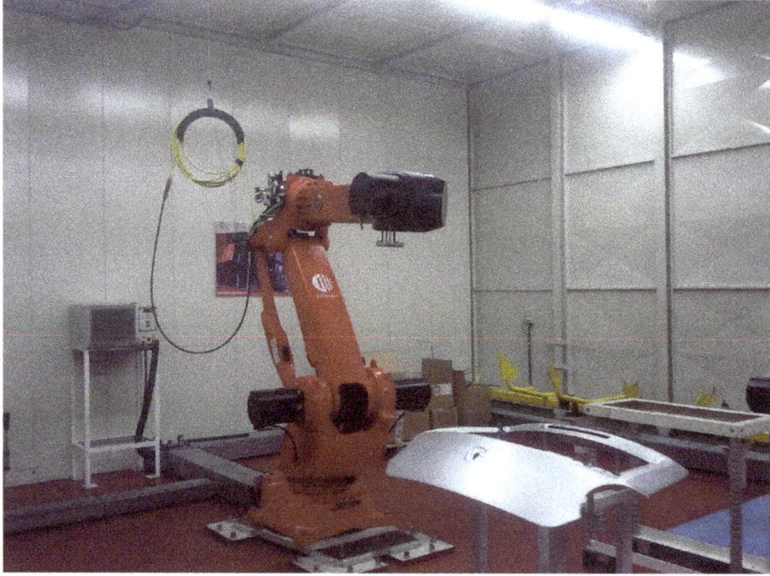

Figure 12. Experimental setting for RLW welding.

This section analyzes the results of all experiments and compares these with each other. The estimation results for energy consumption and the errors in the estimated energy profile were investigated. First of all, the estimated profile was generated with a linear formulation by using a regression model for fluctuation. A first order polynomial was used for drawing the profile. Each cooling and launching cycle was divided into decrease direction and increase direction, respectively. Linear regression was applied to find the coefficients of the exponential function on laser launch and of the log function on the cooling operation.

The second approach is the mean to utilise the neural network model to estimate energy profile with hyperbolic tangential activation function. The overall structure of single network is shown in Figure 13. the variety of the neural network which are generated iteratively from one layer to three layers with increasing of each layer from 4 nodes to 12 nodes. As mentioned in the feature engineering section, laser mode, starting time and starting power are added into the training data by derivation from the time-series power data of the training set. To compare the effectiveness of each aspect, the following four conditions are investigated.

- (1) Linear formulation generated by the chain of linear regressions cooling and launching
- (2) Profile generated by neural network trained only using mode and time
- (3) Profile generated by neural network trained with additional data, which are Starting power and starting time (mode 1)
- (4) Profile generated by neural network trained with the additional conditions on cooling and launching (mode 2)

Condition (2) did not lead to convergence of the loss function. The comparison between (1) and (3) is shown in Figures 14 and 15. The features of exponential and log functions caused the high peaks of each stitch. In the comparison between Mode 1 and Mode 2, the 5×10 two-layered network showed the highest accuracy in Mode 1, while the $10 \times 15 \times 5$ network was better in Mode 2, as shown in Figures 16 and 17 and Table 1. In the table, MAPE (Mean Absolute Percentage Error) showed the gap between the estimated value and the measured value with the following equation.

$$MAPE(Mean\,Absolute\,Percentage\,Error) = \frac{1}{N}\sum_{i=1}^{N}\frac{|E_{measure}(i) - E_{standard}(i)|}{E_{standard}(i)} \tag{2}$$

The detail of this function was described by Wang et al. [24] Both networks showed high performance in terms of drawing an accurate profile; three layers showed a fine graph, then the one of high RMS and R squared.

Table 1. Comparison with the set of neural network models.

Category	Mode 1 (5, 10)	Mode 2 (5, 10)	Mode 1 (10, 15, 5)	Mode 2 (10, 15, 5)	Measured
Energy (KWh)	5.355	5.393	5.384	5.393	5.407
Accuracy (%)	87.94	99.27	97.98	99.26	100.0
MAPE (%)	1.072	0.226	0.509	0.254	0.000

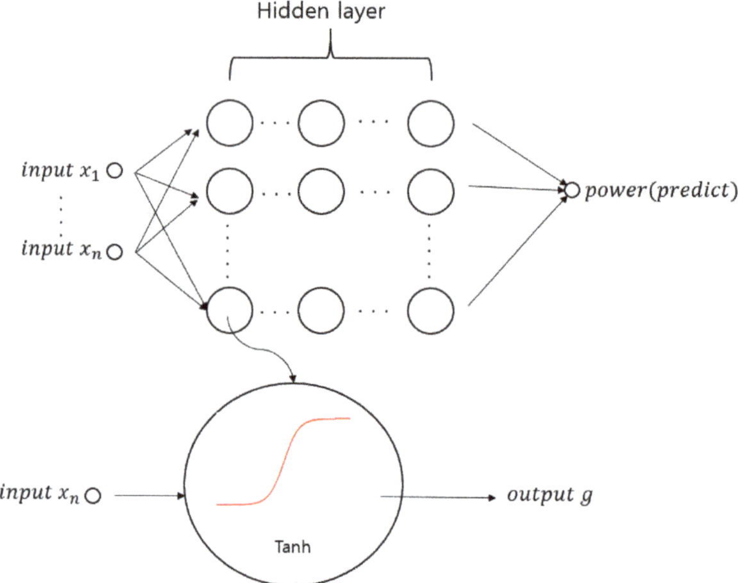

Figure 13. Neural network model.

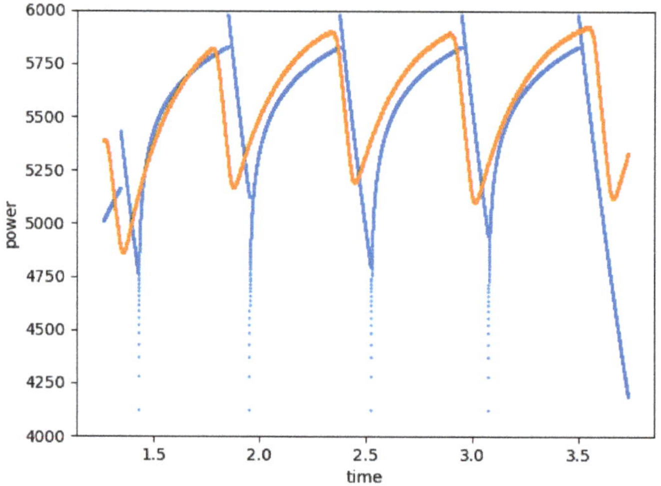

Figure 14. Profile estimated by the llinear regression.

In terms of the energy efficiency, welding speed cannot be changed because it is optimized for the material characteristics and shape of the parts being welded as a precondition. Rapid movement is the only option for changing the process parameters. Figure 18 shows the estimated energy profile generated by three rapid movements using the linear regression model. Scan speed was ten-times faster than joint motion. If rapid movements were only done by scanning, then the total efficiency could be increased. Rapid movement had a tremendous influence with low speeds, but for high speeds, the variation was small. The optimal speed of rapid movement was related to cooling time to get steady state energy. A longer time than this for movement was not necessary because it caused meaningless energy loss. If the cooling time was long, it caused long steady state and increased energy, while shorter time was better because cooling dropped exponentially.

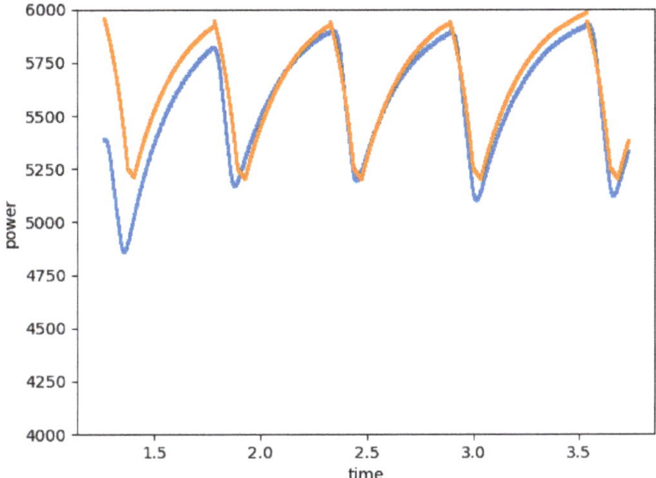

Figure 15. Profile estimated by the neural network with Mode 1.

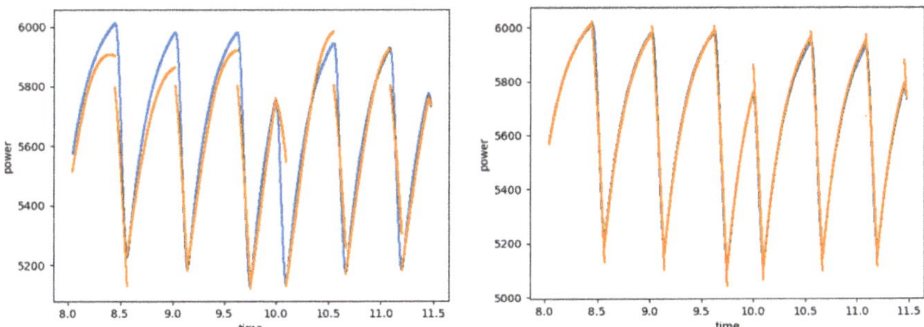

Figure 16. Comparison with Mode 1 (**left**) and Mode 2 (**right**) in two-layer networks.

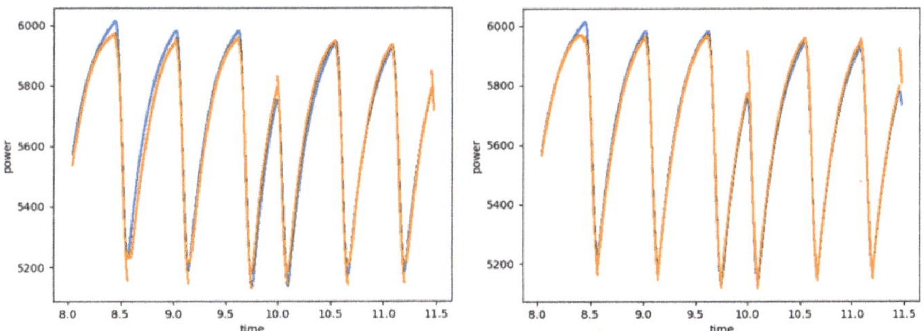

Figure 17. Comparison with Mode 1 (**left**) and Mode 2 (**right**) in three-layer networks.

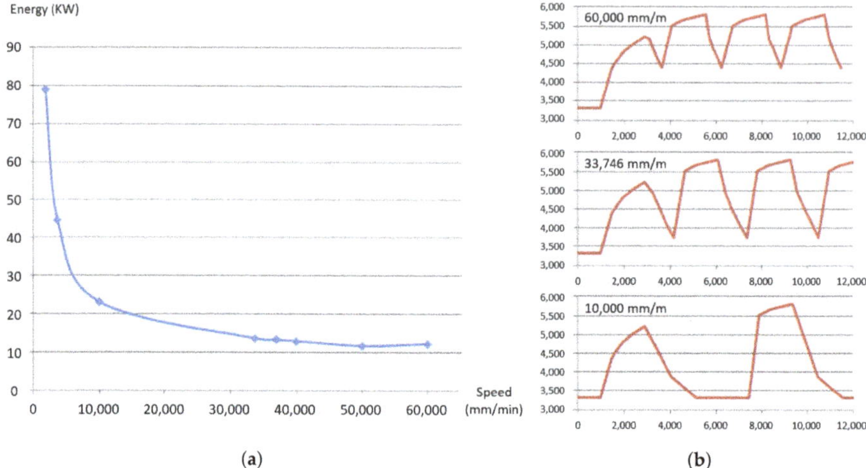

Figure 18. (**a**) Relationship between welding speed and energy use; (**b**) energy profiles for varying speeds. Comparison of the energy profile in different operating conditions.

5. Discussion

This section summarizes the major aspects of the proposed estimation model for RLW. The proposed energy estimation model involved component selection by functional analysis and machine learning compensation by using process parameters and experimental data. To conclude this paper, we highlight those aspects of the proposed energy estimation model that can produce energy profiles for a given welding program and a small amount of experimental data.

5.1. Provision of the Correlation with Laser Power and Process Time

This paper has formalized the effects of functional specification and process parameters in an energy consumption model. This is based on the fact that the laser source determines energy efficiency in the RLW processes. By providing a functional analysis and process analysis, this paper shows how energy profiles were measured and the determination of key factors that influence total energy consumption.

As shown, laser power and process time affected energy consumption in the RLW process. Both analyses showed that the laser source and the chiller used more electrical energy than the robot motors. Laser power is a significant criterion for estimating the energy profile. It is noted that low power laser welding leads to low energy efficiency. However, door materials determine the laser power and welding speed. Hence, rapid movements between welding stitches should be reduced. The integrated approach showed the correlation between laser power and process time in a single dimension. For this reason, the proposed approach is beneficial for optimization of path planning in terms of process time and energy use.

The integration of both analyses is an interesting illustration. In some areas, for example in "energy use distribution", the laser machine had a higher rate of energy consumption than the rate of the industrial robot. This was the opposite to the rate of metal cutting. The steady state energy consumption was relatively lower than for machine tools. In the "reduction of process time", the shortest rapid movement time was not the best solution to minimize the total energy use. Very long continuous welding caused an unstoppable increase in the temperature of the laser disks and, hence, an increase in cooling.

5.2. Machine Learning

Linear regression and neural networks were applied in the energy profile estimation model. The three major process parameters of metal cutting were not given in laser welding. In the laser controller part of the program, unit blocks were separated into steady state and laser launching and cooling. The unit block was the minimum criterion for training in the machine learning approach, and the starting power and starting time were counted from each laser launch to improve the accuracy of the neural network approach. The given training dataset showed the highest accuracy for a two-layered neural network. Linear regression was possible if all stitch profiles were divided into cooling and launching of the laser. However, the recurrent neural network approach showed over-fitting and was rarely trained. The profile data included small fluctuations causing disruptions in the results of the loss functions. Program-based parameters were required in the case of small datasets and were applied to the training data as time-series information to improve the learning rate. Even though front and end training data showed highly accurate results, due to the small amount of training data, a high learning rate was required. Increasing the size of training data is difficult for users of RLW. Mode 2 did not show high improvement compared with Mode 1 even if more features were applied. The dis-connectivity of each launch and cooling caused inaccuracy in finding the starting point of the shift to cooling and to launching.

6. Conclusions

This paper proposed an integrated approach merging a functional model and a process model with machine learning-based compensation. The research was done in the domain of remote laser welding, an assembly technique that has benefits for assembly in terms of energy savings. The characteristics of remote laser welding were applied to the features of the machine learning model and showed improved learning performance. This lead to optimization of performance for the process.

As described in the Discussion Section, the correlation between laser power and process power had been formalized and the use of machine learning described. The formalization led to a better understanding of the parameter settings in planning, as well as evaluation of the benefits of new equipment. This is important in order to allow manufacturers to appreciate the benefits of new equipment and processes, as well as quantifying energy savings. The machine learning aspects showed an example of the application of artificial intelligence technology to overcome the lack of formal understanding of the problem domain. The Discussion Section described the benefits and shortcomings encountered in this application domain. The paper also described the differences between the RLW application methods and those applied to metal cutting processes.

Future work is intended to adapt more features that have two cases: domain knowledge-based feature (model driven) and data science-based feature (data driven). Treatment of data was done from two perspectives. The first was the classical approach for data, which is called the data-driven method. The second method used deep learning and is powerful also for use in other fields.

Author Contributions: Conceptualization, Methodology–J.U.; Funding acquisition, Investigation, Project administration, Resources, Supervision–I.A.S.; Software, Validation & Visualization–J.U. and Y.-k.P.; Writing—original draft–J.U.; Writing—review & editing–J.U. and I.A.S.

Funding: This work has been supported by EU FP7 grant RLW Navigator No. 285051 (Remote Laser Welding System Navigator for Eco and Resilient Automotive Factories, FoF-ICT-2011.7, No. 285051) and by a grant from Kyung Hee University in 2018 (No. 20182219).

Conflicts of Interest: The authors declare no conflict of interest.

References

1. Um, J.; Stroud, I.A. Total energy estimation model for remote laser welding process. In Proceedings of the 46th CIRP Conference on Manufacturing Systems, Sesimbra, Portugal, 29–31 May 2013.
2. Sebastian, T. *Energy Efficiency in Manufacturing Systems*; Springer: Berlin, Germany, 2012.

3. Zhou, L.; Li, J.; Li, F.; Meng, Q.; Li, J.; Xu, X. Energy consumption model and energy efficiency of machine tools: A comprehensive literature review. *J. Clean. Prod.* **2016**, *112*, 3721–3734. [CrossRef]

4. Pagone, E.; Salonitis, K.; Jolly, M. Energy and material efficiency metrics in foundries. In Proceedings of the 15th Global Conference on Sustainable Manufacturing, Haifa, Israel, 25–27 September 2017.

5. Gutowski, T.; Dahmus, J.; Thiriez, A. Electrical energy requirements for manufacturing processes. In Proceedings of the 13th CIRP International Conference on Life Cycle Engineering, Leuven, Belgium, 31 May–2 June 2006.

6. Gutowski, T. The Carbon and Energy Intensity of Manufacturing. In Proceedings of the 40th CIRP International Manufacturing Systems Seminar, Liverpool, UK, 30 May–2 June 2007.

7. Gutowski, T.; Branham, M.; Dahmus, J.; Jones, A.; Thiriez, A. Thermodynamic analysis of resources used in manufacturing processes. *Environ. Sci. Technol.* **2009**, *43*, 1584–1590. [CrossRef] [PubMed]

8. Gontarz, A.; Weiss, L.; Wegener, K. Evaluation approach with function-oriented modeling of machine tools. In Proceedings of the 1st International Conference on Sustainable Intelligent Manufacturing, Leiria, Portugal, 29 June–1 July 2011.

9. Winter, M.; Herrmann, C. Eco-Efficiency of alternative and conventional cutting fluids in external cylindrical grinding. In Proceedings of the 21st CIRP Conference on Life Cycle Engineering, Trondheim, Norway, 18–20 June 2014.

10. Winter, M.; Thiede, S.; Herrmann, C. Influence of the cutting fluid on process energy demand and surface roughness in grinding—A technological, environmental and economic examination. *Int. J. Adv. Manuf. Technol.* **2015**, *77*, 2005–2017. [CrossRef]

11. Ma, F.; Zhang, H.; Hon, K.K.B.; Gong, Q. An optimization approach of selective laser sintering considering energy consumption and material cost. *J. Clean. Prod.* **2018**, *199*, 529–537. [CrossRef]

12. Zhu, Y.; Peng, T.; Jia, G.; Zhang, H.; Xu, S.; Yang, H. Electrical energy consumption and mechanical properties of selective-laser-melting-produced 316L stainless steel samples using various processing parameters. *J. Clean. Prod.* **2019**, *208*, 77–85. [CrossRef]

13. Gregory, J.; Olivares, A.; Staffetti, E. Energy-optimal trajectory planning for robot manipulators with holonomic constraints. *Syst. Control Lett.* **2012**, *61*, 279–291. [CrossRef]

14. Avram, O.I.; Xirouchakis, P. Evaluating the use phase energy requirements of a machine tool system. *J. Clean. Prod.* **2011**, *19*, 699–711. [CrossRef]

15. Duflou, J.R.; Sutherland, J.W.; Dornfeld, D.; Herrmann, C.; Jeswiet, J.; Kara, S.; Hauschild, M.; Kellens, K. Towards energy and resource efficient manufacturing: A processes and systems approach. *CIRP Ann.* **2012**, *61*, 587–609. [CrossRef]

16. Newman, S.T.; Nassehi, A.; Imani-Asrai, R.; Dhokia, V. Energy efficient process planning for CNC machining. *CIRP J. Manuf. Sci. Technol.* **2012**, *5*, 127–136. [CrossRef]

17. Marvuglia, A.; Messineo, A. Using recurrent artificial neural networks to forecast household electricity consumption. In Proceedings of the 2011 2nd International Conference on Advances in Energy Engineering, Bangkok, Thailand, 27–28 December 2011.

18. Biswas, M.R.; Robinson, M.D.; Fumo, N. Prediction of residential building energy consumption: A neural network approach. *Energy* **2016**, *117*, 84–92. [CrossRef]

19. Um, J.; Gontarz, A.; Stround, I. Developing energy estimation model based on Sustainability KPI of machine tools. *Procedia CIRP* **2015**, *26*, 217–222. [CrossRef]

20. Giesen, A.; Speiser, J. Fifteen years of work on thin-disk lasers: Results and scaling laws. *IEEE J. Quantum Electron.* **2007**, *13*, 598–609. [CrossRef]

21. Reinhart, G.; Munzert, U.; Vogl, W. A programming system for robot-based remote-laser-welding with conventional optics. *CIRP Ann. Manuf. Technol.* **2008**, *57*, 37–40. [CrossRef]

22. Du, P.K.; Warren, A.; Lin, S.M. Improved peak detection in mass spectrum by incorporating continuous wavelet transform-based pattern matching. *Bioinformatics* **2006**, *22*, 2059–2065. [CrossRef]

23. Erdős, G.; Kardos, C.; Kemény, Z.; Kovács, A.; Váncza, J. Process planning and offline programming for robotic remote laser welding systems. *Int. J. Comput. Integr. Manuf.* **2016**, *29*, 1287–1306. [CrossRef]
24. Wang, S.; Liang, Y.C.; Li, W.D.; Cai, X.T. Big Data enabled Intelligent Immune System for energy efficient manufacturing management. *J. Clean. Prod.* **2018**, *195*, 507–520. [CrossRef]

Article

Energy Saving Operation of Manufacturing System Based on Dynamic Adaptive Fuzzy Reasoning Petri Net

Junfeng Wang [1,*], Zicheng Fei [1], Qing Chang [2] and Shiqi Li [1]

[1] Department of Industrial and Manufacturing System Engineering, Huazhong University of Science and Technology, Wuhan 430074, China; feizicheng_hust@163.com (Z.F.); sqli@hust.edu.cn (S.L.)

[2] Department of Mechanical and Aerospace Engineering, University of Virginia, Charlottesville, VA 22904 USA; qc9nq@virginia.edu

* Correspondence: wangjf@hust.edu.cn; Tel.: +86-(0)27-8754-1034

Received: 6 May 2019; Accepted: 10 June 2019; Published: 11 June 2019

Abstract: The energy efficient operation of a manufacturing system is important for sustainable development of industry. Apart from the device and process level, energy saving methods at the system level has attracted increasing attention with the rapid growth of the industrial Internet of things technology, which makes it possible to sense and collect real-time data from the production line and provide more opportunities for online control for energy saving purposes. In this paper, a dynamic adaptive fuzzy reasoning Petri net is proposed to decide the machine energy saving state considering the production information of a discrete stochastic manufacturing system. Fuzzy knowledge for energy saving operations of a machine is represented in weighted fuzzy production rules with certain values. The rules describe uncertain, imprecise, and ambiguous knowledge of machine state decisions. This makes an energy saving sleep decision in advance when a machine has the inclination of starvation or blockage, which is based on the real-time production rates and level of connected buffers. A dynamic adaptive fuzzy reasoning Petri net is formally defined to implement the reasoning process of the machine state decision. A manufacturing system case is used to demonstrate the application of our method and the results indicate its effectiveness for energy saving operation purposes.

Keywords: knowledge representation; fuzzy reasoning Petri net; energy efficient operation; manufacturing system

---

1. Introduction

It has been shown that 37% of the energy consumption and 17% of carbon dioxide (CO_2) emissions come from the industrial sector [1]. Moreover, the energy consumption of industries has an annual growth rate of 1.3% from 2013 to 2025 [2]. The manufacturing industry is responsible for 38% of energy-related CO_2 emissions in China [3]. Its sustainable development requires industries to improve their energy efficiency eagerly [4,5]. The International Energy Agency reported that approximately 18–26% of the total energy consumption in manufacturing industries, i.e., 25-37 EJ, can be saved if proper actions are taken [6]. The energy-saving potential in manufacturing industries worldwide is estimated to be 20% until 2050 [7]. Therefore, energy management decisions are more and more important for manufacturing industries all over the world [8].

It is known that idle states are common in the shift time of non-bottleneck machines. The idle period took over 16% of the production time in an aircraft small-parts supplier and there was a turn-off opportunity to save at least 13% of the total energy consumption [9]. Apparently, the status of machines in a manufacturing system should be continuously monitored to make a decision. An OFF/SLEEP action could be taken for energy saving when a machine becomes idle (or tends to be idle). Later, it should be switched on or woken up to recover the production at an appropriate time slot without sacrificing the system throughput.

The development of the industrial internet of things (IoT) has endowed modern manufacturing systems with the sensible ability and real-time production data which can be collected, and provides more opportunities for energy saving operations. Currently, many new computerized numerical control (CNC) machines and robotics have energy-saving modes. However, most manufacturing execution systems have no module or function for energy saving decisions. From the system level, many works focus on the scheduling of jobs to minimize non-processing energy consumption [10]. Research on the dynamic decision of machine states for energy efficient manufacturing have gradually received significant attention from academics and industries.

The paper is organized as follows: Some related literatures are reviewed in Section 2. Section 3 describes the energy saving operations of a manufacturing system and the knowledge representation of machine state reasoning based on weighted fuzzy production rules are proposed. Section 4 proposes a dynamic adaptive fuzzy reasoning Petri net for machine state inferences to save energy considering the real-time production information. In Section 5, a serial manufacturing system is studied to demonstrate the effectiveness of our method and some discussions are also provided. Conclusions and future works are discussed in Section 6.

2. Related Works

In order to improve the energy efficiency of manufacturing systems, energy consumption control should not sacrifice the system throughput. Due to the complexity resulting from the interactions among machines and buffers, the real-time energy saving operation of manufacturing systems mainly focuses on policy methods and analytical models.

The energy saving action of machines can be defined in policies or rules. Mouzon et al. [11] gave several switch-off dispatching policies to minimize the energy consumption in a one-buffer-one-machine system. A simple policy is, e.g., machines were switched into the lower power idle mode when the idle time exceeded the defined threshold value, was used to reduce the energy consumption of idle status in [12]. In a pallet constrained flow shop, Mashaei et al. [13] described a switched-off policy to reduce energy consumption considering design constraints and two idle modes with deterministic warm-up durations. For a single machine with stochastic inter arrival times of parts, several policies, such as N-policy, upstream policy, downstream policy and upstream and downstream policy, were defined to switch off the machine [14]. Jia et al. [15] set both a higher and a lower threshold value of the buffer level to regulate the switch-on/off operations for reserve energy in Bernoulli serial lines. It is obvious that some knowledge of energy efficient production cannot be easily described quantitatively in polices. The parameters of the above control policies are hard to determine for different manufacturing systems.

A quantitative prediction of energy saving opportunities is also a method for energy efficient operation of manufacturing systems. By defining the shutdown time length of a machine without affecting the system throughput as an opportunity window, Chang et al. [16] predicted the switch-off period of a machine based on an approximate analytical model and a real-time control algorithm considering random downtime events. For serial manufacturing systems, Sun and Li [17] presented an analytical energy control opportunity estimation considering the utilization of buffers and stochastic failure of machines. They also proposed a Markov decision process model, which was solved by approximate dynamic programming to choose an optimal machine state [18]. Li et al. [19,20] estimated the energy saving window according to an event-based analysis, and a supervisory method was developed to execute opportunity windows periodically. Zou et al. [21] proposed a stochastic analytical model to calculate the machine shutdown time and recovery time based on a discrete-time Markov chain. They presented a distributed feedback production control method to improve the overall system profit and energy efficiency [22]. Hibino et al. [23] proposed an idle-time prediction model and transition model to decide a proper idle state. Analytical models for energy saving operations at a system level are usually useful to the appointed type of manufacturing systems and the reliability mode of machines. Some corrections must be made to the calculated energy opportunity window for unreliable manufacturing systems.

Fuzzy logic is an ideal method to describe human knowledge and has been widely used in the real-time control of production scheduling, planning and process control [24–26]. The fuzzy method applies to more expert production knowledge and relies less on mathematical models compared with conventional methods [27]. By decomposing a manufacturing system into three basic modules, Wang et al. [28,29] presented a fuzzy logic method to switch on/off machines considering real-time production data.

Petri nets (PNs) and its diverse variants are powerful tools for graphically/mathematically process modeling and formal verification in analysis and control of discrete manufacturing systems [30]. The PNs were used to model and evaluate energy consumption of machine tools and flexible manufacturing systems. Based on generalized stochastic Petri nets, Xie et al. [31] proposed an energy consumption model and an analysis method in order to assess the production time and energy consumption of machine tools. Pang and Le [32] built a weighted p-timed Petri net model to schedule a flexible manufacturing system and generate near-optimal energy schedules. For evaluating energy consumption, Wang et al. [33] presented an energy consumption model for machining systems based on colored timed Petri nets, where the uncertainty of task assignment and the volatility of operation time were treated by Petri net functions. Li et al. [34] used a colored timed object-oriented Petri net to model and predict hybrid energy behaviors of flexible machining systems. A small amount of literature has been found for real-time energy consumption control of machines based on the PNs method. Fei et al. [35] proposed a fuzzy Petri net model for machine ON/OFF reasoning and decisions, where the elements of the fuzzy rules were described as places and transitions. However, the fuzzy rules are static and cannot be adjusted according to real-time production conditions.

Fuzzy reasoning Petri nets (FRPNs) are extensions of the classical PNs for dealing with vague, imprecise or fuzzy information. FRPNs have been used to represent fuzzy production rules (FPRs) and formulate fuzzy rule-based reasoning automatically [36]. The main features of a FRPN are its graphical representations and dynamic processing abilities to model knowledge-based systems. By representing the operation knowledge in weighed fuzzy production rules (WFPRs), this paper presents a dynamic adaptive fuzzy reasoning Petri net (DAFRPN) to represent and infer machine states for energy saving decisions considering real-time buffer levels and production rates. Machine states are monitored and decision cycles switched on to decrease the duration of idle states and the total system energy consumption.

3. Problem Statement and Fuzzy Operation Knowledge Representation

3.1. System Description and Assumptions

For a serial discrete manufacturing system (Figure 1) in this paper, the assumptions are made in time (0,T] as follows.

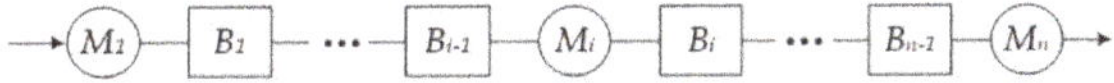

Figure 1. A multistage serial manufacturing system.

1. Machine M_i, $i = 1, 2, \ldots, n$, has a constant cycle time τ_i. The cycle time of machines can be the same or different, which means the system can be synchronous or asynchronous lines.
2. Buffer B_i, $i = 1, 2, \ldots, n-1$, has a finite buffer capacity L_i. The buffer level of B_i at time t is represented as $b_i(t)$ at the beginning of each discrete time slot.
3. The reliability models of machines are assumed to follow a known but not definitive probability distribution. The time dependent failure mode is used in this study.

4. At the beginning of a discrete time slot, M_i is starved if it is powered on and its upstream buffer B_{i-1} is empty. It is assumed that the first machine is never starved.
5. At the beginning of a discrete time slot, M_i is blocked if it is powered on and its downstream buffer B_i is full. It is assumed that the last machine is never blocked.
6. A power-on machine consumes materials, such as electricity, fuel, gas and compressed air. All consumed materials will be converted into energy.
7. In this study, a machine has following states: Processing (PS), starvation (ST), blockage (BL), failure (FL), and sleep for energy saving (SL).

In a time unit, a machine in a processing state consumes the processing energy $E_{ps,i}$. The energy consumption of traditional idle states, i.e., starvation and blockage, is described as $E_{id,i}$. A machine consumes energy $E_{sl,i}$ in the energy saving state, i.e., sleep state. It is assumed that $E_{sl,i} < E_{id,i} < E_{ps,i}$. A machine in a failure state does not consume energy. Buffers do not consume energy. The consumed energy E_i of a machine M_i in time (0,T] is calculated as:

$$E_i(T) = E_{ps,i} \times T_{ps,i} + E_{id,i} \times T_{id,i} + E_{sl,i} \times T_{sl,i}, \tag{1}$$

where $T_{ps,i}$, $T_{id,i}$, $T_{sl,i}$ are the time length of the corresponding machine state.

The overall energy consumption of a manufacturing system E_{sys} during time (0,T] is the summation of the energy consumption of all machines as follows:

$$E_{sys}(T) = \sum_{i=1}^{n} E_i(T), \tag{2}$$

Based on E_{sys}, the total energy cost of machines in a manufacturing system can be achieved. In this paper, the total energy cost of machines, the system throughput and the energy consumption per part are the main energy performance indicators of a manufacturing system.

3.2. Knowledge Representation of Machine Energy Saving Operations

The machine idle states are the main non-valuable operational states with higher energy consumption. Reducing or eliminating the idle periods by switching a machine into the energy saving state for some reasonable duration, and then waking it up, is the main objective of our work. Regarding a power-on machine, the decision of whether or not to switch it into the sleep state is evaluated and executed between two consecutive time slots based on production data. If a sleep operation is decided, the machine enters the sleep state during the subsequent time period. If a running decision is made, the machine is woken up from the sleep state or keeps its running state.

Traditionally, an exact buffer level was assigned as the threshold value for switching ON/OFF a machine that approaches either starvation (due to upstream buffer) or blockage (due to downstream buffer) [14,15]. It is difficult to decide a precise level of all buffers for energy control purposes in different manufacturing systems due to the combinatorial explosion of the buffer levels. Fuzzy production rules (FPRs) are a good tool for processing uncertain, imprecise, ambiguous real-world knowledge, and the uncertainty of the fulfillment of the conditions in rules [27]. Fuzzy knowledge representations for energy saving machine operation based on weighted FPRs (WFPRs) is presented in this study to switch a machine into the energy saving state in advance, if the machine has an inclination of starvation or blockage based on the real level of its upstream buffer, downstream buffer and production rate.

Let R_x be a set of WFPRs for machine state reasoning. Based on the definition of a general WFPR [37], a specific WFPR for the energy saving operation of a power-on machine is described in this paper as follows:

$$R_x : \text{IF } b_i(t) \text{ is } UB \text{ AND } b_{i+1}(t) \text{ is } DB \text{ THEN } s_i(t) \text{ is } SR \ (cf_i(t) = \mu_x), \ \omega_{ub,x}, \ \omega_{db,x}, \tag{3}$$

where

- $b_i(t)$ is the level of upstream buffer of M_i at time point t.
- $b_{i+1}(t)$ is the level of downstream buffer of M_i at time point t.
- UB and DB are the fuzzy sets of buffer levels which are described in 3 linguistic variables {Low, Medium, High} in this paper.
- $s_i(t)$ is the decided machine state, i.e., SR = {Sleep, Running}, at time t.
- The parameter $\mu_x = \{\mu_s, \mu_r\}$ defined in the universe of discourse [0,1] is the certainty factor (CF) representing the strength of the certainty of WFPRs when $s_i(t)$ is {sleep} and {running} respectively.
- $w_{ub,x}$ and $w_{db,x}$ are the set of weights, which are defined in the universe of discourse [0,1] and assigned to the first and second antecedent propositions in the rules. The weights show the relative importance of each antecedent proposition to the consequence proposition. The sum of the proposition weights equals to 1.

All the knowledge defined in WFPRs for energy saving state decisions of power-on machine are shown in Table 1 and explained as following.

Table 1. Weighted adaptive fuzzy production rules for machine state decisions.

No.	$b_i(t)$	$\omega_{ub,x}$	$b_{i+1}(t)$	$\omega_{db,x}$	$s_i(t)$	$cf_i(t)$
1	Low	0.8	Low	0.2	Sleep	μ_s
2	Low	0.6	Medium	0.4	Sleep	μ_s
3	Low	0.5	High	0.5	Sleep	μ_s
4	Medium	0.4	Low	0.6	Running	μ_r
5	Medium	0.5	Medium	0.5	Running	μ_r
6	Medium	0.3	High	0.7	Sleep	μ_s
7	High	0.5	Low	0.5	Running	μ_r
8	High	0.6	Medium	0.4	Running	μ_r
9	High	0.2	High	0.8	Sleep	μ_s

In the antecedent propositions of the WFPRs, six abbreviations are used as {UBL, UBM, UBH, DBL, DBM, DBH} in this paper. For example, UBL means the level of the upstream buffer is low and DBM means the level of the downstream buffer is medium. Each linguistic variable has a membership function. The triangle membership function of the buffer level is adopted in Figure 2 according to [28] and the optimization of membership function is not the focus of this paper. At each decision time point, the real level of buffers is sampled and the degree of membership is calculated. Different weighted fuzzy production rules can be activated for energy saving decisions based on the PN reasoning in the next section.

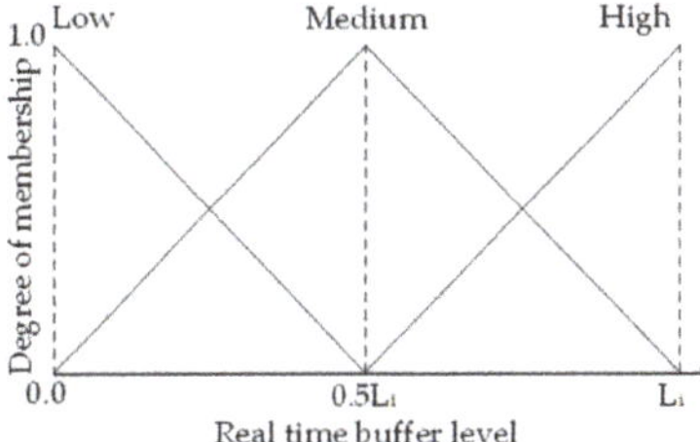

Figure 2. Membership functions of a buffer level.

Each antecedent proposition has a weight, i.e., $\omega_{ub,x}$ and $\omega_{db,x}$, which means the relative degree of importance of a proposition contributing to the consequent proposition. For example, if the level of the upstream buffer $b_i(t)$ is sampled as {Low} and the level of downstream buffer $b_{i+1}(t)$ is also sampled as

{Low} at a time point, the machine has a high-risk of starvation but a low-risk of blockage. For the energy saving purpose, the high-risk of starvation has a more important degree on a sleep decision comparing to a low-risk of blockage. It is reasonable to assign a larger weight to antecedent proposition $\{b_i(t)$ is Low$\}$ and a smaller weight to proposition $\{b_{i+1}(t)$ is Low$\}$ in the first rule shown in Table 1.

The CF in a WFPR indicates the degree of certainty of a rule. The larger the value of a certainty factor, the more the consequent proposition of a rule is believed. The traditional CF is a constant determined by an expert. Our objective is to sleep the machine and save more energy while not sacrificing the machine/system throughput. The decision rules should dynamically balance energy saving operations and product throughput between decision cycles. In this study, the CF is used to dynamically change the truth of a WFPR rule based on the production rate of a machine during the production. For example, when the production rate of a machine is low during a decision cycle, the sleep decision of a WFPR has small truth and the running decision has large truth in order to improve the machine/system throughput. The rules are defined to infer the CF of WFPRs in a decision cycle as follows:

$$R_y : \text{IF } r_i(t) \text{ is } PR \text{ THEN } \mu_s \text{ is } CVS \text{ AND } \mu_r \text{ is } CVR, \tag{4}$$

where $r_i(t)$ is the production rate of a machine within a decision cycle, PR is a fuzzy set, CVS and CVR are the certainty values of the sleep decision and the running decision respectively.

The $r_i(t)$ of a machine is zero when no parts are processed during a decision cycle. The maximum production rate will be $\frac{1}{\tau_i}$ when the machine always processes parts in a decision cycle. The production rate is calculated as follows:

$$r_i(t) = \frac{TP_i(t) - TP_i(t - dc)}{dc} \tag{5}$$

where dc is the decision cycle, $TP_i(t)$ and $TP_i(t\text{-}dc)$ is the throughput of machine i.

The PR is represented with a fuzzy term set {Low, Medium, High}. This study uses a symmetrical triangle membership function $\left[0, \frac{1}{2\tau_i}, \frac{1}{\tau_i}\right]$ for PR shown in Figure 3a. CVS and CVR take the fuzzy term set {Small, Middle, Big} with the triangle membership function $[0, 0.5, 1]$ as shown in Figure 3b,c. The rules for CF reasoning are shown in Table 2. The Mandani scheme is used during the reasoning process of μ_s and μ_r. The knowledge of rules in Table 2 is straightforward. If the production rate of a machine is higher in a decision cycle, a bigger CF for a sleep decision and a smaller CF for a running decision will be assigned to the WFPRs aiming to compromise the energy saving behavior and the machine throughput in the next decision cycle.

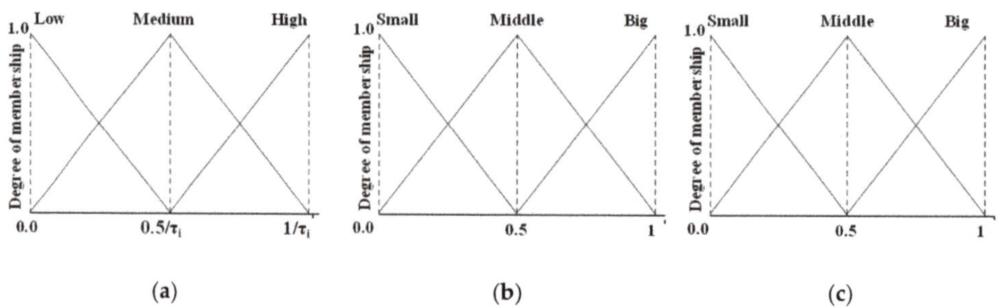

Figure 3. Membership function: (**a**) Production rate of a machine; (**b**) Certainty value of sleep decision; (**c**) Certainty value of running decision.

Table 2. The fuzzy rules for certainty factor reasoning.

No.	r_i	μ_s	μ_r
1	Low	Small	Big
2	Medium	Middle	Middle
3	High	Big	Small

4. Dynamic Adaptive Fuzzy Reasoning Petri Net for Energy Efficient Operation

4.1. Formal Definition of Dynamic Adaptive Fuzzy Reasoning Petri Net

As a graphical and mathematical modeling tool, a Petri net is widely used to describe and study information processing systems. For a rule-based reasoning process, the graphical description of a PN can visualize the structure of a rule-based system and make the model relatively simple and legible. The algebraic equations of a PN can provide a mathematical foundation to express the dynamic behavior of a system. In this section, a dynamic adaptive fuzzy reasoning PN is proposed to implement machine state reasoning for energy saving operations in a manufacturing system for the first time.

A dynamic adaptive fuzzy reasoning Petri net (Figure 4) is built for the machine states decision based on the fuzzy knowledge (Tables 1 and 2) and can be defined as a 9-tuple:

$$DAFRPN = (P,\ TN,\ I,\ O,\ D,\ W,\ \mu,\ \alpha,\ \beta) \tag{6}$$

where

- $P = \{p_1, p_2, \ldots, p_m\}$ is a finite set of propositions or called places, $P = PS \cup PT$, including a set of starting places $PS = \{p \in P |^* p = \varnothing\ and\ p^* \neq \varnothing\}$, a set of terminating places $PT = \{p \in P(^*p \neq \varnothing\ and\ p^* = \varnothing\}$. Here, $^*p$ is the set of all input transitions of p, and p^* is the set of all output transitions of p. Variable m is the number of propositions in the rules. In this paper, there are 8 places, i.e., $m = 8$. $|PS| = 6$, which represent three linguistic levels of upstream buffers and three linguistic levels of downstream buffers. $|PT| = 2$ represents the two decision states {sleep, running} of a machine.
- $TN = \{tn_1, tn_2, \ldots, tn_x\}$ is a finite set of transitions, representing a set of fuzzy inference rules. Variable x is the number of rules. In this paper, there are 9 rules and $x = 9$. Transitions $tn_1 \rightarrow tn_9$ correspond to rules $R_1 \rightarrow R_9$ shown in Table 1.
- $D = \{d_1, d_2, \ldots, d_m\}$, a finite set of propositions in the fuzzy knowledge base. $|P| = |D|$. $P \cap T \cap D = \varnothing$. For example, the proposition $d_1 = \{UBL\}$ describes the fuzzy level of upstream and the proposition $d_7 = \{Sleep\}$ is the decision result of the machine at the next decision cycle.
- $I : PS \rightarrow TN$, the input function, defining the directed arcs from starting places to transitions.
- $O : TN \rightarrow PT$, the output function, defining the directed arcs from transitions to terminating places.
- $W : P \times TN \rightarrow [0, 1]$, an input function whose elements are the weights of input places, which indicates that how much an input place (or corresponding antecedent proposition) impacts on its transition. The sum of the antecedent proposition weights of a transition is equal to 1.
- $\mu : TN \times P \rightarrow [0, 1]$, an output function whose elements are the certainty factors of the transitions/rules, indicating how much a transition/rule impacts on its output places, i.e., the decided machine state. μ is adjusted automatically based on the changing production conditions.
- $\alpha : P \rightarrow [0, 1]$, a correlation function, representing mapping from a set of places to a set of real values in [0,1]. $\alpha(P_i)$ is the fuzzy value of a starting place, indicating the truth degree of proposition d_i.
- $\beta : P \rightarrow D$, a correlation function, represents a bidirectional mapping between a set of places and a set of propositions. For example, $\beta(p_1)$ means the mapping between d_1 and {UBL}.

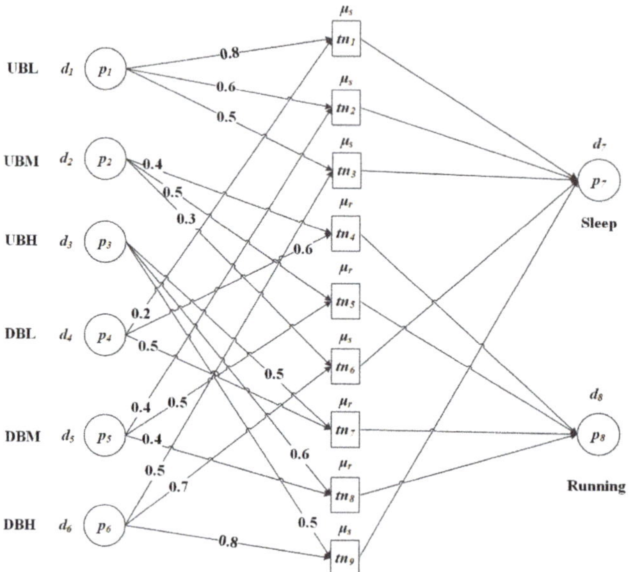

Figure 4. Dynamic adaptive fuzzy reasoning Petri net for machine states decision.

4.2. Knowledge Reasoning for Energy Saving State Decisions of A Machine

In our model, each starting place has a token. At each decision time point, the real level of the upstream and downstream buffer is collected from the manufacturing system by the IoT. The actual buffer level is converted into fuzzy linguistic values which have different fulfillment degrees, i.e., $\alpha(p_i)$, based on membership functions. Let $(tn) = \{p_i\}, i = 1, 2, \ldots 6$, $O(tn) = \{p_k\}$, $k = 7, 8$, certainty factors $\mu(tn) = \{\mu_j\}$, $j = 1, 2, \ldots, 9$, the weights of input places are represented as $w_{11}, w_{12}, \ldots, w_{ij}$.

- Enabling criterion

  $\forall tn \in TN$, transition tn is enabled for $p_i \in I(tn)$,

$$\alpha(p_i) > 0, \tag{7}$$

- Firing criterion

  When tn is fired, a token in the starting place is copied and a token with an equivalent truth value is put into the output place.

  If $\left|{}^*p\right| = 1$, the output place has only one input transition. The equivalent fuzzy value of an output place is defined by:

$$\alpha(p_k) = \mu_j \times \sum \left[\alpha(p_i) \times w_{ij}\right], \ k = 7, 8 \tag{8}$$

  If $\left|{}^*p\right| > 1$, the output place has more than one input transition. The equivalent fuzzy value of an output place is determined by the input transition with the maximum equivalent fuzzy value as follows:

$$\alpha(p_k) = max\{\mu_j \times \sum \left[\alpha(p_i) \times w_{ij}\right]\}, \ k = 7, 8 \tag{9}$$

  The reasoning process of machine states can be realized in a parallel way by matrix equation as in [38] and is not described here.

- Decision Making

After the firing of all transitions, the conclusion proposition with the larger fuzzy truth value is chosen as the decided machine state at a current decision time point. During the production process, the fuzzy reasoning process based on DAFRPN will execute at each decision time point. The machine state will be switched to running or sleep, and the idle period of a machine can be reduced to achieve energy saving operations.

5. Numerical Experiments and Discussions

5.1. A Serial Automotive Powertrain Line

In order to illustrate the feasibility and effectiveness of our proposed method, a 6M5B serial manufacturing system is taken as a case study for comparison with the control method in [22,29]. This system is a simplified version of a real automotive powertrain production line and parameters are mocked up for confidence. The reliability models of machines are assumed as exponential distribution. The power rate is assumed to be 0.2 \$/kWh. The simulation is repeated twenty times and all simulations last three weeks (30240 minutes). Tables 3 and 4 show the system parameters.

Table 3. Parameters of machines in a serial manufacturing system.

Machine	M_1	M_2	M_3	M_4	M_5	M_6
MTBF (min)	5422	6301.2	11872.2	5440.2	6412.8	6250.8
MTTR (min)	130.8	208.2	409.8	279.6	205.2	250.8
Cycle time (min)	3.5	4.3	2.7	9.4	1.1	5.9
Working energy (kWh)	450	300	240	288	660	360

Table 4. Parameters of buffers in a serial manufacturing system.

Buffer	B_1	B_2	B_3	B_4	B_5
Capacity	120	150	160	50	150
Initial level	70	30	50	40	50

This study conducted simulation experiments in three scenarios: Baseline without machine control (S1); only one machine is controlled (S2), e.g., M_2 or M_5; multi-machines are controlled (S3), i.e., M_1, M_2, M_3, M_5, except the bottleneck machine and the last machine for maximizing the system throughput. First, the baseline scenario is simulated and the system bottleneck is identified as M_4 based on the machine blockage and starvation data using the method in [39]. The decision cycle of each machine in controlled scenarios S2 and S3 is set as five times as much as its cycle time. The system performances of three scenarios are compared in Table 5.

Table 5. Comparison of system performances in three scenarios.

Scenario	Sleep Time(min) [95% CI]	Throughput [95% CI]	Total Energy Cost [95% CI]	Energy Cost Per Part (\$)
S1	-	3168.45 [3149.72,3187.18]	225727.80 [224839.67,226615.94]	71.24
S2 (M_2)	14021.62 [13877.85,14165.38]	3168.45 [3149.72,3187.18]	211707.56 [210800.94,212614.18]	66.82
S2 (M_5)	15468.38 [15346.97,15589.78]	3167.05 [3148.41,3185.69]	191700.72 [190945.86,192455.59]	60.53
S3	73722.71 [73323.99,74121.42]	3166.55 [3147.81,3185.29]	123747.56 [123055.63,124439.49]	39.08

From Table 5, the machines have notable sleep time in S2 and S3. For single machine control, the sleep time of M_2 (14021.62 min) and M_5 (15468.38 min) is about 46.37% and 51.15% of their total work time (30240 min). When multi-machines are controlled simultaneously, the total sleep time of M_1, M_2, M_3, M_5 (73722.71 min) is 60.95% of the four controlled machines' work time. The system throughput has negligible loss in all control scenarios with distinct energy cost savings. Compared with energy cost in S1, the system energy cost reduces by 6.21%, 15.07%, 45.18% in S2(M_2), S2(M_5) and S3 when different machines are controlled. Regarding the energy cost per part, all controlled scenarios also have observable cost savings. The energy cost per part decreases from 71.24 $ in S1 to 39.08 $ in S3, which has a 45.14% cost reduction. The results show that the controlled scenarios are effective energy saving operations of the manufacturing system.

5.2. Discussions

The purpose of machine control is to sleep the machine when it has the tendency for starvation or blockage without sacrificing the machine and system throughput. The performances of M_2 and M_5 are shown in Figure 5 for the three scenarios. The throughput of M_2 only slightly reduces approximately by 0.16% and 1.8% in S2 and S3. The energy cost of M_2 dramatically reduces from 29497.95 (S1) to 15476.69 (S2) and 14043.98 (S3), which save 47.53% and 52.39% of the energy cost respectively. The M_5 has negligible throughput loss, i.e., 0.04% and 0.05% of S1, in controlled scenarios, which means the control method does not affect the productivity of M_5. The machine M_5 achieves a 52.34% energy cost reduction in S2 and a 65.73% energy cost reduction in S3 compared with S1. From the single machine aspect, the control method has a remarkable effect for energy saving operations.

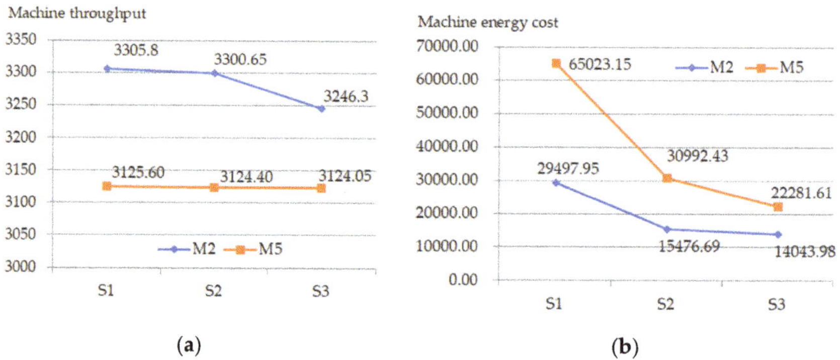

Figure 5. The performances of machine M_2 and M_5 in three scenarios: (**a**) Throughput of machines; (**b**) Energy cost of machines.

Table 6 shows the time length of different states, which are throughput loss (TPL) and energy cost reduction (ECR) of each machine in S1 and S3. It can be seen that the processing time of each machine in S3 has a slight reduction compared with S1, which results in a corresponding loss of the machine throughput. When multi-machines are controlled, the blockage and starvation time of machines are obviously decreased and are changed into sleep time. In S1, the total idle time of M_3 is 20997.81 minutes. In S3, machine M_3 has the longest sleep time, i.e., 20676.43 minutes, which reaches 98.47% conversion of the idle time and accounts for 68.37% of the total simulation time. Among the controlled machines, M_1 has the largest throughput loss percentage (2.29%) in S3. The throughput of bottleneck M_4 statistically keeps the same, which indicates the multi-machine control has no influence on the bottleneck machine M_4. Each controlled machine has gained over 50% of energy cost savings, with only 0.06% system throughput loss.

Table 6. Time length of machine states and machine performances comparison in S1 and S3.

Time Length	Scenario	M_1	M_2	M_3	M_4	M_5	M_6
PS	S1	11744.60	14214.94	8615.97	29008.40	3438.16	18693.86
	S3	11475.28	13959.09	8563.86	29007.46	3436.46	18682.65
BL	S1	17894.49	15283.01	20841.81	6.26	0.00	0.00
	S3	0.00	84.89	373.36	7.20	0.00	0.00
ST	S1	0.00	0.00	156.00	0.00	26117.82	10641.95
	S3	0.00	0.00	0.00	0.00	6691.55	10655.03
FL	S1	600.91	742.05	626.21	1225.34	684.02	904.19
	S3	600.15	742.21	626.35	1225.34	684.10	902.33
SL	S3	18164.58	15453.81	20676.43	0.00	19427.90	0.00
TPL	%	2.29%	1.80%	0.60%	0.00%	0.05%	0.06%
ECR	%	61.28%	52.39%	69.82%	0.00%	65.73%	0.01%

During the simulation, this study recorded the main parameters at each decision time point. Figure 6 is a randomly sampled experiment time phase in S3 from 22512 minutes to 24591 minutes, which shows the state changing, the production rate and the level of related buffers for M_2.

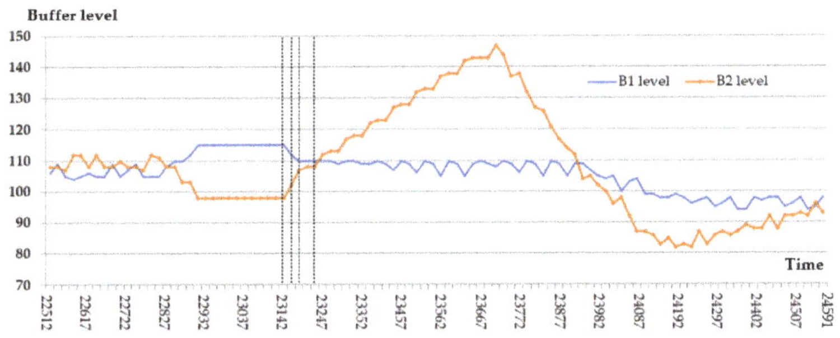

(**a**)

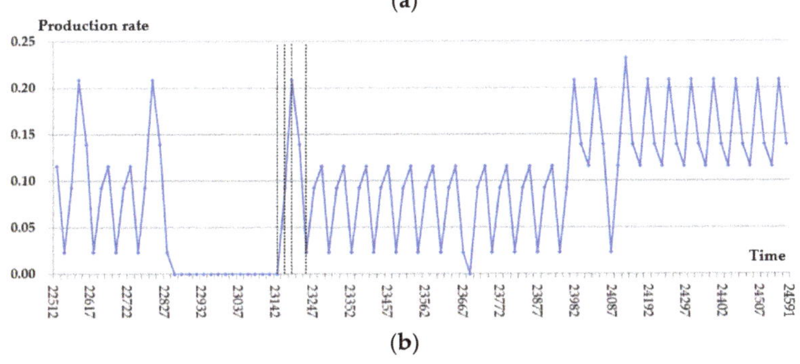

(**b**)

Figure 6. *Cont.*

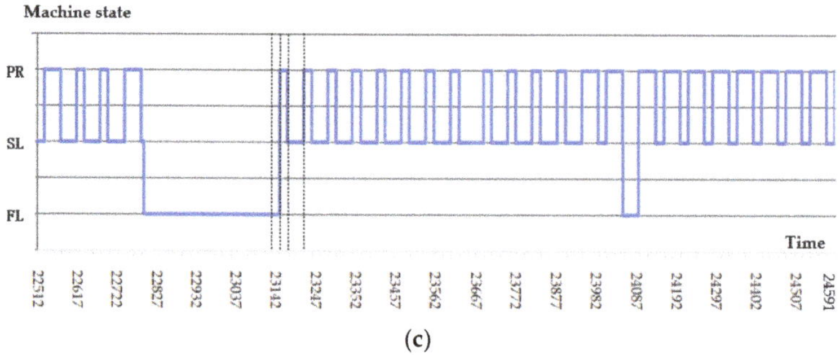

Figure 6. A random sampled simulation in S3 at each decision time point: (**a**) Level of buffer before/after M_2; (**b**) Processing rate of M_2; (**c**) State change of M_2.

From Figure 6a, it can be concluded that the buffer, B_1 and B_2, are not empty or full when multi-machines are controlled in S3. Consequently, the blockage/starvation time of M_1, M_2 and M_3 are greatly decreased as shown in Table 6. Due to the sleep and failure times, the production rate of M_2 is changed accordingly in Figure 6b. Based on the buffer level and production rate, the state of M_2 is decided using the DAFRPN. For example, at time point 23142, the machine is still at the failure state with zero production rate and buffer level $\{b_1 = 115, b_2 = 98\}$. Twenty-one minutes later, i.e., 23163 minutes, the production rate is 0.0930 and the buffer level is $b_1 = 112$, $b_2 = 102$. The decided state of M_2 is running, as shown in Figure 6c. The machine continually processes parts until the next decision time point, 23184. The production rate is 0.2093 and the buffer level is $b_1 = 110$, $b_2 = 107$. The state of M_2 will be switched as sleep based on the DAFRPN. After two decision intervals, i.e., 23226 minutes, the machine will be woken up with the production rate 0.0233 and buffer level $\{b_1 = 110, b_2 = 108\}$. It reveals that M_2 is turned into a sleep state if it has the tendency of blockage or starvation with higher production rates.

In order to investigate the performance of our dynamic adaptive fuzzy reasoning Petri net method, the results of the mathematical method of energy saving opportunity windows [22] and the fuzzy control method [29] for the above case line are compared in Table 7. Our method has the least throughput loss when a single machine and multi-machines are controlled. The energy cost reduction in S3 of our method is higher than [22] but lower than [29]. Compared with the fuzzy control method in [29], our method has less parameters to be set for control. There are many threshold values that must be adjusted for better control performance [29], which is time consuming and not practical in actual application. The mathematical method in [22] assumed that there was a slow machine in the line which meant their method is only applicable to asynchronous lines. Our method has no restriction on machine cycle time and can be used for both synchronous and asynchronous manufacturing systems. If it is assumed the profit per part produced is 300 $ and 600 $ respectively as in [22], the profit increase percentage of different methods considering throughput loss and energy cost can be calculated (Table 7). It is obvious that the increase percentage of a system profit will decrease owing to the reduced throughput with higher prices for parts. As the method in this study has the least throughput loss, it can be seen that the profit increase percentage of this method has less reduction when produced parts are more valuable.

Table 7. System performance comparison of different methods.

Methods	[22]	M_5		M_1, M_2, M_3, M_5	
		[29]	Our	[29]	Our
Throughput loss (%)	2.70%	0.69%	0.04%	0.23%	0.06%
Energy cost reduction (%)	27.00%	25.48%	15.07%	51.76%	45.18%
Energy cost reduction per part (%)	24.98%	24.95%	15.03%	51.66%	45.14%
Increase of system profit (%) (300 $)	5.15%	7.07%	4.64%	15.90%	13.99%
Increase of system profit (%) (600 $)	0.68%	2.67%	1.98%	6.74%	6.02%

The system performances of different decision cycles in S3 are compared in Table 8 with the same WFPR weights. Decision cycles are set as two times, five times and ten times as much as the machine cycle time respectively. From Table 8, it can be found that the system throughput do not have an obvious change under different decision cycles. When a longer decision cycle is adopted, the total sleep time gets smaller, and the total energy cost and the energy cost per part becomes larger. A longer decision interval means a less frequent data collection from the workshop, which makes it harder to catch the transient state of the system, thus missing some energy saving chances to reduce idle time. A smaller decision interval can catch much more energy saving opportunities but with a frequent switch between a sleep state and processing state. Based on the cycle time of machines and the results of simulation experiments, an appropriate decision cycle interval should be determined.

Table 8. System performance comparison of different decision cycles in S3.

Decision Cycle (Times)	Sleep Time (min) [95% CI]	Throughput [95% CI]	Total Energy Cost [95% CI]	Energy Cost Per Part ($)
2	78560.86 [78103.31,79018.40]	3166.85 [3148.16,3185.54]	113726.48 [113025.77,114427.20]	35.91
5	73722.71 [73323.99,74121.42]	3166.55 [3147.81,3185.29]	123747.56 [123055.63,124439.49]	39.08
10	66122.12 [65619.01,66625.22]	3166.75 [3148.06,3185.44]	136515.89 [135625.89,137405.89]	43.11

The selection of different input weight combinations will influence the effects of the energy saving method. Many groups of input weights have been adopted in our simulation experiments, from which three groups chosen as examples (G1, G2, G3) and the system performances of these three groups are shown in Table 9. The decision cycle is five times as much as the cycle time in these three experiments. It can be seen that different input weight groups all obtain the energy cost reduction and have certain effects on system performances. The optimal combination of input weights can be determined based on the results of simulation experiments and the actual situation of the shop floor in the future work.

Table 9. System performance comparison of input weights group in S3.

Input Weights Group [1]	Sleep Time (min) [95% CI]	Throughput [95% CI]	Total Energy Cost [95% CI]	Energy Cost Per Part ($)
G1	73722.71 [73323.99,74121.42]	3166.55 [3147.381,3185.29]	123747.56 [123055.63,124439.49]	39.08
G2	73667.62 [73266.93,74068.31]	3166.55 [3147.81,3185.29]	123793.75 [123097.26,124490.24]	39.09
G3	60671.83 [60280.04,61063.62]	3167.35 [3148.63,3186.07]	152435.80 [151310.40,153561.20]	48.13

[1] G1 use the input weights in Table 1. In G2, the input weight $(\omega_{ub,x}, \omega_{db,x})$ of rule No. 1,6,9 in Table 1 is re-assigned as (0.7,0.3),(0.4,0.6), (0.3,0.7). In G3, the input weight $(\omega_{ub,x}, \omega_{db,x})$ of rule No. 2,4,8 in Table 1 is re-assigned as (0.7,0.3),(0.3,0.7), (0.7,0.3).

In this paper, the authors do not consider the warm-up energy from a sleep state to a running state in order to compare the results with the literature [22,29]. The proposed method can be used when a warm-up process is taken into account. In our method, the elements of the DAFRPN, such as the amount of places and arcs, can be varied if the granularity of the buffer level in linguistic values and the number of decided machine states are changed. For example, the buffer level can be described in five grads {Very low, Low, Medium, High, Very high} and the machine can have three decided states {Light sleep, Deep sleep, Running}. The DAFRPN can be adjusted conveniently based on the newly defined WFPRs of the energy saving operation knowledge, which will have twenty-five rules. There will be ten starting places and three terminating places in the DAFRPN. This makes the DAFRPN method more flexible for different control granularity.

6. Conclusions

Energy saving operations of manufacturing systems is a popular research topic in academia and industry for sustainable development. Due to the stochastic and dynamic properties of manufacturing systems, the real-time energy saving operations considering production conditions is a difficult problem at a system level. A control method based on dynamic adaptive fuzzy reasoning Petri nets is proposed to make a decision on machine states for reducing idle times. Weighted fuzzy production rules with certain values are used to describe fuzzy knowledge of machine state decisions considering online production information. The real-time production rate and buffer levels are formatted as linguistic fuzzy sets to represent the imprecise knowledge of machine operations. The fuzzy knowledge describes the purpose for energy saving operation, i.e., turning the machine to sleep if the machine has an inclination of starvation or blockage during the production process. A dynamic adaptive fuzzy reasoning Petri net is formally defined in this paper to implement reasoning processes of machine state decisions. A serial automotive powertrain line is taken to conduct simulation experiments. The results of three scenarios show the effectiveness and flexibility of our method for energy saving manufacturing system operations. For future works, the energy saving operation method based on dynamic adaptive fuzzy reasoning Petri nets will be extended and applied to manufacturing systems with parallel, assembly or disassembly structures.

Author Contributions: The authors contributed equally to this work.

Funding: This research was funded by National Natural Science Foundation of China grant number No. 71571075.

Acknowledgments: Authors thank the anonymous reviewers for the valuable comments.

Conflicts of Interest: The authors declare no conflicts of interest.

Abbreviations and Nomenclature

Abbreviations	Description
PNs	Petri nets
FRPNs	Fuzzy reasoning Petri nets
FPRs	Fuzzy production rules
WFPRs	Weighed fuzzy production rules
DAFRPN	Dynamic adaptive fuzzy reasoning Petri net
PS	Processing state
ST	Starvation state
BL	Blockage state
FL	Failure state
SL	Sleep state
CF	Certainty factor
UBL	The level of the upstream buffer is low

DBM	The level of the downstream buffer is medium
CI	Confidence interval
TPL	Throughput loss
ECR	Energy cost reduction
PS	Starting places
PT	Terminating places
τ_i	Cycle time of M_i,
L_i	Buffer capacity of B_i
$b_i(t)$	Buffer level of B_i at time t
$E_{ps,i}$	Energy consumption of processing state
$E_{id,i}$	Energy consumption of idle state
$E_{sl,i}$	Energy consumption of sleep state
$T_{ps,i}$	Time length of processing state
$T_{id,i}$	Time length of idle state
$T_{sl,i}$	Time length of sleep state
$E_i(T)$	Energy consumption of M_i in time (0,T]
$E_{sys}(T)$	Energy consumption of the system during time (0,T]
$b_i(t)$	Level of the upstream buffer of M_i at time point t
$b_{i+1}(t)$	Level of the downstream buffer of M_i at time point t
$s_i(t)$	Decided machine state
μ_s	Certainty factor of WFPR when $s_i(t)$ is sleep
μ_r	Certainty factor of WFPR when $s_i(t)$ is running
$w_{ub,x}$	Weight of the first antecedent proposition in the rule
$w_{db,x}$	Weight of the second antecedent proposition in the rule
$r_i(t)$	Production rate of M_i within decision cycle
dc	Decision cycle
$TP_i(t)$	Throughput of M_i at time point t

References

1. Park, C.W.; Kwon, K.S.; Kim, W.B.; Min, B.K.; Park, S.J.; Sung, I.H.; Seok, J. Energy consumption reduction technology in manufacturing–a selective review of policies, standards, and research. *Int. J. Precis. Eng. Manuf.* **2009**, *10*, 151–173. [CrossRef]
2. Energy Information Administration. Available online: http://www.eia.gov/outlooks/aeo/pdf/0383(2015).pdf (accessed on 20 December 2016).
3. Zhang, X.P.; Cheng, X.M. Energy consumption, carbon emissions, and economic growth in China. *Ecol. Econ.* **2009**, *68*, 2706–2712. [CrossRef]
4. Thomas, Z.; Charlotte, M. Evaluating the management system approach for industrial energy efficiency improvements. *Energies* **2016**, *9*, 774.
5. Lawrence, A.; Thollander, P.; Andrei, M.; Karlsson, M. Specific energy consumption/use (SEC) in energy management for improving energy efficiency in industry: Meaning, usage and differences. *Energies* **2019**, *12*, 247. [CrossRef]
6. International Energy Agency. Available online: http://indiaenvironmentportal.org.in/files/Indicators_2008.pdf (accessed on 20 October 2008).
7. Gutowski, T.G.; Allwood, J.M.; Herrmann, C.; Sahni, S. A global assessment of manufacturing: Economic development, energy use, carbon emissions, and the potential for energy efficiency and materials recycling. *Annu. Rev. Environ. Resour.* **2013**, *38*, 81–106. [CrossRef]
8. Patrik, T.; Jenny, P. Industrial energy management decision making for improved energy efficiency—strategic system perspectives and situated action in combination. *Energies* **2015**, *8*, 5694–5703.
9. Twomey, J.; Yildirim, M.; Whitman, B.L.; Liao, H.; Ahmad, J. *Energy Profiles of Manufacturing Equipment for Reducing Energy Consumption in A Production Setting. Working Paper*; Wichita State University: Wichita, KS, USA, 2008.
10. Peng, C.; Peng, T.; Zhang, Y.; Tang, R.; Hu, L. Minimising non-processing energy consumption and tardiness fines in a mixed-flow shop. *Energies* **2018**, *11*, 3382. [CrossRef]

11. Mouzon, G.; Yildirim, M.B.; Twomey, J. Operational methods for minimization of energy consumption of manufacturing equipment. *Int. J. Prod. Res.* **2007**, *45*, 4247–4271. [CrossRef]
12. Prabhu, V.V.; Jeon, H.W.; Taisch, M. Modeling green factory physics-an analytical approach. In Proceedings of the 8th IEEE International Conference on Automation Science and Engineering, Seoul, Korea, 20–24 August 2012.
13. Mashaei, M.; Lennartson, B. Energy reduction in a pallet-constrained flow shop through on–off control of idle machines. *IEEE Trans. Autom. Sci. Eng.* **2013**, *10*, 45–56. [CrossRef]
14. Frigerio, N.; Matta, A. Energy-efficient control strategies for machine tools with stochastic arrivals. *IEEE Trans. Autom. Sci. Eng.* **2015**, *12*, 50–61. [CrossRef]
15. Jia, Z.; Zhang, L.; Arinez, J.; Xiao, G. Performance analysis for serial production lines with Bernoulli machines and real-time WIP-based machine switch-on/off control. *Int. J. Prod. Res.* **2016**, *54*, 1–17. [CrossRef]
16. Chang, Q.; Xiao, G.; Biller, S.; Li, L. Energy saving opportunity analysis of automotive serial production systems. *IEEE Trans. Autom. Sci. Eng.* **2013**, *10*, 334–342. [CrossRef]
17. Sun, Z.; Li, L. Opportunity estimation for real time energy control of sustainable manufacturing systems. *IEEE Trans. Autom. Sci. Eng.* **2013**, *10*, 38–44. [CrossRef]
18. Li, L.; Sun, Z. Dynamic energy control for energy efficiency improvement of sustainable manufacturing systems using Markov decision process. *IEEE Trans. Syst. Man Cybern. Syst.* **2013**, *43*, 1195–1205. [CrossRef]
19. Li, Y.; Chang, Q.; Ni, J.; Brundage, M.P. Event-based supervisory control for energy efficient manufacturing systems. *IEEE Trans. Autom. Sci. Eng.* **2018**, *15*, 92–103. [CrossRef]
20. Li, Y.; Wang, J.; Chang, Q. Event-based production control for energy efficiency improvement in sustainable multistage manufacturing systems. *ASME J. Manuf. Sci. Eng.* **2018**, *141*. [CrossRef]
21. Zou, J.; Arinez, J.; Chang, Q.; Lei, Y. Opportunity window for energy saving and maintenance in stochastic production systems. *Trans. ASME J. Manuf. Sci. Eng.* **2016**, *138*, 121009. [CrossRef]
22. Zou, J.; Chang, Q.; Arinez, J.; Xiao, G.X. Data-driven modeling and real-time distributed control for energy efficient manufacturing systems. *Energy* **2017**, *127*, 247–257. [CrossRef]
23. Hibino, H.; Yanaga, K. Decision support for energy-saving idle production facility operations in a production line based on an M2M environment. *Procedia CIRP* **2017**, *61*, 399–403. [CrossRef]
24. Azadegan, A.; Porobic, L.; Ghazinoory, S.; Samouei, P.; Kheirkhah, A.S. Fuzzy logic in manufacturing: A review of literature and a specialized application. *Int. J. Prod. Econ.* **2011**, *132*, 258–270. [CrossRef]
25. Tsourveloudis, N.C.; Doitsidis, L.; Ioannidis, S. Work-in-process scheduling by evolutionary tuned fuzzy controllers. *Int. J. Adv. Manuf. Technol.* **2007**, *34*, 748–761. [CrossRef]
26. Tamani, K.; Boukezzoula, R.; Habchi, G. Application of a continuous supervisory fuzzy control on a discrete scheduling of manufacturing systems. *Eng. Appl. Artif. Intell.* **2011**, *24*, 1162–1173. [CrossRef]
27. Precup, R.E.; Hellendoorn, H. A survey on industrial applications of fuzzy control. *Comput. Ind.* **2011**, *62*, 213–226. [CrossRef]
28. Wang, J.; Fei, Z.; Chang, Q.; Li, S.; Fu, Y. Multi-state decision of unreliable machines for energy-efficient production considering work-in-process inventory. *Int. J. Adv. Manuf. Technol.* **2019**, *102*, 1009–1021. [CrossRef]
29. Wang, J.; Fei, Z.; Chang, Q.; Fu, Y.; Li, S. Energy saving operation of multi-stage stochastic manufacturing systems based on fuzzy logic. *Int. J. Simul. Model.* **2019**, *18*, 138–149. [CrossRef]
30. Liu, H.C.; You, J.X.; Li, Z.W.; Tian, G. Fuzzy petri nets for knowledge representation and reasoning: A literature review. *Eng. Appl. Artif. Intell.* **2017**, *60*, 45–56. [CrossRef]
31. Xie, N.; Duan, M.; Chinnam, R.B.; Li, A.; Xue, W. An energy modeling and evaluation approach for machine tools using generalized stochastic petri nets. *J. Clean. Prod.* **2016**, *113*, 523–531. [CrossRef]
32. Pang, C.K.; Le, C.V. Optimization of total energy consumption in flexible manufacturing systems using weighted p-timed petri nets and dynamic programming. *IEEE Trans. Autom. Sci. Eng.* **2014**, *11*, 1083–1096. [CrossRef]
33. Wang, Q.; Wang, X.; Yang, S. Energy modeling and simulation of flexible manufacturing system based on colored timed Petri net. *J. Ind. Ecol.* **2014**, *18*, 558–566. [CrossRef]
34. Li, Y.; He, Y.; Wang, Y.; Wang, Y.; Yan, P.; Lin, S. A modeling method for hybrid energy behaviors in flexible machining systems. *Energy* **2015**, *86*, 164–174. [CrossRef]

35. Fei, Z.; Li, S.; Chang, Q.; Wang, J.; Huang, Y. Fuzzy petri net based intelligent machine operation of energy efficient manufacturing system. In Proceedings of the 14th IEEE Conference on Automation Science and Engineering (CASE), Munich, Germany, 20–24 August 2018; pp. 1593–1598.
36. Gao, M.; Zhou, M.C.; Huang, X.; Wu, Z. Fuzzy reasoning Petri nets. *IEEE Trans. Syst. Man Cybern. A* **2013**, *33*, 314–323.
37. Yeung, D.S.; Tsang, E.C.C. Weighted fuzzy production rules. *Fuzzy Sets Syst.* **1997**, *88*, 299–313. [CrossRef]
38. Liu, H.C.; Lin, Q.L.; Mao, L.X.; Zhang, Z.Y. Dynamic adaptive fuzzy petri nets for knowledge representation and reasoning. *IEEE Trans. Syst. Man Cybern. S* **2013**, *43*, 1399–1410. [CrossRef]
39. Li, J.; Meerkov, S.M.; Zhang, L. Production systems engineering: Problems, solutions, and applications. *Annu. Rev. Control.* **2010**, *34*, 73–88. [CrossRef]

Article

Life-Cycle and Energy Assessment of Automotive Component Manufacturing: The Dilemma Between Aluminum and Cast Iron

Konstantinos Salonitis, Mark Jolly *, Emanuele Pagone and Michail Papanikolaou

Manufacturing Theme, Cranfield University, Cranfield, Bedford, MK43 0AL, UK
* Correspondence: m.r.jolly@cranfield.ac.uk; Tel.: +44 (0)1234 758347

Received: 6 June 2019; Accepted: 30 June 2019; Published: 3 July 2019

Abstract: Considering the manufacturing of automotive components, there exists a dilemma around the substitution of traditional cast iron (CI) with lighter metals. Currently, aluminum alloys, being lighter compared to traditional materials, are considered as a more environmentally friendly solution. However, the energy required for the extraction of the primary materials and manufacturing of components is usually not taken into account in this debate. In this study, an extensive literature review was performed to estimate the overall energy required for the manufacturing of an engine cylinder block using (a) cast iron and (b) aluminum alloys. Moreover, data from over 100 automotive companies, ranging from mining companies to consultancy firms, were collected in order to support the soundness of this investigation. The environmental impact of the manufacturing of engine blocks made of these materials is presented with respect to the energy burden; the "cradle-to-grave approach" was implemented to take into account the energy input of each stage of the component life cycle starting from the resource extraction and reaching to the end-of-life processing stage. Our results indicate that, although aluminum components contribute toward reduced fuel consumption during their use phase, the vehicle distance needed to be covered in order to compensate for the up-front energy consumption related to the primary material production and manufacturing phases is very high. Thus, the substitution of traditional materials with lightweight ones in the automotive industry should be very thoughtfully evaluated.

Keywords: Manufacturing; energy efficiency; life-cycle assessment; aluminum; cast iron

---

1. Introduction

Over the years, the material selection for modern car components changed a lot. As a reference, in the 1970s, a design engineer would have to select from four to five sheet forming grades, whereas today there are more than 50 options [1]. A number of material selection criteria need to be considered including corrosion and wear resistance, crashworthiness, and manufacturability. At the same time, legislation pushes for lighter vehicles, on the basis that lighter cars result in lower fuel consumption. Since 1995 in Europe, the average car CO_2 emission requirement dropped from 186 g/km to 161 g/km in 2005, and it is expected to further reduce to 95 g/km in 2021 [2]. For achieving these requirements, automotive manufacturers opt to use aluminum alloys in vehicles as a "lightweight" material. The average usage of aluminum (Al) in a passenger car varies from 12% to 60% depending on the vehicle. With regard to cast Al alloys, these are mostly used for engine blocks, cylinder heads, and wheels, although they are increasingly used for nodes in the chassis structure and can potentially reduce weight by 40%.

Substituting with lower-density materials leads to lower tailpipe emissions; however, this does not consider the CO_2 footprint of the materials used in the manufacturing of vehicles. The CO_2 footprint of any material is related to its embodied energy, which is a synonym of the "track record" of a material and the way it is produced. In every production phase, energy is needed for changing the phase, geometry, and properties of the material. This energy is, thus, virtually embodied in the material. Ashby et al. [3] presented the embodied energy of producing components for the automotive industry and discussed the contribution of each life-cycle phase. According to their investigation, the energy involved during the use phase of a vehicle is much larger than that during the material extraction and manufacturing phase. Similar conclusions were reached by Sorger et al. [4] who investigated the effects of substituting an aluminum cylinder block with a newly developed one made of cast iron (CI). Their results showed that the CI engine block presents some significant advantages with respect to cost, energy savings, and CO_2 emissions.

Manufacturing process efficiency can obviously have a great impact on the energy consumption during that life-cycle phase of the vehicle. Salonitis and Ball [5] highlighted the importance of energy efficiency for both manufacturing processes and systems. One of the most energy-consuming manufacturing processes is casting (when considering all sub-processes such as melting, holding, finishing), and a lot of research is undertaken on how to improve its energy efficiency [6–10]. The casting process is used in the automotive sector for the manufacturing of a number of components both in the powertrain and in the body in white (BIW). A couple of attempts were also reported on the use of different materials for the casting of automotive components [11,12].

The objective of the present investigation is to establish a methodology for the environmental impact assessment of substitution of materials in the automotive sector so as to improve the current decision-making practices in the automotive sector. The discussion is on whether Al alloys are a better option than cast iron (CI), when the total energy burden is considered (and not only the tailpipe emissions). For assessing the energy required, an extensive literature review was undertaken, and over 100 experts from the automotive supply chain, such as original equipment manufacturers (OEMs), engine design consultancy firms, foundries, mining companies, primary alloy producers, and recycling companies, as well as machining, heat treatment, and impregnation companies, were contacted. The case study selected was the engine block, as it is the single heaviest component in most passenger cars.

2. Methodology: Assessment Approach

Focusing only on the use phase, or only on the manufacturing phase for the assessment of the overall environmental impact of a product does not allow for a full understanding of the whole picture. The "cradle-to-grave" approach aims to include the energy consumption that occurs due to resource extraction and processing, component and product assembly, use, and end-of-life processing of a vehicle (Figure 1). The evaluation of the overall impacts that a product has on the environment through all of these life-cycle stages would give a complete picture of the lightweighting shift validity.

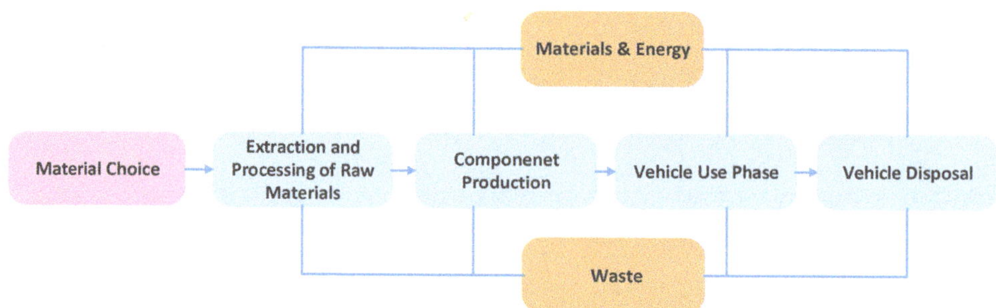

Figure 1. "Cradle-to-grave" approach.

For assessing the energy required and the CO_2 emissions in each stage of the life cycle, an extensive literature review was undertaken. The present study focused on all the processes, from cradle to grave, in the production of passenger vehicle engine blocks, such as mining, smelting and electrolysis, melting, holding, casting, fettling, heat treatment, machining, impregnation, and recycling.

3. Embodied Energy in Materials Due to Primary Production

The starting point is the calculation of the primary production energy for each type of material. For the calculation of embodied energy, the methodology proposed by Brimacombe et al. [13] is used.

3.1. Primary Aluminum Production

The production of primary aluminum requires a number of steps. Allwood and Cullen [14] suggested that, for primary aluminum, the energy required is of the order of 170 GJ/ton. The literature review indicated that energy ranges from 50 to 100 GJ/ton. Due to the ambiguity in these figures, the energy requirements were calculated theoretically; Figure 2 shows that, for one ton of primary aluminum, 98 GJ of energy is required. In the paragraphs below, the calculation of these figures is explained.

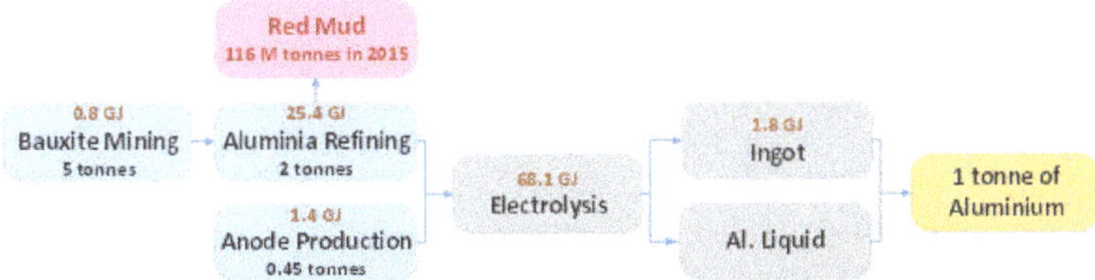

Figure 2. Primary aluminum production steps with associated energy content for producing one ton of material.

Primary production of aluminum starts with the mining of dry bauxite, which requires 0.17 ± 0.08 GJ/t. This figure was calculated after reviewing a number of reported energy figures in the literature review as listed in Table 1.

Table 1. Bauxite mining energy per ton of bauxite.

Source	Energy (GJ/t)
[15]	0.145
[16]	0.150
[17]	0.150
[18]	0.153
[19]	0.188
[20]	0.210

Alumina is refined from bauxite through the Bayer process, where the main steps are digestion, clarification, precipitation, and calcination [21]. Firstly, dry bauxite is crushed in large mills and blended with liquor to form slurry. Then, lime and caustic soda are added, mixed, and poured into the digester, where a solution of hot caustic soda dissolves the alumina. During the digestion, impurities drop to the bottom and form a solid waste residue called red mud. In order to separate the alumina from the red mud, the mix is moved to clarification. By cooling, aluminum hydroxide is precipitated from the caustic soda and then washed. The last step is calcination, where the water content in hydroxide is removed, and the alumina white powder is produced [22]. The energy consumption in this process varies in a range where the calculated average is 13.2 ± 6.4 GJ/t of alumina (Table 2).

Table 2. Alumina refining energy per ton of alumina.

Source	Energy (GJ/t)
[19]	13.17
[17]	12.52
[23]	10.65
[15]	12.77
[16]	14.20
[18]	17.90
[24]	15.00
[25]	13.80
[23]	10.95
[26]	10.65
[20]	13.82

Red mud is highly alkaline (pH = 13), having great environmental impact; thus, it is very difficult to dispose of. It represents a major problem in the primary aluminum production. Red mud disposal covers vast areas which consequently cannot be built or farmed on, even after red mud is dried after several years. The most common ways to dispose of it is by land storage in the form of lagoons, dry stacking, or dry cake [23]. Two or more tons of red mud is produced for every ton of aluminum.

The key process for producing Al is electrolysis. Alumina is dissolved in a molten cryolite to decrease the melting point of alumina. The process, known as the Hall–Heroult process after the inventors, passes an electric current through the molten alumina to dissociate it into aluminum and oxygen. The oxygen reacts with the carbon anode to produce CO_2, whilst molten aluminum remains and is tapped off periodically into teapot ladles [22]. In terms of process consumables, carbon anodes are used. A mix of calcined petroleum coke, recycled anodes butts, and coal are baked at 1150 °C to produce anodes, consuming 3.1 GJ per ton of anode. Depending on the anode use, the produced Al can be differentiated. The two main technologies are prebake (anodes are baked in ovens and then consumed in the electrolysis cells) and Soderberg (anodes are baked directly in the electrolysis cell) [23]. Furthermore, the carbon anode is totally consumed in Soderberg technologies, while, in prebake technologies, 80% is consumed and the other 20% is used again in the anode production process. In Europe, most of the electrolysis facilities use prebake technology with the exception of two Soderberg smelters placed in Spain. By calculating the average from the range of energy consumptions required for electrolysis, the process consumes approximately 54.4 ± 4.5 GJ/t of produced Al (Table 3). Also, if 80% of the total amount of carbon anode is converted into carbon dioxide, an extra energy of 14 GJ/t aluminum is added to the process [26], ending up with a total energy consumption of 68 GJ/t aluminum in the electrolysis process.

Table 3. Electrolysis energy per ton of aluminum.

Sources	Energy (GJ/t)
[27]	56
[24]	52
[28]	66
[21]	54
[29]	53
[16]	55
[20]	47
[23] (95% prebaked and 5% Soderberg)	53.6
[23] (89% prebaked and 11% Soderberg)	55.0
[24]	50
[18]	55
[26]	56

Afterward, the molten aluminum is poured into molds to solidify in different shapes, which are shipped as ingots. In some cases, liquid aluminum is transported in insulated ladles by road depending on the proximity of the foundry [18]. The average energy consumption for ingot casting from the range collected from the literature review is 1.81 ± 0.17 GJ per ton of aluminum (Table 4). Finally, by adding up all the energy consumed in all the different processes, the production of one ton of primary aluminum requires 98 GJ.

Table 4. Cast ingot energy per ton of aluminum.

Sources	Energy (GJ/t)
[26]	2.00
[18]	1.77
[20]	1.67

3.2. Pig Iron Production

Similarly, for primary iron/steel, the energy required for the production of pig iron, according to the literature review, ranges from 20 to 40 GJ/ton. Revisiting the process and calculating the energy per phases theoretically indicated that the energy content of one ton of primary iron is 17 GJ (Figure 3). In the paragraphs below, the calculation of these figures is explained.

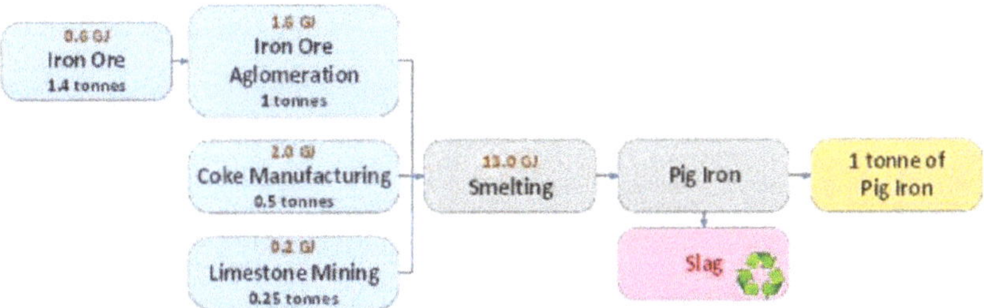

Figure 3. Primary iron production steps with associated energy content for producing one ton of material.

According to Moll et al. [30], the main raw material in pig iron production is iron ore, consuming an average energy of 0.44 ± 0.2 GJ/t of iron ore mined (Table 5). Fine iron ores are converted into lump ores before charging into the blast furnace, in a process known as iron ore agglomeration. There are two different processes of agglomeration which are used in industry: sintering and pelletizing. Sintering plants are usually located near the blast furnace site, while pelletizing plants are situated near the mines [31]. From the range of data collected, the average energy required for this process is 1.59 ± 0.36 GJ/t of iron agglomerate (Table 6).

Table 5. Iron ore mining and concentration energy per ton of iron ore.

Sources	Energy (GJ/t)
[32]	0.153
[33]	0.142
[30]	0.177
[27]	0.956
[34]	0.750

Table 6. Iron ore agglomeration per ton of iron ore agglomerated.

Sources	Energy (GJ/t)
[35]	1.70
[33]	1.50
[27]	1.37
[36]	1.60
[30] (pelletizing)	1.33
[30] (sintering)	1.55
[37] (pelletizing)	0.82
[37] (sintering)	1.54
[38] (sintering)	2.25
[34] (sintering)	1.75
[31] (pelletizing)	2.10
[31] (sintering)	1.60

Coal is converted at high temperatures to produce coke, which will provide permeability, heat, and gases which are required to reduce and melt the iron ore, pellets, and sinter [39]. The energy consumed to produce one ton of coke is approximately 3.98 ± 1.1 GJ (Table 7). In some countries like Brazil, charcoal is commonly used in the production of pig iron instead of coke.

Table 7. Coke manufacturing energy per ton of coke.

Sources	Specific Country	Energy (GJ/t)
[27]		2.19
[40]		3.70
[34]	Germany 2003	3.70
	Japan 2002	3.50
	China 2004	4.20
[35]		4.30
[37]		4.45
[22]		3.59
[36]		5.80
[38]		2.40
[31]		6.00

Finally, limestone is added in order to remove the impurities [33]. Similar to iron ore, limestone also has to be extracted from the earth, in a process that consumes close to 0.9 ± 0.5 GJ per ton (Table 8).

Table 8. Energy consumption per ton of limestone.

Sources	Energy (GJ/t)
[41]	0.964
[27]	0.848

The iron ore (lump, sinter, and/or pellets), along with additives such as limestone and reducing agents (coke), is put into the blast furnace in order to smelt. Then, a hot air blast is injected into the blast furnace. The limestone is melted to remove the sulfur and other impurities, originating a residue known as slag. This process, known as smelting, is the most energy-consuming in the production of pig iron, accounting for 13 GJ (Table 9) of the total 17.4 GJ per ton of pig iron.

Table 9. Energy consumption per ton of limestone.

Sources	Specifics	Energy (GJ/t)
[22]		16.90
[38]		13.6–16.2
[42]	Blast furnace	12.3
[40]	Blast furnace	10.4
[27]	Raw iron manufacturing	12.8
[31]	Blast furnace	13–14.1
[43]	Blast furnace	12.7–18.6
[44]		12.0
[34]	blast furnace	10.4
[45]		12.2
[36]	blast furnace	10.4
[37]		13.63

3.3. Outcome

In Figure 4, the various stages and their energy consumption for the production of one ton of pig iron and primary aluminum are shown. The difference in the total energy consumed to produce one ton of primary aluminum when compared to the production of the same amount of pig iron sums up to roughly 80 GJ.

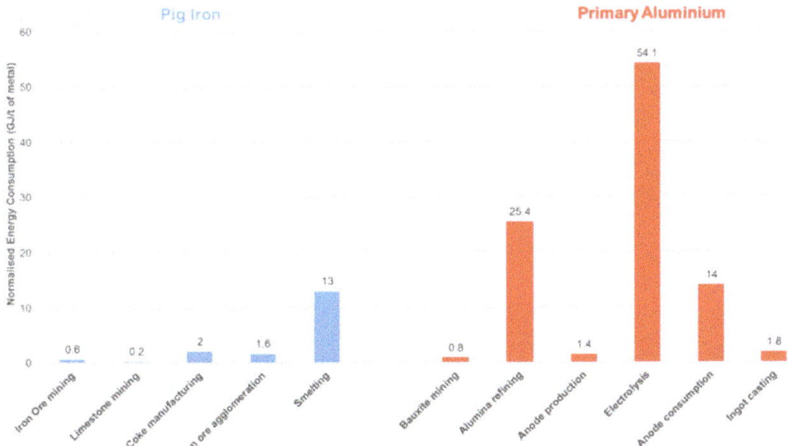

Figure 4. Energy consumption for the production of pig iron and primary aluminum.

Furthermore, red mud is a by-product of the primary aluminum production at a rate of two tons per ton of aluminum (120 million tons per year) and, at this moment, there are no solutions for it. On the other hand, the slag from the smelting process is easily recycled into road and cement making. Finally, the electrolysis of alumina consumes four times more energy than the whole production of pig iron.

4. Case Study: Engine Block

The heaviest single component in a passenger vehicle is the cylinder block. Over the last 10 years, the most significant transformation in engines was the capacity to provide more power with a lower displacement. This is a result of one of the most significant engine trends: downsizing. Comparing 2001 with 2013, engine power increased 20% while engine displacement decreased by 10% [46]. Furthermore, the top-selling vehicle models worldwide follow this trend. According to Reference [46], the engine displacement of the most sold vehicles is between 0.8 and 2.0 L, except for the United States of America (USA) and Canada, where engines with more power and displacement are highly valued.

The four-cylinder blocks were selected as a case study in the present study, as they represent approximately 71% of the total blocks manufactured worldwide [47]. For the reasons mentioned in the previous paragraph, the present investigation focuses on in-line four-cylinder 1.6-L engine blocks. These can be found in both diesel and petrol versions and in both CI and Al-alloy materials. Al-alloy engine blocks are lighter than CI engine blocks as illustrated in Figure 5. However, due to the fact that CI is stronger than Al alloys, Al-alloy engine blocks need thicker walls between cylinder bores, making them longer. As a result, the volume of CI required is considerably less, being in the region of 55% of that of the equivalent Al-alloy block, and CI engines are considerably more compact. As illustrated in Figure 5, the weight differentials between the petrol and diesel engines made of Al alloy and CI are 9 and 11 kg, respectively. However, more compact engines lead to an even smaller weight difference in the fully assembled engine, as a result of smaller ancillary components. Thus, our calculations are based on an on-the-road weight differential for the engine of 7 kg and 9 kg for petrol and diesel, respectively, which was substantiated by a number of design consultancy firms and OEMs.

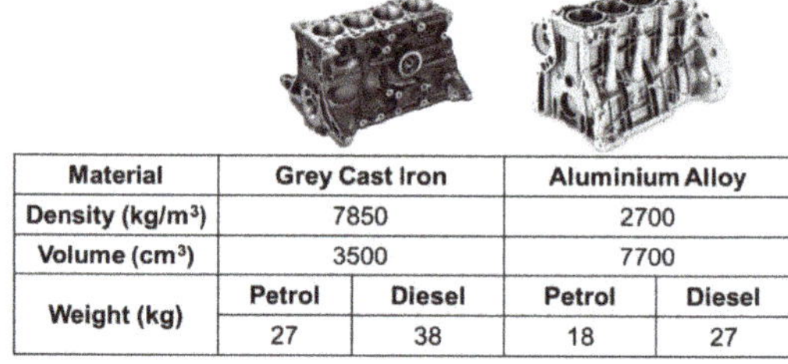

Material	Grey Cast Iron		Aluminium Alloy	
Density (kg/m³)	7850		2700	
Volume (cm³)	3500		7700	
Weight (kg)	Petrol	Diesel	Petrol	Diesel
	27	38	18	27

Figure 5. Weight difference between cast iron and aluminum-alloy engine blocks according to the fuel consumed by the vehicle for 1.6-L engines.

In Figures 6 and 7, the process flow for manufacturing the engine blocks from CI and Al alloys, respectively, is presented. The key difference between the two process flows is the need for heat treatment in the case of Al-alloy engine blocks and the use of liners.

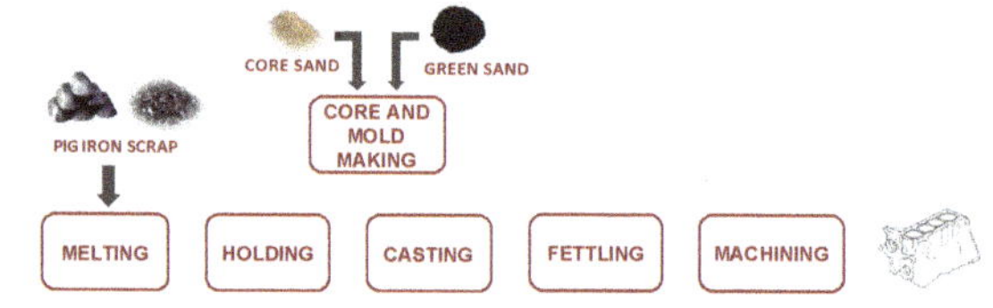

Figure 6. Process flow for cast iron (CI) engine block manufacturing.

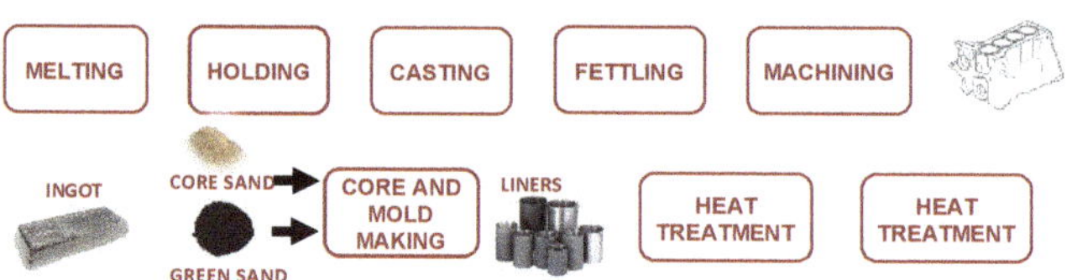

Figure 7. Process flow for Al-alloy engine block manufacturing.

4.1. CI Engine Blocks

Producing engine blocks from cast iron requires casting to a near net shape and machining to the exact dimensions. For collecting the required data (material use and energy consumption), three casting foundries were visited. These three foundries are responsible for the production of more than 60% of the world's cast-iron engine blocks.

4.1.1. Melting Stage

The casting temperatures for CI and Al vary around 1500 °C and 730 °C, respectively. This normally occurs in a melting furnace which can differ from foundry to foundry and/or for different metals. Normally, two types of furnaces are used: cupola and induction. By a number of foundries, it was verified that they only use cupola furnaces to produce CI engine blocks. Cupola furnaces use coke as an energy source, and their thermal efficiency ranges between 20 and 30%. The main inputs in these furnaces are pig iron (4.8%), ferrosilicon 75% Si (4%), and steel and/or CI scrap (91.2%). Unrecoverable metal losses, mainly due to oxidation, are reported by foundries, to an average of 2%. In total, three CI foundries were audited, and the energy per ton of liquid metal was measured to be 3.9 ± 0.1 GJ (Figure 8).

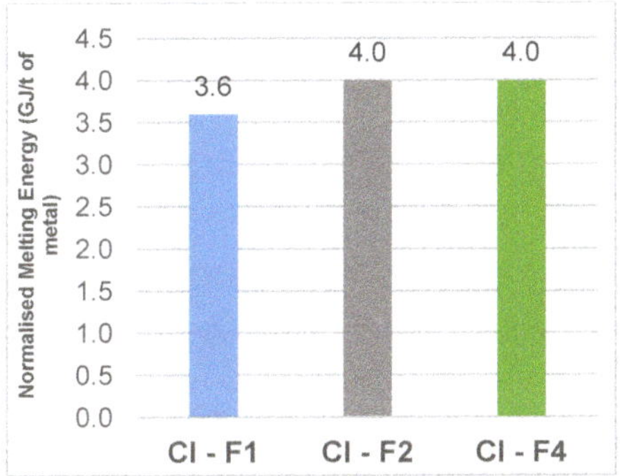

Figure 8. Melting energy per ton of liquid metal in three different cast-iron foundries.

4.1.2. Holding Stage

After melting, to keep the metal at casting temperature and with a consistent composition, it is transferred and kept in the holding furnace as a buffer due to different production rates. The energy per ton of liquid metal was measured to be 0.2 ± 0.1 GJ in two foundries (Figure 9). The holding furnaces in both foundries were electricity powered. One of the biggest factors in the energy consumption during the holding process is the holding time. This changes from foundry to foundry according to the production rate, casting method, and of course the type of metal. In the holding process, the foundries reported an unrecoverable metal loss of 2%.

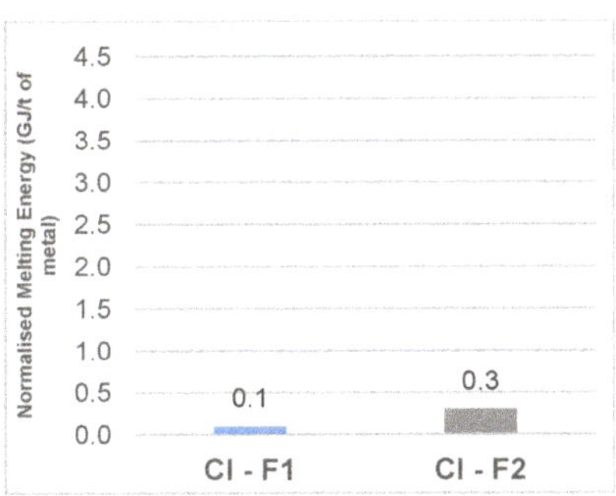

Figure 9. Holding energy per ton of liquid metal in two different cast-iron foundries.

4.1.3. Core- and Mold-Making Stage

In engine block castings, cores are used to form the complex internal geometry of the block. Cores are made from silica sand using the cold box method, where a binder system is used to cure the sand and resin to form the core. The design of the core varies depending on the material to be casted, and, for CI engine blocks, the reported core weight is 42.6 ± 4 kg (Figure 10a). The process of core-making also consumes a significant amount of energy, as the cores are normally coated and baked before use. Three foundries reported average energy needed for core-making to be 0.97 ± 0.3 GJ per ton of core sand (Figure 10b). In addition to the cores, a sand mold is used to form the outer limits of the casting. It is also used to support the core package, which together form the core package system. The weight of the sand mold, according to one of the foundries, is approximately 180 kg. For the formation of the mold, machining is used that is reported to consume 0.16 ± 0.2 GJ (Figure 10c) per ton of green sand.

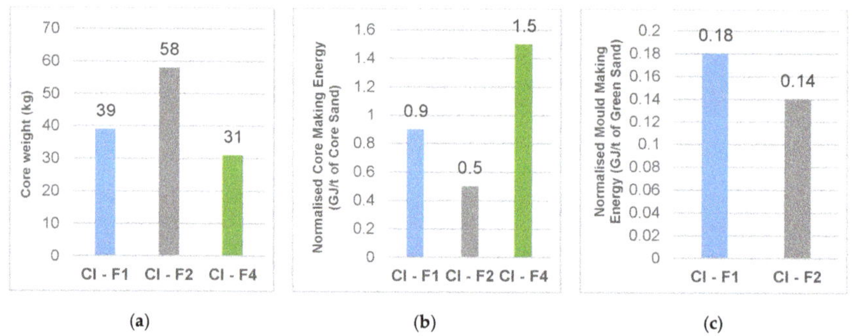

Figure 10. Mold- and core-making: (**a**) core weight, (**b**) core-making energy, and (**c**) mold-making energy.

4.1.4. Casting Stage

For the casting of CI into engine blocks, all visited foundries reported that only gravity sand casting is used, using green sand molds and a core package. In gravity sand casting, liquid metal is poured into a cavity that is formed by a monolithic sand mold, as explained previously. The pouring of the metal can be fully automated, semi-automated, or completely manual. Flow rates of the metal may vary from the beginning to the end of a casting campaign as the pouring ladle empties. Metal flow velocities should be adequate to avoid turbulence and achieve a good-quality casting. Sand castings have a low cooling rate because of the sand insulating mass surrounding the casting.

4.1.5. Fettling Stage

Following the casting process and the removal of the solid block from the sand mold, it has to be roughly machined to remove secondary cavities, risers, runners, and gates (also known as fettling). This excess material is usually re-melted. The mold yield reported from all three foundries was 75 ± 1% (Figure 11a). The energy consumed during the process varies significantly per foundry, and the reported values range from 0.1 to 1.4 GJ per ton of liquid metal (Figure 11b).

4.1.6. Machining Stage

Castings are produced volumetrically larger than required. Surfaces such as cylinder bores, deck faces, crankshaft bores, etc. are casted with an excess material of 2–3 mm that allows later dimensional corrections. A large number of holes must be drilled for oil circulation, bolts, etc. The main machining operations in an engine block are cubing, boring, drilling, and threading. Machining performance and, consequently, machining energy consumption may vary according to the machining parameters used. The energy can be significantly reduced by arranging for casting feeders to be located on areas which are to be machined. The approach used to quantify the energy requirements during machining is based on an analytical model provided by MAG Manufacturing Technology [48]. The model used is based on real machining energy measurements and has the capability to aggregate all the ancillary energy requirements (air, coolant supply, etc.) into each operation. Cycle times for each operation were obtained from machining outsourcing houses for two in-line four-cylinder blocks. The total energy consumption calculated for machining one cast-iron block is 61 MJ, i.e., 1.6 GJ/ton of cast-iron block. Usually, 10 kg of material is removed, which represents 20% of the block.

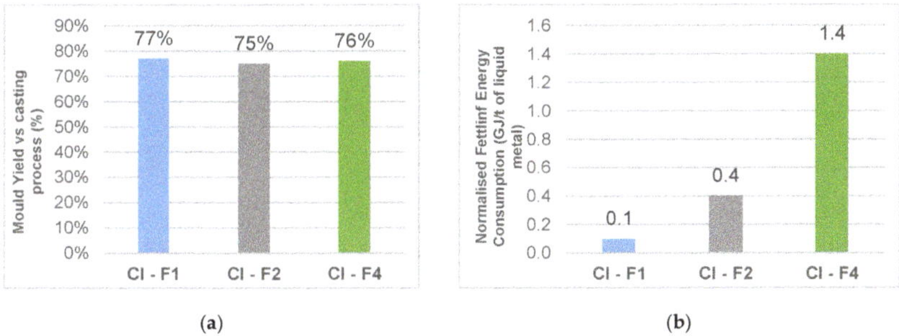

(a)   (b)

Figure 11. Fettling process: (**a**) mold yield in different casting processes; (**b**) fettling energy consumption.

4.1.7. Ancillary Processes

Miscellaneous energy is related to the facility operation and other ancillary processes like heating, lighting, etc. The energies included in each foundry for the miscellaneous processes vary widely. In the case of the three CI foundries, the reported energy ranged from 0.1 to 3.8 GJ per ton of good casting.

4.1.8. Inspection Stage

Quality inspection is undertaken throughout the casting process. Foundries aim to minimize their internal rejection rate to increase their efficiency by applying strict internal inspection standards in order to not ship and transport bad product. CI foundries reported an average of 3% internal scrap and 0.5% external scrap. Internal scraped CI blocks are re-melted directly.

4.1.9. Material Recycling

In all foundries, material is recycled. The furnace charge that foundries are using for engine block manufacturing comes from two different sources—external recycling (new scrap, old scrap, turnings, and dross) and in-house recycling (Figure 12). According to foundry practices, the ratio between the two differs. The dominant production route for steel made from scrap is electric arc furnace, while the energy needed is equal to on average 7 GJ/ton (Table 10). The most common route for primary steel production is basic oxygen furnace that converts pig iron into steel. The energy for this step on average is equal to 0.8 GJ/ton. Together with pig iron production energy, a full steelmaking process is equal to 18.2 GJ per ton.

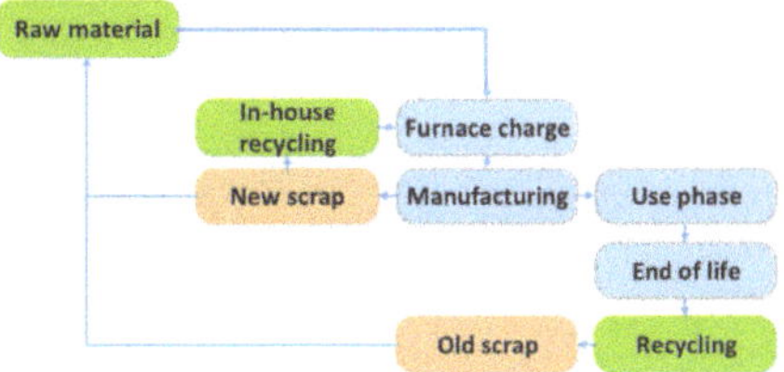

Figure 12. Material flow diagram of the recycling.

Table 10. Energy for steel recycling with electric arc furnace.

Source	Energy (GJ/t)
World Steel Association (2015)	5.3–8.7
[49]	6–15
[50]	8.1–9.0
[13]	10
[22]	5.5
[43]	5.3
[44]	5.5

Because the history of the scrap that is used as a furnace charge is not known [13], it is necessary to consider all the stages that the material might go through, from initial manufacture to final disposal. Based on the number of product cycles, the embodied energy in the material can be estimated by calculation. The total energy content for the chosen number of cycles can be calculated as follows [13]:

$$X = \left(X_{pr} - X_{re}\right)\left[\frac{(1-r)}{(1-r^n)}\right] + X_{re}.$$

(1)

According to Equation (2), the energy burden for multiple recycling, where the material is recycled indefinitely, can be obtained by calculating the following [13]:

$$X = X_{pr} - r\left(X_{pr} - X_{re}\right),$$

(2)

where X_{pr} stands for energy for manufacturing one ton of material via the primary route, X_{re} is the energy for manufacturing one ton of material via the recycling route, r is the overall recycling efficiency over one life cycle ($r = RR{\cdot}Y$), RR is the scrap recovery rate (%), and Y stands for the efficiency of the recycling process (%). Figure 13 represents the embodied energy for steel scrap after recycling. For the electric arc furnace (EAF) route and steel scrap processing, the average overall recycling efficiency (r) includes the furnace yield and the efficiency of recovering the steel at the end-of-life ($r = 0.89$) [13].

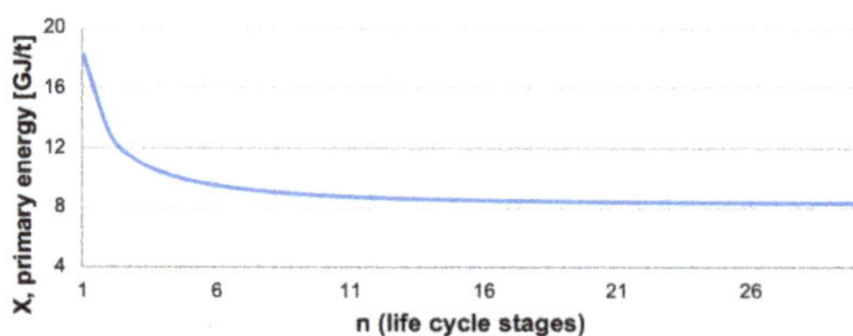

Figure 13. Steel scrap embodied energy ($X = 8.2$) for electric arc furnace (EAF) recycling route.

However, the above analysis considers only the once-through product system. To undergo a full energy analysis, the influence of recycling and reusing material in the casting process should also be considered [14]. As a result, the multiple life-cycle method needs to be adopted. The residue metal that can be again re-melted comes from fettling (in the form of runners and feeders), rarely machining (swarf), and internal inspection. Apart from metal, other process materials like core sand and green sand can also be recycled (via thermomechanical or thermal sand reclamation) or reused [51]. The embodied energies for cast-iron, core sand after reclamation and green sand are illustrated in Figure 14a,b,c respectively.

The alloying and treatment materials need to be considered as well. For CI, ferrosilicon is added to enhance the grain structure and metallurgy of the finished component. The energy content to produce one ton of ferrosilicon master alloy is just over 30 GJ. However, the addition rate into the iron is such that this contributes 1.6 GJ/ton of CI engine blocks.

Figure 15 shows the Sankey diagram representation of the energy and materials flows. Using this, the largest areas of energy input, recycling loops, and material losses are shown.

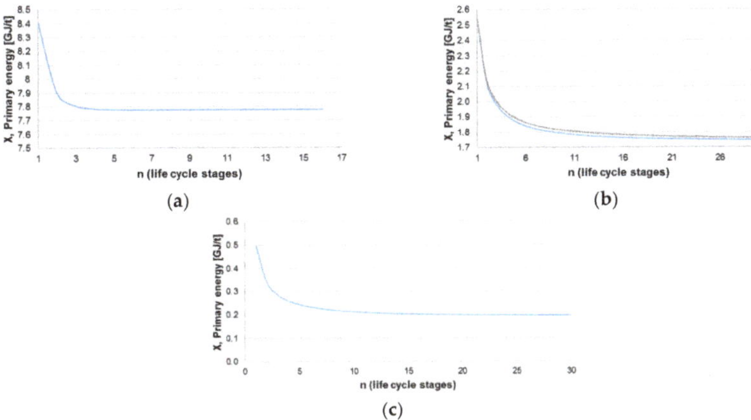

Figure 14. (**a**) Energy embodied in metal collected from the production stage and re-melted in-house in the cast-iron foundries (assumed 2% of the embodied energy for pig iron addition), (**b**) energy embodied in core sand after reclamation process, and (**c**) energy embodied in green sand for its multiple reuse.

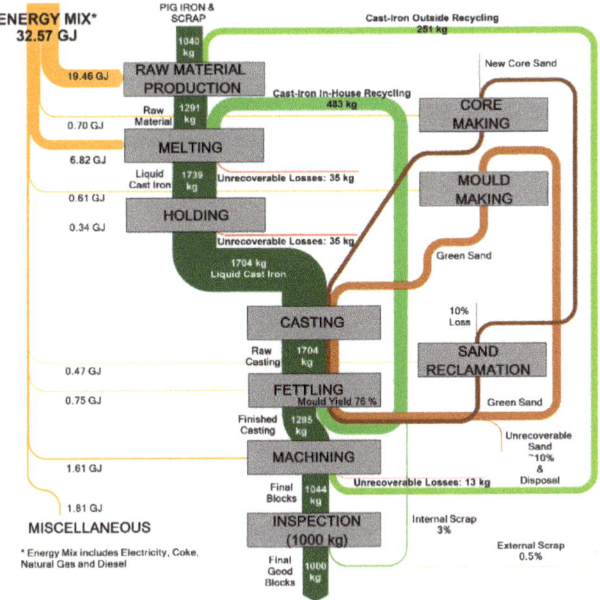

Figure 15. Energy and material flow in CI sand casting, showing that 1000 kg of good castings require the melting of 1739 kg of CI and 32.57 GJ.

4.2. Al-Alloy Engine Blocks

Figure 7 shows the process flow for Al-alloy engine block manufacturing. Compared to CI engine blocks, the process is slightly more complicated, as there is need for the use of liners as explained later on, as well as heat treatment of the cast components. Furthermore, the casting processes to be used vary from company to company. Three different casting processes can be identified that are widely used for the manufacturing, namely high-pressure die casting (HPDC), low-pressure die casting (LPDC), and low-pressure sand casting (LPSC; also known as the Cosworth process). Overall, 70% of aluminum-alloy engine blocks are casted by HPDC, while the other 30% are casted through a combination of the other methods [23].

The LPDC process consists of a dosing furnace which is pressurized, forcing liquid aluminum to enter the mold from the bottom. The mold consists of steel dies combined with internal sand cores. The repeatable rising and falling of the metal through the delivery tube may introduce oxide layers which eventually are delivered to the casting. LPDC is used for medium to long series of casting runs, where better mechanical properties are required when compared to HPDC. In HPDC, the alloy is inserted into a cold chamber and a hydraulic piston squeezes the metal into a steel die mold at extremely high speed (up to 80 m/s) and pressure (3500 tons). No sand cores can withstand the high pressure; thus, the HPDC block designs are limited to open-deck blocks.

Similar to cast-iron green sand casting, aluminum gravity sand casting also uses core packages. In the LPSC (Cosworth process), the metal is usually pumped into the sand mold from the bottom by an electrical pump. The difference from LPDC is that the metal in the pump never drops back to the level of the metal and, consequently, the level of oxide generated is potentially lower than in a gravity system [52]. Data were collected from a number of foundries that employ such processes.

4.2.1. Melting and Holding

In Al-alloy engine block foundries, tower furnaces are most commonly used [6]. The unrecoverable metal losses are of the same order of magnitude as CI. The foundries contacted reported an average energy consumption of 6.5 ± 3 GJ per ton of liquid metal (Figure 16). With regard to the holding of the liquid metal, the holding time varies between foundries. In HPDC and LPDC, the holding time is around four hours, while for LPSC it is 13 hours because of the additional time required for refining of the metal. The foundry using the Cosworth process used holding as a refining step to allow unwanted trace elements to settle out of the liquid Al alloy and oxides to float to the surface. Figure 17 shows the holding energy in GJ per ton of liquid metal.

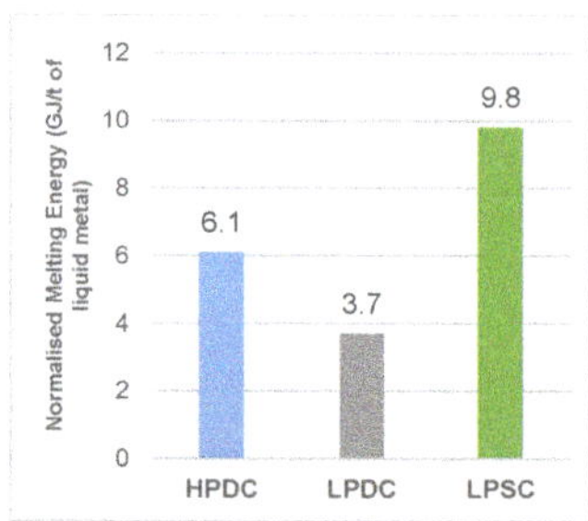

Figure 16. Melting energy per ton of liquid metal in three different Al foundries.

4.2.2. Core- and Mold-Making

The material and the process used for the core- and mold-making depend on the type of the casting process to be used. In LPSC foundries, cores are made from silica sand using the cold box method, where a binder system is used to cure the sand and resin to form the core. In HPDC, sand cores are not used due to the high-pressure injection of the metal which would destroy the cores. The core weight also varies for the different metals. The cores in cast iron sand casting are much heavier than in aluminum LPDC. This is because it includes the whole core package (cores + core shells). The energy required for making cores and the mold is quite similar with cast-iron sand casting (CISC), with the exception of when dies are used.

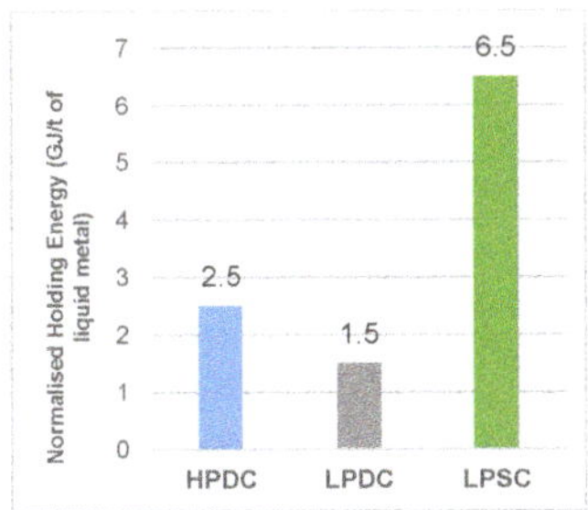

Figure 17. Holding energy per ton of liquid metal in three different Al foundries.

4.2.3. Casting

The four different casting processes were presented already. As per CI, the energy consumed during the casting process is negligible with the exception of HPDC. In HPDC, automatic spray-up for lubrication and robotic casting removal after solidification also consume a lot of energy. The dies are usually monolithic and contain cooling and heating channels. Due to these extra energies in HPDC, a casting energy is accounted for in this casting method only (1.2 GJ per ton of casting). HPDC parts are near net shape, and less fettling and machining operations are required. Due to the nature of metal filling, HPDC castings are often non-heat-treatable but might go through a stress-relieving thermal cycle.

4.2.4. Fettling

Once the cast engine block is removed from the sand mold or the die, fettling is required as per the CI process as well. In the case of Al-alloy engine blocks, the reported mold yield is lower compared to CI and is approximately 65 ± 2%. The material removed from fettling aluminum-alloy engine blocks can be re-melted directly in the foundry or sold to an external recycling company to be transformed again into aluminum alloys. The second case relates to aluminum LPDC and, therefore, the calculations in this study, for LPSC, are based on outside recycling. The energy consumed during the fettling process was reported in all three foundries to be 0.6 GJ per ton of liquid metal.

4.2.5. Heat Treatment

A key difference in the CI process flow is the need for heat treatment. Al–Si alloys used to produce Al-alloy engine blocks usually require T6 and T7 heat treatments which are used to improve both mechanical and wear properties [53]. Foundries also reported that T5 is the most common heat treatment process used in HPDC. The average energy consumption per casting can be calculated when temperature and holding times are known.

Considering a treatment efficiency of 100%, the average energy consumption for heat treatments T6 and T7 can be calculated to be 3.2 GJ/ton of finished casting. For T5, the average energy consumption is calculated to be 1.0 GJ/t. For the case of engine block casting, 20% heat treatment efficiency is required; thus, the values considered were scaled accordingly.

4.2.6. Impregnation

Casting introduces porosities during the solidification of the liquid metal. Turbulent metal flow, gas entrapment, and metal shrinkage are the main factors that introduce voids in the casting. Porosity is more pronounced in aluminum-alloy castings because of its higher volumetric shrinkage and hydrogen content. The three main forms of porosity are full enclosed, blind, and through porosity. Such porosity could result in leaking under pressure, and would, thus, require the block to be scrapped. an impregnation process that introduces a polymer sealant in the pores and cracks of castings is used for this reason. The most commonly used impregnation process is the vacuum dry process. The castings are stashed into a basket and inserted in a series of chambers until a full impregnation cycle is achieved.

Around 90% of the energy in an impregnation cycle is consumed heating up the water at around 90 °C, and the remaining 10% is used for circulation pumps, vacuum pumps, rotational mechanisms, and other ancillary systems. The energy involved in the process was ascertained to be around 7.2 MJ/engine block.

4.2.7. Machining

Using the MAG analytical model [48], the total energy consumption for machining is 51 MJ, of which 13 MJ is for the initial machining of the cylinder liners.

4.2.8. Liner Casting

For the aluminum-alloy in-line four-cylinder blocks, for all casting processes, cast-iron cylinder liners are cast in the block. The wear and mechanical properties of hypoeutectic alloy sliding surfaces are not adequate to withstand the friction of the moving piston in the cylinder bore. Cast-in CI liners are used for the tribological system "cylinder–piston–piston ring". The liners are centrifugally cast and the induction pre-heated prior to casting at around 375 °C to achieve better bonding with the liquid Al, ending up with a total energy of 188 MJ/engine block. For the fettling of the solid casting system, the yield ratio is approximately 67% with a total energy consumption of 0.6 GJ per ton of liquid metal.

4.2.9. Material Recycling

As with the CI foundries, Al foundries charge their furnaces with recycled material as well. The process is quite similar; however, Figure 12 needs to be updated in order to include secondary smelter and the production of ingots. Figure 18 illustrates the common processes for the material flow of the recycling model for the case of Al alloys.

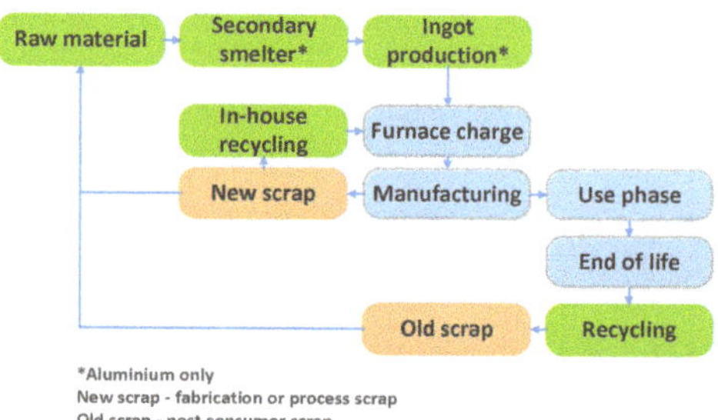

Figure 18. Material flow diagram of the recycling.

Al-alloy engine blocks are usually made from secondary ingot. The alloy used is A383 or A380 for LPDC and HPDC, and A319 for LPSC. The process of recycling Al scrap to form the alloys is by refining, a process that uses a combination of rotary and reverberatory furnaces [54]. The recycled Al can have similar properties to primary Al. However, in a course of multiple recycling, more and more alloying elements are introduced into the metal cycle. Secondary alloys have relatively high levels of impurities, especially iron, which is detrimental to many properties. The multiple life-cycle method is, thus, used (as in the CI recycling) for calculating an average energy consumption.

Figures 19–21 present the Sankey diagrams for the LPSC, LPDC, and HPDC cases, respectively.

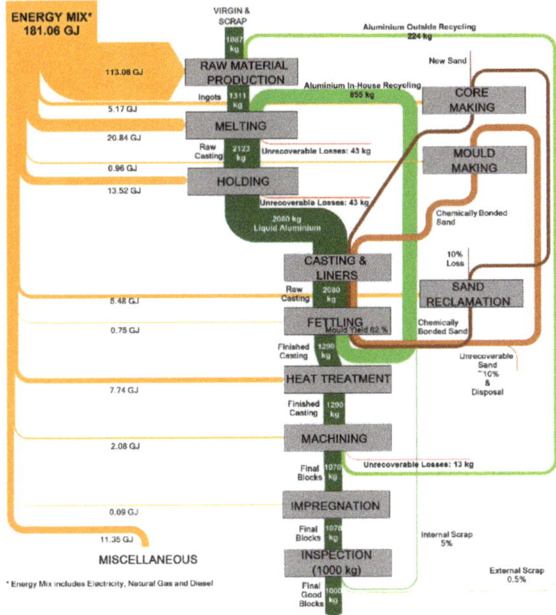

Figure 19. Energy and material flow in low-pressure sand casting (LPSC), showing that 1000 kg of good castings require the melting of 2123 kg of Al and 181.06 GJ.

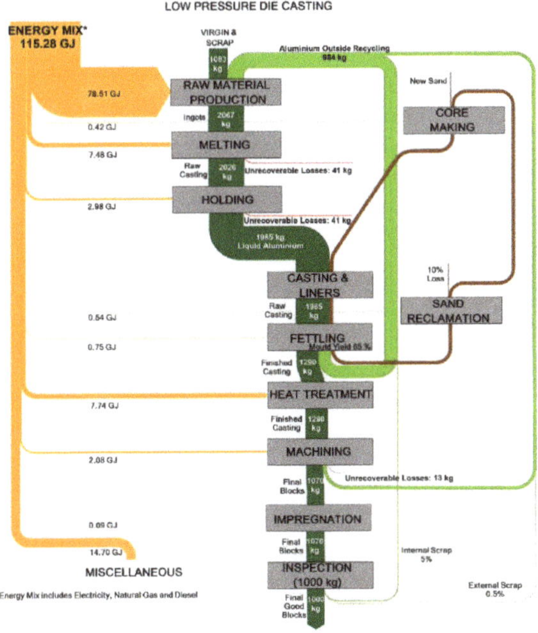

Figure 20. Energy and material flow in low-pressure die casting (LPDC), showing that 1000 kg of good castings require the melting of 2067 kg of Al and 115.28 GJ.

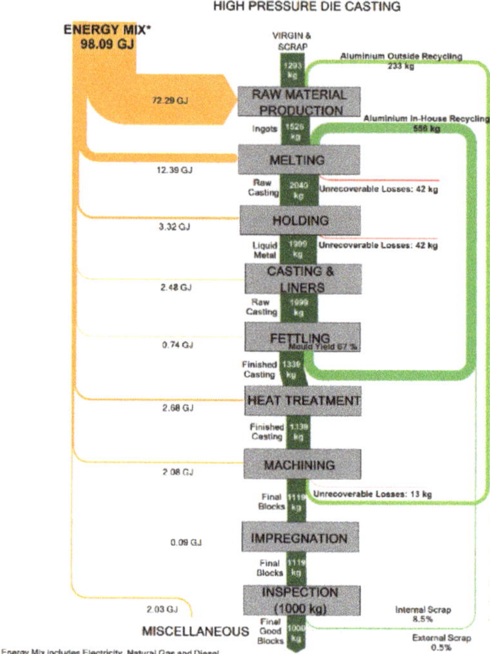

Figure 21. Energy and material flow in high-pressure sand casting (HPDC), showing that 1000 kg of good castings require the melting of 2040 kg of Al and 98.09 GJ.

5. The Answer to the Dilemma between Al Alloys and CI

Figure 22 shows the energy breakdown in each material source and indicates that ingot and external scrap represent the highest embodied energy of the charge and feedstock for Al-alloy and CI engine blocks. Figure 23 demonstrates the process energy breakdown for each casting. It is obvious that the CI engine block requires considerably less energy. The excess energy spent for the manufacturing of Al-alloy engine blocks should be compensated for by the fact that the vehicle is lighter and, thus, consumes less energy during its use.

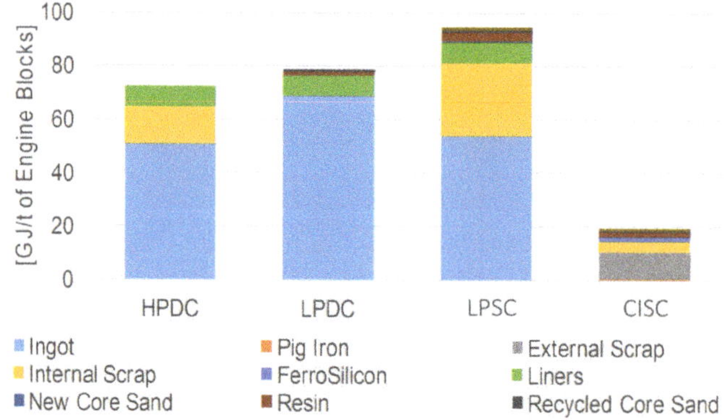

Figure 22. Embodied material energy per ton of engine blocks.

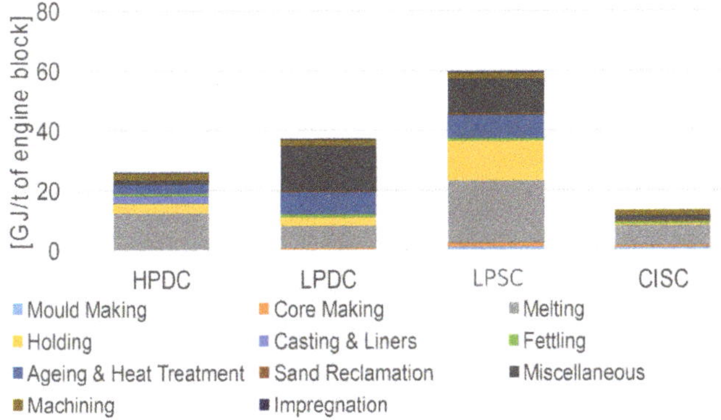

Figure 23. Process energy per ton of engine blocks.

Figures 22 and 23 provide information about the embodied and process energy per ton of engine block. However, it is equally significant to represent the data above using a single block as a functional unit. The values of process, embodied, and total energy, which is equal to the sum of the embodied and process energy, required for the production of each single engine block via the four manufacturing processes, are listed in Table 11.

Table 11. Total energy per engine block. HPDC—high-pressure die casting; LPDC—low-pressure die casting; LPSC—low-pressure sand casting; CISC—cast-iron sand casting.

	HPDC		LPDC		LPSC		CISC	
	Diesel	**Petrol**	**Diesel**	**Petrol**	**Diesel**	**Petrol**	**Diesel**	**Petrol**
Process energy (GJ/t)	25.8	25.8	36.78	36.78	59.12	59.12	13.11	13.11
Embodied energy (GJ/t)	72.37	72.37	78.63	78.63	114	114	19.46	19.46
Weight of single block (kg)	27	18	27	18	27	18	38	27
Process energy (GJ/block)	0.64	0.41	0.91	0.58	1.46	0.93	0.5	0.35
Embodied energy (GJ/block)	1.79	1.14	1.94	1.24	2.81	1.79	0.74	0.53
Total energy (GJ/block)	2.43	1.54	2.85	1.81	4.28	2.72	1.24	0.88

The embodied energy due to manufacturing and use is illustrated in Figure 24 (shown for the case of diesel engines; similar results were attained for petrol engines). The starting values of the embodied energy correspond to the total energy of the manufacturing process (Table 11). It is evident that the vehicle would have to be driven more in order for the lightweighting to yield benefits. This is due to the much higher embodied energy of Al alloys compared with CI as a result of the huge energy content during both the electrolysis and bauxite conversion stages of the production of aluminum.

The distance needed to be covered by a vehicle in order to compensate for the additional energy due the manufacturing and primary production of the engine block is estimated using the break-even distance (*BED*) according to

$$BED = \frac{\Delta PEB}{\left(\delta F_s \cdot E_f \cdot \Delta M\right)} \cdot 10^4, \tag{3}$$

where ΔPEB (*MJ*) is the difference in the process energy burden between the manufacturing process with the lowest total energy (CISC) and the rest of the processes, δF_s is the fuel savings, E_f is the energy content of the fuel, and ΔM is the engine block weight differential (Table 12). The values of the break-even distance for the two types of engine blocks (diesel and petrol) and the various manufacturing processes under examination are summarized in Table 13.

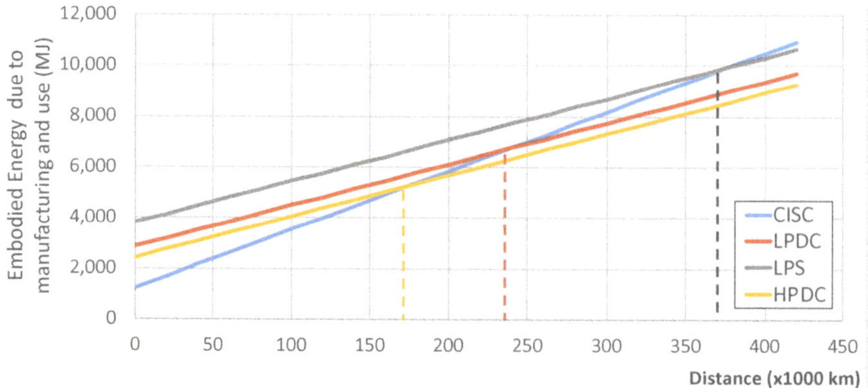

Figure 24. Break-even distance for paying back the lightweight material (for a diesel automotive vehicle of 1200 kg with an average consumption of 7l,100 km).

Table 12. Parameters for the break-even distance (BED) calculation.

	Diesel	Petrol
$\delta F_s\left(\frac{L}{100\ km\ \times\ 100\ kg}\right)$	0.2	0.25
$E_f\left(\frac{MJ}{L}\right)$	38.6	34.2
$\Delta M(kg)$	9	7

Table 13. BED (km) vs. CISC for various types of engine blocks and manufacturing processes.

	Diesel	Petrol
HPDC	170,889	110,611
LPDC	232,141	155,809
LPSC	369,221	256,960

6. Conclusions

Currently, the legislation around the automotive industry is focused on the reduction of tailpipe emissions of vehicles and does not consider the production phase of automotive components. Automotive companies are compelled to pursue a lightweighting and engine-downsizing design strategy to comply with the steadily more stringent targets in emission standards. The objective of this investigation was to perform a thorough life-cycle analysis of an automotive component (engine block) made of two different materials, CI and Al alloy, in order to review the potential energy savings of lightweighting.

The "cradle-to-grave" approach was adopted to calculate the overall energy requirements, including the energies for the production of the raw materials, while acknowledging the embodied energy from the initial manufacture up to the final disposal. Our results indicate that the energy required for the primary production and manufacture of CI engine blocks is much lower compared to the Al-alloy engine case. On the other hand, Al-alloy blocks are more lightweight and contribute to the increase in fuel savings during the use phase of the particular component.

In order to evaluate the effects of lightweighting on the overall energy consumption during the component's life cycle, the weighted average break-even distance (required to compensate for the extra energy consumption in Al-alloy engine blocks) was estimated and found to be around 175,000 km. The breakeven distance fluctuated between 175,000 and 370,000 km for a diesel and between 115,000 km and 260,000 km for a petrol engine block. The conclusion drawn is that, compared to an average passenger vehicle life of 200,000 km, for the LPDC and LPSC processes, the vehicle will never recover the extra energy in the Al-alloy engine blocks while being on the road. Therefore, the substitution of materials, traditionally used in the automotive industry, with lighter ones should be very carefully considered.

Author Contributions: Conceptualization, M.J. and K.S.; methodology, M.J. and K.S.; validation, E.P. and M.P.; formal analysis, all; investigation, all; writing—original draft preparation, K.S.; writing—review and editing, M.P.; visualization, E.P.; supervision, M.J. and K.S.

Funding: This research was partially funded by the UK EPSRC projects "Small Is Beautiful" and "Energy Resilient Manufacturing 2: Small Is Beautiful Phase 2 (SIB2)" for funding this work under grants EP/M013863/1 and EP/P012272/1, respectively.

Acknowledgments: The authors would like to acknowledge the UK EPSRC projects "Small Is Beautiful" and "Energy Resilient Manufacturing 2: Small Is Beautiful Phase 2 (SIB2)". Earlier versions of this manuscript were presented at the 38th International Vienna Motor Symposium, Vienna, Austria in 2017 [11] and the TMS Annual Meeting and Exhibition in 2018 [12].

Conflicts of Interest: The authors declare no conflicts of interest.

References

1. Davies, G. *Materials for Automobile Bodies*; Butterworth-Heinemann: Oxford, UK, 2012.
2. ACEA European Automobile Manufacturers' Association. *Reducing CO_2 Emissions from Cars and Vans*; ACEA European Automobile Manufacturers' Association: Brussels, Belgium, 2015.
3. Ashby, M.; Coulter, P.; Ball, N.; Bream, C. *The CES EduPack Eco Audit Tool—A White Paper 2009*; Granta Publications: London, UK, 2009.
4. Sorger, H.; Schöffmann, W.; Wolf, W.; Steinberg, W. Lightweight design of cast iron cylinder blocks. *MTZ Worldw.* **2015**, *76*, 22–27. [CrossRef]
5. Salonitis, K.; Ball, P. Energy efficient manufacturing from machine tools to manufacturing systems. *Procedia CIRP* **2013**, *7*, 634–639. [CrossRef]
6. Salonitis, K.; Zeng, B.; Mehrabi, H.A.; Jolly, M. The challenges for energy efficient casting processes. *Procedia CIRP* **2016**, *40*, 24–29. [CrossRef]
7. Salonitis, K.; Jolly, M.R.; Zeng, B.; Mehrabi, H. Improvements in energy consumption and environmental impact by novel single shot melting process for casting. *J. Clean. Prod.* **2016**, *137*, 1532–1542. [CrossRef]
8. Salonitis, K.; Jolly, M.; Zeng, B. Simulation based energy and resource efficient casting process chain selection: A case study. *Procedia Manuf.* **2017**, *8*, 67–74. [CrossRef]
9. Pagone, E.; Jolly, M.; Salonitis, K. The development of a tool to promote sustainability in casting processes. *Procedia CIRP* **2016**, *55*, 53–58. [CrossRef]
10. Pagone, E.; Salonitis, K.; Jolly, M. Energy and material efficiency metrics in foundries. *Procedia Manuf.* **2018**, *21*, 421–428. [CrossRef]
11. Jolly, M.; Salonitis, K. Primary manufacturing, engine production and on-the-road CO_2: How can the automotive industry best contribute to environmental sustainability. In Proceedings of the 38th International Vienna Motor Symposium, Vienna, Austria, 27–28 April 2017.
12. Gonçalves, M.; Jolly, M.R.; Salonitis, K.; Pagone, E. Resource Efficiency Analysis of High Pressure Die Casting Process. In Proceedings of the TMS Annual Meeting & Exhibition, Phoenix, AZ, USA, 11–15 March 2018; pp. 1041–1047.
13. Brimacombe, L.; Coleman, N.; Honess, C. Recycling, reuse and the sustainability of steel. *Millenium Steel* **2005**, *446*, 3–7.
14. Allwood, J.M.; Cullen, J.M.; Carruth, M.A.; Cooper, D.R.; McBrien, M.; Milford, R.L.; Moynihan, M.C.; Patel, A.C.H. *Sustainable Materials: With Both Eyes Open*; UIT Cambridge: Cambridge, UK, 2012.
15. Balomenos, E.; Panias, D.; Paspaliaris, I. Energy and exergy analysis of the primary aluminum production processes: A review on current and future sustainability. *Miner. Process. Extr. Metall. Rev.* **2011**, *32*, 69–89. [CrossRef]
16. Association, A. *Aluminum: The Element of Sustainability*; A North American Aluminum Industry Sustainability Report; The Alumininum Association: Arlington, VA, USA, 2011; p. 33.
17. World Aluminium—Alumina Production. Available online: http://www.world-aluminium.org/statistics/alumina-production/ (accessed on 1 June 2019).

18. Saur, K. *Life Cycle Assessment of Aluminium: Inventory Data for the Worldwide Primary Aluminium Industry*; International Aluminium Institute: London, UK, 2003.

19. Green, J.A.S. *Aluminum Recycling and Processing for Energy Conservation and Sustainability*; ASM International: Materials Park, OH, USA, 2007.

20. PE Americas. *Life Cycle Impact Assessment of Aluminum Beverage Cans*; Aluminum Association, Inc.: Washington, DC, USA, 2010.

21. Alcoa. *Transforming Annual Report*; Alcoa Corporate Center: Pittsburgh, PA, USA, 2014.

22. Margolis, N.; Sousa, L. *Energy and Environmental Profile of the US Aluminum Industry*; US Department of Energy: Washington, DC, USA, 2017.

23. European Aluminium Association. *Environmental Profile Report for the European Aluminium Industry*; European Aluminium Association: Brussels, Belgium, 2013.

24. International Aluminium Institute. *Aluminium for Future Generations*; International Aluminium Institute: London, UK, 2009.

25. Scarsella, A.A.; Noack, S.; Gasafi, E.; Klett, C.; Koschnick, A. Energy in alumina refining: Setting new limits. In *Light Metals 2015*; Springer: Cham, Switzerland, 2015; pp. 131–136.

26. Moya, J.A.; Boulamati, A.; Slingerland, S.; Van Der Veen, R.; Gancheva, M.; Rademaekers, K.M.; Kuenen, J.J.P.; Visschedijk, A.J.H. *Energy Efficiency and GHG Emissions: Prospective Scenarios for the Aluminium Industry*; European Commission, Joint Research Centre, Institute for Energy and Transport: Brussels, Belgium, 2015.

27. Wurtemberg, J.M.V. *Lightweight Materials for Automotive Applications*; Sintercast: Pully, Switzerland, 1994.

28. Rio Tinto Minerals. *2012 Sustainable Development Report*; Rio Tinto Minerals: Greenwood Village, CO, USA, 2013.

29. Alcoa Canada. *Outlook on Sustainability 2013*; Alcoa Canada: Montréal, QC, Canada.

30. Moll, S.; Acosta, J.; Schütz, H.; Ritthoff, M. *Iron and Steel—A Materials System Analysis*; European Topic Centre on Resource and Waste Management: Copenhagen, Danmark, 2005.

31. Hasanbeigi, A. *Emerging Energy-Efficiency and Carbon Dioxide Emissions-Reduction Technologies for the Iron and Steel Industry*; Berkeley Lab: Berkeley, CA, USA, 2013.

32. Norgate, T.; Haque, N. Energy and greenhouse gas impacts of mining and mineral processing operations. *J. Clean. Prod.* **2010**, *18*, 266–274. [CrossRef]

33. De la Torre de Palacios, L. Natural resources sustainability: iron ore mining. *Dyna* **2011**, *78*, 227–234.

34. *Tracking Industrial Energy Efficiency and CO_2 Emissions*; International Energy Agency: Paris, France, 2007.

35. Bettinger, D. *Energy Efficiency in Iron & Steelmaking*; SusChem: Linz, Austria, 2011.

36. Fruehan, R.J.; Fortini, O.; Paxton, H.W.; Brindle, R. *Theoretical Minimum Energies to Produce Steel for Selected Conditions*; Carnegie Mellon University: Pittsburgh, PA, USA, 2000.

37. U.S. Department of Energy. *Bandwidth Study on Energy Use and Potential Energy Saving Opportunities in U.S. Iron and Steel Manufacturing*; U.S. Department of Energy: Washington, DC, USA, 2015.

38. De Paula, G.M. *Produção Independente de Ferro-Gusa ("Guseiros")*; Núcleo de Estudos de Economias de Baixo Carbono: Ribeirão Preto, Brazil, 2014.

39. Americal Iron and Steel Institute Steel Production. Available online: https://www.steel.org/steel-technology/steel-production (accessed on 1 June 2019).

40. International Energy Agency. *Tracking Clean Energy Process 2015*; International Energy Agency: Paris, France, 2015.

41. University of Tennessee—Center for Clean Products. *Limestone Quarrying and Processing: A Life-Cycle Inventory*; Natural Stone Council: Hollis, NH, USA, 2008.

42. IEA. *Global Industry Dialogue and Expert Review Workshop*; International Energy Agency: Paris, France, 2014.

43. Price, L.; Phylipsen, D.; Worrell, E. *Energy Use and Carbon Dioxide Emissions in the Steel Sector in Key Developing Countries*; Lawrence Berkeley National Laboratory, University of California: Berkeley, CA, USA, 2001.

44. Hasanbeigi, A.; Jiang, Z.; Price, L. *Analysis of the Past and Future Trends of Energy Use in Key Medium-and Large-Sized Chinese Steel Enterprises, 2000–2030*; Berkeley Lab.: Berkeley, CA, USA, 2013.

45. Worrell, E.; Price, L.; Neelis, M.; Galitsky, C.; Nan, Z. *World Best Practice Energy Intensity Values for Selected Industrial Sectors*; Berkeley Lab.: Berkeley, CA, USA, 2007.

46. Mock, P. *EU CO_2 Standards for Passenger Cars and Light-Commercial Vehicles*; International Council on Clean Transportation: Berlin, Germany, 2014.

47. Trechow, P. *Windstrom treibt Erdgasfahrzeuge an*; VDI Nachrichten: Düsseld, Germany, 2011.

48. MAG—Manufacturing Technology. Available online: http://www.mag-ias.com/web/en/index.php (accessed on 28 May 2019).
49. Sverdrup, H.U.; Ragnarsdottir, K.V.; Koca, D. Aluminium for the future: Modelling the global production, market supply, demand, price and long term development of the global reserves. *Resour. Conserv. Recycl.* **2015**, *103*, 139–154. [CrossRef]
50. CES Edupack. Available online: https://grantadesign.com/education/ces-edupack/ (accessed on 1 June 2019).
51. Fenyes, M. Maximising sand recovery in the foundry. In *Transactions of 58th IFC*; Ahmedabad, India, 2010; pp. 49–54.
52. Jolly, M. 1.18 - Castings. In *Comprehensive Structural Integrity*; Milne, I., Ritchie, R.O., Karihaloo, B., Eds.; Pergamon: Oxford, UK, 2003; pp. 377–466. ISBN 978-0-08-043749-1.
53. Ye, H. An overview of the development of al-si-alloy based material for engine applications. *J. Mater. Eng. Perform.* **2003**, *12*, 288–297. [CrossRef]
54. Boin, U.M.J.; Bertram, M. Melting standardized aluminum scrap: A mass balance model for europe. *JOM* **2005**, *57*, 26–33. [CrossRef]

Article

Impact of Structural Changes on Energy Efficiency of Finnish Pulp and Paper Industry

Satu Kähkönen [1,*], Esa Vakkilainen [1] and Timo Laukkanen [2]

[1] Laboratory of Sustainable Energy Systems, School of Energy Systems, LUT University, 53851 Lappeenranta, Finland; esa.vakkilainen@lut.fi
[2] Department of Mechanical Engineering, School of Engineering, Aalto University, 00076 Espoo, Finland; timo.laukkanen@aalto.fi
* Correspondence: satu.kahkonen@lut.fi

Received: 20 August 2019; Accepted: 23 September 2019; Published: 26 September 2019

Abstract: A key challenge in prevention of global warming is how to increase energy efficiency, to be able to deal with increased fossil CO_2 emissions from rising energy usage. Increasing energy efficiency will decrease energy usage and is in a key role in emission mitigation. The focus is the pulp and paper industry, which is energy-intensive. Development of industrial energy efficiency has been studied before but the role of industrial transformation is still mostly unknown. The knowledge must be improved, to be able to predict future developments in the most effective way. In this research, impact of various production unit closures and start-ups on energy efficiency of the Finnish pulp and paper industry were studied utilizing statistical analysis. Results indicate that about 20% of the Finnish pulp and paper industry energy efficiency improvement between 2011 and 2017 is caused by the major structural changes. The rest, 80% of the progress, was mainly due to improved technology and more optimal operational modes. Additional findings suggest that modern mill start-ups have a significantly greater potential to reduce energy consumption than old mill closures.

Keywords: energy efficiency; pulp; paper; energy consumption; structural change

---

1. Introduction

World energy usage increases as a result from global population growth and increasing level of wellbeing [1]. Energy supply security, reduction of greenhouse gas emissions, and efforts to reduce global warming are the main issues acting as a driving force for changing our present energy usage [2]. Energy efficiency improvement has an important role in the cost-effective energy saving and in the sustainable development [3]. Energy efficiency means an ability to produce a high number of products with as low an amount of energy as possible. In addition to environmental advantages, enhancing energy efficiency lowers operating costs and consequently improves competitiveness [4]. Of the various sectors, industry was the largest energy consumer in the world in 2016: its share was about 30% of total consumed energy [5]. Therefore, significant results can be reached by improving industrial energy efficiency [6,7].

In Finland, industry and construction consumed 47% of total electricity in 2015 [8]. Inside the field of Finnish industry, forest industry was clearly the major consumer [9]. Production of pulp and, subsequently, paper requires significant amounts of energy. A high number of factors affect the energy consumption of pulp and paper industry mills, such as the type, size, age, and location of the mill, type of products, raw materials, and processes, as well as operational choices [10]. Even if energy consumption is measured to sufficient accuracy, evaluating the role of various changes to energy efficiency is challenging. Many factors have an influence on energy efficiency, but in many cases the size of the impact is unknown. By enhancing awareness about affecting factors, it is possible to develop mills and processes in an efficient way towards energy efficient operation. Several studies have been done relating energy efficiency measurement and improvement on industry [11–17]. Mostly, they concentrate on finding relevant energy intensity values [18,19]. Less attention has been paid to the effect of structural changes, i.e., retirements and additions of capacity.

The aim of this work is to study the impact of structural changes on energy efficiency of Finnish pulp and paper industry. EU (European Union) and IPPC (Integrated Pollution Prevention and Control) have called for intensification of energy usage [9]. At the same time, the business environment of pulp and paper industry is rapidly changing, and consequently pulp and paper industry has gone through a large structural change. Pulp and paper industry has been forced to implement notable changes to operate in a profitable way despite the challenging business environment. The demand for printing and writing paper is decreasing in EU, whereas an increasing amount of packaging materials is needed [20]. Thus, production grades are changing. Many mills have or are planning to enlarge their product portfolio towards new bioproducts, such as biofuels and biomaterials [21]. The major changes seen in Finland have been old unit closures, new unit start-ups and conversions of paper mills to products with increased demand, such as paperboard. It can safely be assumed that neither closed and nor started units have an average energy efficiency. Therefore, this study focuses on the effect of closures and start-ups on the pulp and paper industrial energy efficiency.

This paper consists of five parts. Section 2 introduces Finnish pulp and paper industry, especially developments in its production and energy consumption. Section 3 presents research methods and data gathering process. Results are shown in Section 4. Section 5 is discussion that considers received results. Finland was chosen as the target country of the research. It belongs to the major forest industry countries and the structural change has been significant in the past decade. The study is executed as a scenario analysis. Mill energy consumption data, gathered from various sources, is used for creating three scenarios, which are compared. The study focuses on energy aspects and for example economic measures are excluded.

2. Finnish Pulp and Paper Industry

Two-thirds of Finland is covered by forests [22]. Due to the ample resources Finland is one of the most important countries for the forest industry in the world. Forest industry was the second-largest employer in Finland in 2017 with 42,000 employees and accounting about 20% of both gross value of manufacturing and export [23].

Pulp and paper industry consumes most of the energy the forest industry uses. Mechanical forest industry is much less energy intensive. Its main products are sawn wood and wood-based panels. New forest industry bioproducts such as pellets, biogas, or biofuels are still of small volume. Wood pulp is the main virgin fiber material of paper. Also recycled or non-wood fiber is used. Chemical pulp is produced using sulphate process, sulphite process or other minor pulping processes. These processes are based on separating lignin from wood fibers by using alkaline chemicals in cooking. 80% of pulp produced in the world is produced by sulphate process [24]. In Finland, the sulphite process has not been used since 1992 [22]. Mechanical pulping consists of a grinding and refining processes, which utilizes mechanical energy to separate wood fibers. Many pulping processes accounted for as mechanical pulping are actually combinations of chemical and mechanical stages. Pulps produced by different processes have various properties and manufacturing costs. In addition to chosen process, properties are affected by treatment of fiber after pulping, such as bleaching. Paper is made in a paper machine that consists of forming, pressing, and drying sections. Different finishing processes like calendaring and coating can be utilized to achieve needed paper properties. The major paper grades are packaging papers and paperboards, printing and writing papers, and tissue papers.

2.1. Production

The major products of Finnish forest industry are pulp, paper, and mechanical forest industry products such as sawn wood and wood-based panels. In addition, a wide range of bioproducts is produced. Between 2010 and 2017, production of pulp and mechanical forest industry products has increased 336,000 tons and 2,350,000 m^3, respectively [20,25]. Paper production has been declining and in 2017 paper production was 1,462,000 tons lower than in 2010 [20]. Finland is an important exporter of pulp and paper products [26]. Exported pulp and paper were 3.7 million tons and 10 million tons in 2018, respectively [20].

Finnish forest industry has gone through a large structural change, which has modified product portfolios and production rates. Table 1 shows production volumes at the beginning and end of the decade. Production of printing and writing paper has decreased following the global trend. A main factor is the lowered demand for newsprint and magazine paper as a result from online publications. The production of paperboard has instead increased. As the level of wellbeing, especially in Asia, has risen, the demand for pulp and paper is increasing there [10]. Higher global pulp demand and favorable prices have increased chemical pulp production in Finland [26].

Table 1. Pulp and paper production in Finland in 2010 and 2017. Data from [20].

Grade	2010 (1000 t/a)	2017 (1000 t/a)	Change (%)
Paperboard	2830	3622	33
Other paper	1462	1232	−4
Printing and writing paper	7466	5422	−26
Chemical pulp	6733	7703	14
Mechanical and semi-chemical pulp	3775	3141	−13

Finnish pulp and paper industry consists currently of 17 paper mills, 14 paperboard mills, and 18 pulp mills [23]. Only few stand-alone pulp mills produce dried pulp that is delivered to paper mills. Over 80% of the pulp, as well as over 90% of the paper, is produced in integrated mills, as can be seen in Table 2. Integrated mill means that both pulp and paper are produced at the same site. Integration rate affects significantly the energy usage and the energy efficiency. Integrated mills produce pulp more energy efficiently than stand-alone mills, because they have no need for pulp drying and re-pulping and secondary heat can be used to preheat water needed in the paper machine. Practically all mechanical and recycled pulp units in Finland are integrated with a paper mill.

Table 2. Division of pulp and paper production in Finland. Elaborated from [9].

Category	Share of Pulp Production (%)	Share of Paper Production (%)
Bleached chemi-thermomechanical pulp	4	-
Integrated recycled pulp and paper	2	3
Stand-alone chemical pulp	13	-
Stand-alone paper	-	6
Integrated mechanical pulp and paper	20	35
Integrated chemical pulp and paper	60	55

2.2. Energy Usage

Finnish forest industry consumed 19.7 TWh electricity in 2017, which is about one fifth of Finnish total electricity consumption [8]. Total energy consumption in Finland was 1352.3 PJ in 2017 [27]. Forest industry used 217 PJ fuels [8], which is a significant share of the total fuel usage. Forest industry is renewable-intensive, and it accounts for 45% of production and consumption of Finnish bioenergy [10]. During the 2010s, consumption of primary energy decreased only 0.3% Fuel consumption development towards fossil fuel-free operation is a significant change. The share of biofuels increased from 76% to 85% whereas the share of fossil fuels (natural gas, heavy fuel oil, and coal) decreased from 17% to 10%. Primary energy consumption in 2010 and 2017 is presented in Table 3.

Table 3. Primary energy consumption of Finnish forest industry in 2010–2017. Elaborated from [8].

Primary Energy Usage (1000 TJ)	2010	2017	Change
Biofuels, liquid	13,534	14,984	11%
Biofuels, solid	29,787	35,468	19%
Natural gas	30,820	13,248	−57%
Peat	12,649	7875	−38%
Heavy fuel oil	6217	5454	−12%
Coal	1048	2536	142%
Others	1910	2685	41%
Total	217,774	217,115	−0.3%

Table 4 presents electricity consumption development of Finnish forest industry in 2010–2017. Data utilized in the table is presented in Supplementary material. Pulp and paper industry consumes over five times more electricity than mechanical forest industry and production of bioproducts combined. That is because chemical forest industry utilizes processes with a high energy-intensity. On the other hand, chemical pulp mills produce a lot of energy from the combustion of their sidestreams. Modern pulp mills are able to produce much more heat and over double the electricity than their processes require [28]. The surplus energy can be sold. In addition to electricity consumption, pulp and paper industry is a significant heat consumer. It is challenging to study the heat consumption of Finnish forest industry. Mills do not record or inform their heat consumption precisely because heat is not as valuable a product for mills as electricity. Therefore, this paper does not present development of heat consumption for the lack of data, but heat saving due to structural changes will be calculated.

Table 4. Electricity consumption of Finnish forest industry in 2010 and 2017.

Electricity Consumption (GWh)	2010	2017	Change	Share of Total Consumption in 2017
Bioproducts	382	667	74%	3%
Mechanical forest industry	956	1138	19%	6%
Pulp and paper	20792	17819	−17%	91%
Total	22130	19624	−13%	

Finnish forest industry electricity consumption has been slightly decreasing during the last decade. The annual electricity consumption has decreased during 2010–2017 by 2506 GWh. Reduced production of paper has decreased electricity use. At the same time, the higher amounts of bioproducts, pulp, and mechanical forest industry products have increased the total electricity consumption. In addition, changes in the end products have affected the energy consumption. The part of the electricity consumption that is not attributed to production changes can be assumed to be caused by energy efficiency improvements.

Available high-quality data restricts the reviewed period to 2011–2017. Only electricity consumption statistics, including consumption of the total forest industry, are available [8]. Therefore, the consumption of pulp and paper industry must be calculated. The most accurate way for defining electricity consumption of pulp and paper industry is to subtract electricity consumptions of mechanical forest industry and bioproducts manufacturing from the total consumption. The electricity consumption (Figure 1) is estimated using several sources, which are introduced in Supplementary material. Annual electricity usage was 2336 GWh higher in 2011 than in 2017. A significant reason for electricity consumption decrease is the reduction of production rate and evolved product palette. Even if the chemical pulp production has increased, the paper production has decreased more, and consequently total electricity usage is currently lower than at the beginning of the decade. Based on gathered production data (Supplementary material) and typical electricity consumption values [29], the decrease caused by lower production rates and changed products is 1004 GWh (~5%). Therefore, an approximate

1332 GWh (~7%) decrease must be attributed to other reasons. The major factor for this decrease has been the energy efficiency improvement.

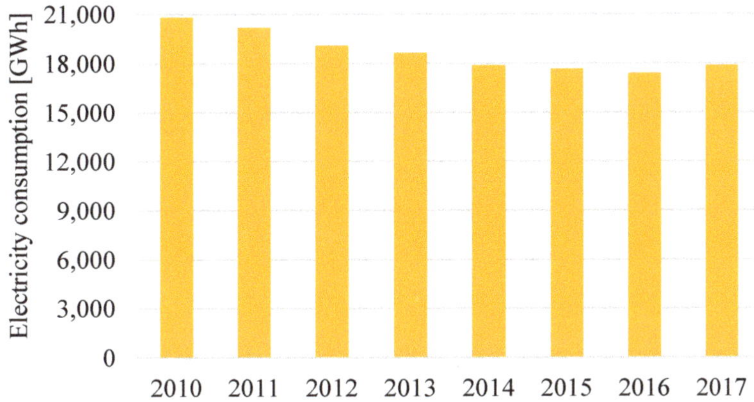

Figure 1. Pulp and paper industry electricity consumption in 2011–2017.

2.3. Structural Changes

Structural changes have been manifested in unit closures, unit start-ups and unit conversions to produce more profitable grades [9,29,30]. Changes in demand have led to significant reductions in pulp and paper production capacity [31]. In addition to whole mill closures and start-ups, major capacity increases and single machine upgrades have been taken into account in this study. In cases of conversions, old product is counted as a closure and new product is counted as a start-up. Table 5 presents closures during reviewed period. Paper production capacity removals have been 2,600,000 tons. Pulp capacity closures account to 530,000 tons. The largest wave of closures has occurred in printing and writing paper mills.

Table 5. Unit closures in Finland in 2011–2017 [9].

Unit	Capacity Decrease (t/a)	Product	Year of Closure
Myllykoski (UPM)	600,000	Magazine paper	2011
Myllykoski (UPM)	213,000	Groundwood	2011
Ääneskoski (M-Real)	200,000	Fine paper	2012
Rauma (UPM)	245,000	Magazine paper	2013
Veitsiluoto (Stora Enso)	190,000	Magazine paper	2014
Lohja (Loparex)	68,000	Other paper	2014
Kaukas (UPM)	270,000	Magazine paper	2014
Jämsänkoski (UPM)	270,000	Magazine paper	2014
Kauttua (Jujo Thermal)	10,000	Other paper	2015
Varkaus (Stora Enso)	285,000	Fine paper	2015
Kotka (Kotkamills)	185,000	Magazine paper	2016
Tervasaari (UPM)	100,000	Other paper	2016
Kyrö (Metsä Board)	105,000	Other paper	2016
Kyrö (Metsä Board)	74,000	Groundwood	2016
Äänekoski (Metsä Fibre)	500,000	Chemical pulp	2017

Table 6 shows the start-ups during reviewed period. During the 2010s, significant forest industry investments have been made in Finland: Reforms mainly concern enlargements in chemical pulp and paperboard production, but also biofuel mills have been built [32]. The increased demand for paperboard can be seen in the table. The number of unit start-ups is lower than the number of unit closures. There are three mills that were converted to produce paperboard, one totally new pulp mill, and three mills with a significant capacity growth. The capacity increase of paper and pulp has been 880,000 tons and 1,570,000 tons, respectively.

Table 6. Unit start-ups in Finland in 2011–2017 [9].

Unit	Capacity Increase (t/a)	Product	Year of Start-Up
Simpele (Metsä Board)	80,000	Paperboard	2011
Imatra (Stora Enso)	20,000	Paperboard	2015
Varkaus (Stora Enso)	380,000	Paperboard	2015
Kymi (UPM)	170,000	Chemical pulp	2015
Kotka (Kotkamills)	400,000	Paperboard	2016
Äänekoski (Metsä Fibre)	1,300,000	Chemical pulp	2017
Kymi (UPM)	100,000	Chemical pulp	2017

3. Methods

In this study, a method for analyzing structural energy efficiency changes was developed. Utilizing the method required gathering a high amount of individual mill energy consumption data from various sources. Both electricity and heat consumption data were gathered, verified, and analyzed.

3.1. Statistical Analysis

The study was executed as a statistical analysis. We derived energy utilization for Finnish pulp and paper sector as reference scenario one. Two additional energy consumption scenarios were made. The second scenario assumes that no units were closed but new ones were started. The third one assumes no new units were started but old ones were closed. Yearly production of all grades in each scenario was kept constant. Used method is presented step by step in Figure 2. Firstly, mills' heat and electricity consumption data, as well as mills' production rates, were collected and validated. To facilitate product division, all mills were categorized to groups introduced in Table 7. Mills' heat and electricity consumptions were decomposed to different products to enable comparing similar products. Energy consumptions of closed and started mills were compared with average existing ones. The average existing mill was defined using consumption values of at least five Finnish pulp and paper industry mills producing similar products as certain changed mill. Changes in energy consumption were calculated by assuming that average mills would replace the production of changed mills. Mills never operate with full capacity during the whole year due to maintenance stoppages etc. Calculations are done utilizing production rates 85% of the maximum capacity. Finally, obtained yearly energy consumption values were compared with statistical values.

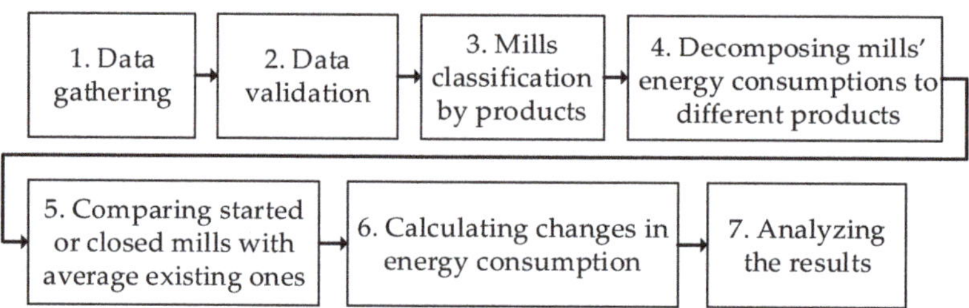

Figure 2. Utilized methodology step by step.

Table 7. Pulp and paper categorization groups.

Pulp	Paper
Chemical pulp	Paperboard
Groundwood	Magazine paper
Refiner	Fine paper
Semi-chemical pulp	Other paper
Chemi-mechanical pulp	
Recycled pulp	

Many Finnish mills are integrated, which makes evaluating energy consumption challenging. Mills reports usually the consumption of the whole mill, and therefore the consumption must be divided to different processes and products. The total energy consumption consists of the sum of consumption of every pulp and paper grade produced in the same mill:

$$E_{tot} = \sum_{i=1}^{n} P_i \cdot e_i \qquad (1)$$

where E_{tot} is total energy consumption of the mill, P_i is production rate of the product and e_i is a weighting factor. The weighting factor is a typical specific energy consumption of the product. To be able to divide the consumption values to different pulp and paper grades, weighting factors presented in Table 8 are utilized.

Table 8. Weighting factors for dividing consumptions for different products. Data from [29].

Grade	Electricity Consumption (GJ/t)	Heat Consumption (GJ/t)
Chemical pulp	2.4	11.2
Groundwood	6.5	0
Refiner	5.5	−2.4
Semi-chemical pulp	1.6	3.6
Recycled pulp	1.4	0.2
Magazine paper	2.5	4.6
Fine paper	3.1	4.9
Fluting	1.2	5
Paperboard	2.4	5.6
Newsprint	2.1	4.7
Other paper	2.5	7.5

3.2. Data Gathering

Verifying and analyzing the results requires gathering reliable heat and electricity consumption data from every Finnish pulp and paper mill. In addition, the products and production rates of every mill must be known. The focus was on closed and started mills and mills similar with changed ones. Finding energy consumption data is difficult because most companies like to keep their consumption and production values as a trade secret for commercial purposes. A high number of sources, mainly environmental reports and permits, university theses dealing with energy use of individual mills, articles, and other publications, were utilized for assigning heat and electricity production and consumption values for each mill and each product. Gathered data and sources are presented in Supplementary material. Most of this work was started in a project for the Finnish Ministry of Environment [30]. Annual production rates for 2011–2017 were obtained from Finnish Forest Industries [33]. Occasionally, there was a high statistical difference with some of the values when compared with other similar mills. Clearly incorrect values were left out of the study or corrected to more reasonable ones.

Gathering statistical data includes known sources of error. Mills do not necessarily measure and report their production and consumption values in a same way and the measurement devices and practices can be different. In addition, specific energy consumption in a certain mill varies significantly, for example due to capacity utilization rate or climate conditions (winter/summer). Some gathered values were reported several years ago which also increases the possibility of errors. However, utilization of individual mill values instead of statistical averages allows one to study the structural changes.

4. Results

Calculated results are shown in Figure 3a,b. Positive values define years with electricity or heat energy savings due to structural changes. Negative values indicate years when average of existing mills have been more energy efficient than started mills or less efficient than closed mills, and therefore the total energy consumption has increased. Figure 3a presents the impacts of unit closures on heat and electricity consumption. Heat is saved every year excluding 2013 when only one, relatively energy-efficient, paper machine was closed. Between 2015 and 2017, heat was saved but electricity consumption increased somewhat. Total heat and electricity savings due to closures were 193 GWh and 109 GWh, respectively. Figure 3b presents the effects of unit start-ups. New start-ups seem to improve the energy efficiency. The most significant savings occurred in 2017 when a modern pulp mill with a high capacity was started. In 2012–2014, no new mills were started. Total heat and electricity savings due to start-ups were 383 GWh and 191 GWh, respectively. Closures and start-ups together saved heat and electricity 577 GWh and 299 GWh, respectively. As many factors, especially the accuracy of data reported by mills, affect the results, they should be viewed as the best estimation with the gathered data and used assumptions.

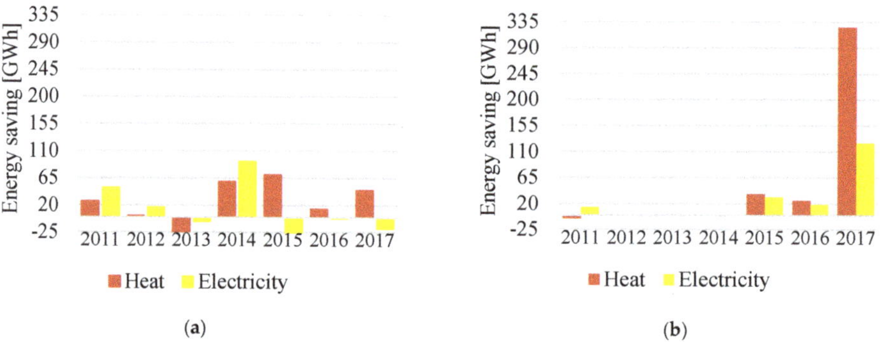

Figure 3. (**a**) Energy savings due to unit closures. (**b**) Energy saving due to unit start-ups.

Figure 4 presents the changes in Finnish pulp and paper electricity consumption. Changed production volumes are the main reason for changes in electricity usage. The black line estimates electricity consumption of chemical pulp and paper industry if the units present in 2011 would have continued to produce each year's production. It can be seen that both unit closures and start-ups have decreased the electricity consumption. Therefore, it seems safe to say they have increased the energy efficiency.

Figure 5 sums up the results of the development of electricity usage over the studied period. Decrease of annual consumption in pulp and paper industry during reviewed period was 2336 GWh. Approximately 1004 GWh was accounted for by decreased production rate and changed production portfolio. Therefore, decrease of 1332 GWh has been due to various energy efficiency improvements. Structural changes have decreased the annual electricity consumption by 299 GWh, and therefore almost 22% of the energy efficiency improvement is accounted by unit closures and start-ups. The respective shares of the start-up and closures were about 8% and 14%. The majority, almost 80%, of the improvement is derived from other energy efficiency improvement projects. It was previously mentioned that production of changed mills is 85% of the total capacity. If the total capacities had been used instead, the structural changes' share of the energy efficiency improvement would have been 26%. These results are valid only with the assumptions used here. If production volumes or the mix of different products change, the results would change.

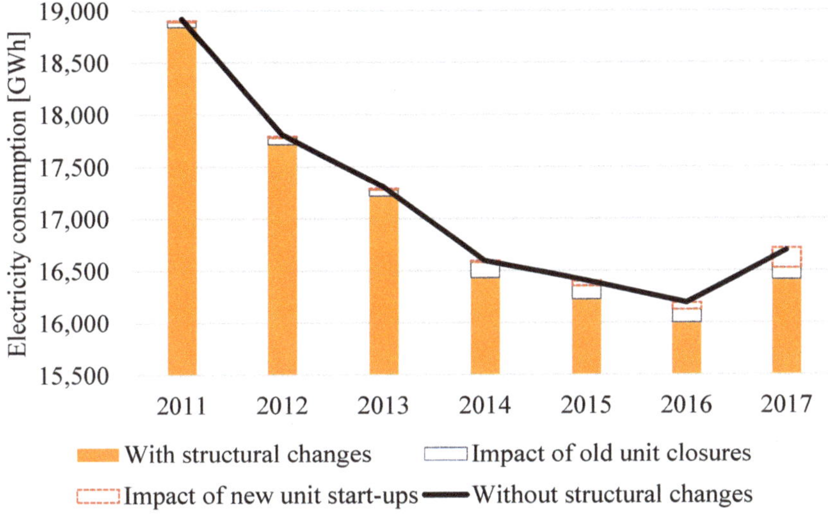

Figure 4. Effect of structural changes on electricity consumption.

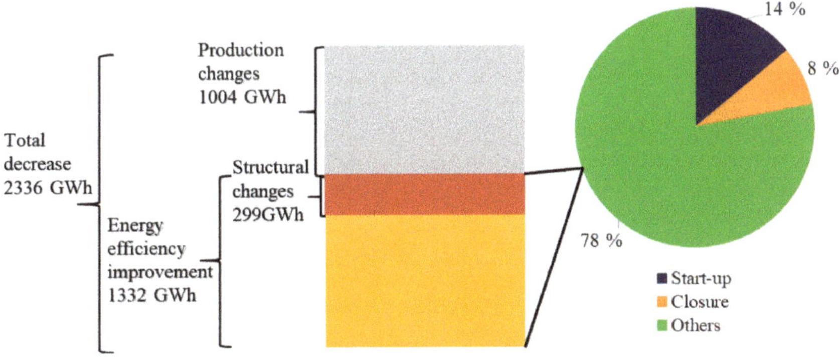

Figure 5. Division of electricity consumption changes in Finnish P&P industry in 2011–2017.

5. Discussion

The results indicate that, considering electricity, the structural changes had about 20% effect on the total energy efficiency improvement. With a moderate to high certainty, it can be stated that structural changes have improved the energy efficiency of Finnish forest industry. However, the structural changes explain only one-fifth of the improvement, and the remaining four-fifths must be explained with other factors. These major factors are changes in used technology and improved modes of operation. These technology changes include both small and large energy efficiency investments. Large investments, like modernizations of individual departments, improved utilization of secondary and waste heat or water cycle closures, can have significant impact on energy consumption.

Small changes, like repairing defective components or correcting operating practices, also have a clear influence on energy efficiency. Improvements in modes of operation consist of education and motivation of personnel as well as ensuring reliable data from processes is collected.

The results suggest that the size of heat savings was almost twice the size of the savings in electricity consumption. That finding meets the current trends. New technologies, for example, improved pressing and drying technologies, have increased the electricity consumption while they have decreased heat consumption [9]. In addition, modern devices for environmental protection have typically higher electricity consumption than older ones [10]. In summary, heat consumption of modern mills should clearly be lower than the heat consumption of older ones, and on the other hand, electricity consumption might be similar or even higher in modern mills in comparison to old ones.

Another interesting finding is that start-ups seem to improve energy efficiency more than closures. Start-ups have saved 70% more heat and electricity than closures, even if the amount of closed capacity was 30% higher than the started one. This finding can be explained by a high-level maintenance and regular upgrades of old mills in Finland. With good care, operating mills are kept in a good shape and the energy efficiency of them is thus not significantly lower than an average Finnish mill. On the other hand, this study indicates that modern mills have a high potential to reduce energy consumption of pulp and paper industry.

This study is valuable for policy makers, legislation, and industries. It indicates that structural changes have only a minor impact on the total energy efficiency and therefore results highlights importance of improving existing mills. Efforts used for energy efficiency improvement during recent decades have realized as significant results. On the other hand, a new large-size mill started in 2017 operates with high energy efficiency but its impact on total energy efficiency is low. The results encourages actors to invest in existing mills' improvement and maintenance. The Finnish pulp and paper industry has changed significantly during 2010s. The change has affected production rates and portfolio as well as energy usage and efficiency. Change will likely continue in the future. It is probable that new bio-based products reach an important position and even more printing and writing paper will be replaced with paperboard [29]. The changes will lead to modifications of pulp and paper making units. The results of this study can be used for estimating energy efficiency changes also in the future.

In further studies, it would be interesting to define the shares of other factors of energy efficiency improvement. Studying this topic would be difficult because a high amount of data about changes done in the mills, operational modes of the mills and energy consumptions should be made available. Also, the impact of structural changes on a global level should be examined. Varying structure of pulp and paper production as well as mill age will lead to changes in the presented results in every country. For example, countries with old mills operating with original processes probably save a high amount of energy by closing these old mills.

6. Conclusions

Finnish pulp and paper industry has gone through a large structural change that has manifested itself as several unit closures and a few new unit start-ups. In addition to changing production rate and product portfolio, structural changes have affected energy efficiency of the Finnish pulp and paper industry. The study was executed by collecting a high amount of mills' operational data and creating three scenarios. With the scenarios, energy efficiency improvement due to structural changes was estimated. Between 2011 and 2017, annual electricity consumption has decreased due to reduced production rate, changed products, and energy efficiency improvements. Energy efficiency improvements consist of several factors, such as structural changes, improved technology and processes, and more optimal operational choices. This study estimates that approximately 20% of the energy efficiency improvement is a result of structural changes. The remaining 80% is a sum of the other factors. The study also indicates that modern mill start-ups have a greater effect on energy efficiency than old mill closures. However, improving existing mills has a higher effect on total energy efficiency than closing old and starting new mills.

Supplementary Materials: The following are available online at http://www.mdpi.com/1996-1073/12/19/3689/s1, Table S1: Products and production rates of Finnish pulp and paper mills. Table S2: Electricity consumption of Finnish pulp and paper mills. Table S3: Heat consumption of Finnish pulp and paper mills. Table S4: Specific electricity consumption of Finnish pulp and paper production. Table S5: Specific heat consumption of Finnish pulp and paper production. Table S6: Production rates of mechanical forest industry products. Table S7: Electricity consumption of mechanical forest industry products. Table S8: Production rates of bioproducts. Table S9: Electricity consumption of bioproducts. Figure S1: Electricity consumption of Finnish forest industry in 2010s.

Author Contributions: S.K. was responsible for the calculations and writing the manuscript. E.V. and T.L. made valuable comments and remarks during the process of writing.

Funding: The project has received funding from the Academy of Finland project "Role of forest industry transformation in energy efficiency improvement and reducing CO_2 emissions" which is gratefully acknowledged.

Conflicts of Interest: The authors declare no conflict of interest.

References

1. International Energy Agency (IEA). Statistics. Available online: https://www.iea.org/statistics/?country=WORLD&year=2016&category=Energy%20consumption&indicator=TFCbySuorce&mode=chart&dataTable=BALANCES (accessed on 24 June 2019).

2. Worrel, E.; Bernstein, L.; Roy, J.; Price, L.; Harnisch, J. Industrial energy efficiency and climate change mitigation. *Energy Eff.* **2009**, *2*, 109–123. [CrossRef]

3. Fracaro, G.; Vakkilainen, E.; Hamaguchi, M.; Melegari de Souza, S.N. Energy Efficiency in the Brazilian Pulp and Paper Industry. *Energies* **2012**, *5*, 3550–3572. [CrossRef]

4. Backman, F. Barriers to Energy Efficiency in Swedish Non-Energy-Intensive Micro-and Small-Sized Enterprises—A Case Study of a Local Energy Program. *Energies* **2017**, *10*, 100. [CrossRef]

5. International Energy Agency (IEA). Wolrd Balance. 2016. Available online: https://www.iea.org/sankey/ (accessed on 13 June 2019).

6. Fleiter, T.; Fehrenbach, E.; Worrell, E.; Eichhammer, W. Energy efficiency in the German pulp and paper industry—A model-based assessment of saving potential. *Energy* **2012**, *40*, 84–99. [CrossRef]

7. Lin, B.; Zheng, Q. Energy efficiency evolution of China's paper industry. *J. Clean. Prod.* **2014**, *140*, 1105–1117. [CrossRef]

8. Finnish Forest Industries. Statistics: Energy and Logistics. Available online: https://www.forestindustries.fi/statistics/energy-and-logistics/ (accessed on 18 June 2019).

9. Kähkönen, S. Energy Efficiency of Finnish Forest Industry. Master's Thesis, LUT University, Lappeenranta, Finland, 2019.

10. Vakkilainen, E.; Kivistö, A. *Energy Consumption Trends and Energy Consumption in Modern Mills in Forest Industry Production*; LUT Energy: Lappeenranta, Finland, 2017; 198p.

11. Blomberg, J.; Henriksson, E.; Lundmark, R. Energy efficiency and policy in Swedish pulp and paper mills: A data envelopment analysis approach. *Energy Policy* **2011**, *42*, 569–579. [CrossRef]

12. Peng, L.; Zeng, X.; Wang, Y.; Hong, G.B. Analysis of energy efficiency and carbon dioxide reduction in the Chinese pulp and paper industry. *Energy Policy* **2015**, *80*, 65–75. [CrossRef]

13. Corcelli, F.; Fiorentino, G.; Vehmas, J.; Ulgiati, S. Energy efficiency and environmental assessment of papermaking from chemical pulp—A Finland case study. *J. Clean. Prod.* **2018**, *198*, 96–111. [CrossRef]

14. Soepardi, A.; Pratikto, P.; Santoso, P.B.; Tama, I.P. Linking of Barriers to Energy Efficiency Improvement in Indonesia's Steel Industry. *Energies* **2017**, *11*, 234. [CrossRef]

15. Solnørdal, M.T.; Foss, L. Closing the Energy Efficiency Gap—A Systematic Review of Empirical Articles on Drivers to Energy Efficiency in Manufacturing Firms. *Energies* **2018**, *11*, 518. [CrossRef]

16. Oh, S.C.; Hildreth, A.J. Estimating the Technical Improvement of Energy Efficiency in the Automotive Industry—Stocastic and Deterministic Frontier Benchmarking Approaches. *Energies* **2014**, *7*, 6199–6222. [CrossRef]

17. Stenqvist, C. Trends in energy performance of the Swedish pulp and paper industry 1984–2011. *Energy Eff.* **2015**, *8*, 1–17. [CrossRef]

18. Worrell, E.; Neelis, M.; Price, L.; Galitsky, C.; Zhou, N. *World Best Practice Energy Intensity Values for Selected Industrial Sectors*; Rev. 2; Publication Number: LBNL-62806; Lawrence Berkeley National Laboratory: Berkeley, CA, USA, 2008; 44p.

19. Laurijssen, J.; Faaij, A.; Worrel, E. Benchmarking energy use in the paper industry: A benchmarking study on process unit level. *Energy Eff.* **2013**, *6*, 49–63. [CrossRef]

20. Finnish Forest Industries. Statistics: Pulp and Paper Industry. Available online: https://www.forestindustries.fi/statistics/pulp-and-paper-industry/ (accessed on 20 June 2019).

21. Hamaguchi, M.; Cardoso, M.; Vakkilainen, E. Alternative Technologies for Biofuels Production in Kraft Pulp Mills—Potential and Prospects. *Energies* **2012**, *5*, 2288–2309. [CrossRef]

22. Seppälä, M.; Klemetti, U.; Kortelainen, V.A.; Lyytikäinen, J.; Siitonen, H.; Sironen, R. *Paperimassan Valmistus (Pulp Production)*; Gummerus Kirjapaino Oy: Jyväskylä, Finland, 2001; 196p. (In Finnish)

23. Finnish Forest Industries. Statistics: Forest Industry. Available online: https://www.forestindustries.fi/statistics/forest-industry/ (accessed on 19 June 2019).

24. European Commission. *Reference Document on Best Available Techniques for Pulp, Paper and Board*; Joint Research Centre, Institute for Prospective Technological Studies: Seville, Spain, 2015; 906p.

25. Food and Agriculture Organization (FAO). Forestry Production and Trade. Available online: http://www.fao.org/faostat/en/#data/FO (accessed on 18 June 2019).

26. Viitanen, J.; Hänninen, R. *Metsäsektorin Suhdannetiedote 2018. (Forest Sector Trends 2018)*; Natural Resources Institute Finland (Luke). Luonnonvarakeskus: Helsinki, Finland, 2018; 24p. (In Finnish)

27. Tilastokeskus. Energia. Available online: https://www.stat.fi/tup/suoluk/suoluk_energia.html (accessed on 18 June 2019).

28. International Renewable Energy Agency (IRENA). *Bioenergy from Finnish Forests*; International Renewable Energy Agency: Abu Dhabi, UAE, 2018; 36p.

29. Pöyry. *Suomen Metsäteollisuus 2015–2035 (Finnish Forest Industry 2015–2035)*; Ministry of employment and the economy: Helsinki, Finland, 2016; 57p. (In Finnish)

30. Kivistö, A.; Ikonen, U.; Tanskanen, T.; Vakkilainen, E.; Ojanen, P.; Pesari, J. *Forest Industry Environmental Strategy (Governance)*; The Forest Industry Development Scenarios for 2020; Reports 96; Centre for Economic Development, Transport and the Environment: Kouvola, Finland, 2013; 70p. (In Finnish)

31. Natural Resources Institute Finland (Luke). Suomen Metsät 2012: Suomen Metsät ja Metsätalous Pähkinän Kuoressa (Finnish Forests 2012: Finnish Forests and Forest Economy in a Nutshell). Available online: http://www.metla.fi/metinfo/kestavyys/SF-1-forest-industry.htm (accessed on 16 September 2019). (In Finnish)

32. Finnish Forest Industries. Metsäteollisuuden Tulevaisuuden Toimintaympäristö (Operating Environment of Forest Industry in the Future). Available online: http://www.teollisuudenmetsanhoitajat.fi/wp-content/uploads/2017/08/Tomi-Salo-Tulevaisuus.pdf (accessed on 16 September 2019). (In Finnish)

33. Finnish Forest Industries. Metsäteollisuuden Ympäristötilastot (Forest Industry Environmental Statistics). 2011–2017. Available online: https://www.metsateollisuus.fi/kategoriat/julkaisut/ymparisto (accessed on 25 June 2019). (In Finnish)

Article

Real Time Energy Performance Control for Industrial Compressed Air Systems: Methodology and Applications [†]

Miriam Benedetti [1], Francesca Bonfà [1], Vito Introna [2], Annalisa Santolamazza [2,*] and Stefano Ubertini [3]

[1]   Energy, New Technology and Environment Agency (ENEA), Rome 00123, Italy;
     miriam.benedetti@enea.it (M.B.); francesca.bonfa@enea.it (F.B.)
[2]   Department of Enterprise Engineering, University of Rome Tor Vergata, Rome 00133, Italy;
     vito.introna@uniroma2.it
[3]   DEIM School of Engineering, University of Tuscia, Viterbo 01100, Italy; stefano.ubertini@unitus.it
*   Correspondence: annalisa.santolamazza@uniroma2.it
†   This paper is an extended version of the conference paper: Bonfá, F.; Benedetti, M.; Ubertini, S.; Introna, V.;
     Santolamazza, A. New efficiency opportunities arising from intelligent real time control tools applications:
     The case of compressed air systems' energy efficiency in production and use. Energy Procedia 2019,
     158, 4198–4203.

Received: 30 July 2019; Accepted: 14 October 2019; Published: 17 October 2019

Abstract: Most manufacturing and process industries require compressed air to such an extent that in Europe, for instance, about 10% of the total electrical energy consumption of industries is due to compressed air systems (CAS). However, energy efficiency in compressed air production and handling is often ignored or underestimated, mainly because of the lack of awareness about its energy consumption, caused by the absence of proper measurements on CAS in most industrial plants. Therefore, any effective energy saving intervention on generation, distribution and transformation of compressed air requires proper energy information management. In this paper we demonstrate the importance of monitoring and controlling energy performance in compressed air generation and use, to enable energy saving practices, to enhance the outcomes of energy management projects, and to obtain additional benefits for non-energy-related activities, such as operations, maintenance management and energy accounting. In particular, we propose a novel methodology based on measured data, and baseline definition through statistical modelling and control charts. The proposed methodology is tested on a real compressed air system of a pharmaceutical manufacturing plant in order to verify its effectiveness and applicability.

Keywords: energy efficiency; compressed air systems; energy data analysis; energy measures; performance control; operations; maintenance; energy accounting

---

1. Introduction

The research presented in this paper is part of a wider research project addressed to the optimization of the energy use of compressed air systems (CAS) in manufacturing plants in the context of Italian energy intensive companies [1–3].

Through the information obtained by more than 15,000 energy audits and data collected through questionnaires, detailed analyses have led to the conclusion that the actual state of the CAS within the Italian industry shows great opportunities for improvement. Indeed, the analyses, on the one hand, confirmed that compressed air production accounts for quite an high portion of the overall Italian industrial sector's electric energy consumption, and, on the other hand, revealed the widespread lack of monitoring systems to measure the energy consumption of CAS [3]. In fact, despite the importance attributed by most research and legislation to the use of reliable measures for energy efficiency and energy savings in industrial sites [4–8], this practice is clearly still quite far from being diffusely implemented. In addition, this is due to the limited knowledge of ensuing energy efficiency and non-energy efficiency benefits. Therefore, there are still huge opportunities for improvement for both technical and managerial aspects. In particular, the improvement of energy performance control seems to be one of the most important interventions to apply as it can provide guidance to the company in the identification of energy waste, in the of optimization of the use of energy, in the improvement of maintenance methodologies and in the assessment and verification of results in economic terms.

Moreover, the increasingly fast innovation in the fields of distributed control, smart metering and machine learning can foster the implementation of a dynamic energy control by companies [9], in an Industry 4.0 perspective. Indeed, the analysis of process data and the relationships among variables enable the acquisition of valuable insight for different applications, such as process monitoring, fault diagnosis, mode clustering, soft sensing of key variables/quality variables, etc. [10].

For example, in operations management, by comparing measured data to a highly accurate and reliable reference, it is possible to exert an efficient and more effective real time control of the performance of the system and an optimization of its operating conditions [11].

Furthermore, from a maintenance point of view, the use of advanced techniques can support the real time performance control during production, implementing faults detection [12–15], fault diagnosis [16,17] and remaining useful life estimation to support the optimization of manufacturing management and maintenance scheduling [18].

Finally, the use of real time control tools can benefit the organization also from a pure economical perspective. Modelling the energy performance of the system, in fact, allows a more accurate estimation of the energy budget, not only enabling a timely discovery of anomalous behaviors but also supporting management in the investigation of their actual causes, thus enacting an efficient control of the overall performance of the system [19].

The aim of the paper is to present a general computing approach applied to energy performance control which can also jointly support the decision-making process of three different non-energy-related activities, namely operations, maintenance and energy accounting.

The approach presented can be applied to any energy use in industry, but is here specifically tailored to CAS. The paper is structured as follows: Section 2 describes the background of the problem, discussing the different methodological approaches identified in literature to impact energy performance control and the main non-energy-related activities mentioned; Section 3 provides a description of the methodology adopted while Section 4 shows the results of the application of the proposed methodology to a real case study: the compressed air system of a pharmaceutical production plant. Finally, Section 5 ends the paper, presenting the main conclusions and implications of the study.

2. Background

In recent years, with the development of new information technologies the paradigm of the fourth technological revolution, namely Industry 4.0, has emerged. This concept is based on innovative technologies such as cyber-physical systems (CPS), the Internet of Things (IoT) and cloud computing and its aim is to achieve higher operational efficiency, productivity and automation, through a continuous integration and analysis of heterogeneous data [20,21]. Considering that some of the most prominent features of this paradigm are digitization, optimization, automation, human machine interaction (HMI), and automatic data exchange and communication, it is clear how Industry 4.0 is an industrial process of both value adding and knowledge management, supported by the use of internet technologies and advanced algorithms [21]. In this context, an important aspect of Industry 4.0 is the ability of manufacturing systems to monitor and analyze the industrial reality at different levels and with different purposes, through the use of a digital copy ("digital twin") of the physical system, process or manufacturing phenomena studied [22,23], thus making possible the enactment of smart decisions achieved through a continuous and synchronized information flow among humans, machines, and sensors alike [24].

In particular, a growing number of publications have been dedicated in the recent years to the study of new control tools based on the use of real time data from measurement systems in the field of energy performance control as well as in the main applications mentioned before: operations, maintenance and energy accounting. While presenting a variety of methodologies and different approaches, the scientific literature in these fields tends to focus on addressing these issues separately instead of analyzing the possibility of tackling them in a joint manner.

In order to examine the implications of the methodology presented in this paper, the main aspects and opportunities arisen in the mentioned applications are here briefly described (to obtain further information about the scientific literature on these topics refer to Appendix A for a deeper analysis of the publications examined).

Regarding the energy domain, various applications can benefit from a more complex data analysis. For instance, most of the different approaches available in literature to monitor and control energy performance in industrial plants share the same main phases [25,26]: measurement plan and data collection, baseline definition, implementation of control over time through comparison between the baseline, and monitored energy consumption. However, various techniques might be used to characterize the baseline energy performance of the system, depending on its complexity and on the availability and quality of data (e.g., air compressors, boilers, pumps, buildings, etc.) [27–33]. Often, for this purpose, statistical regression is used, and more recently—machine learning methodologies, such as artificial neural networks or support vector regression; all of which have been tested. The most common approach is to evaluate the residual between actual energy consumption and the prediction obtained from the baseline model. In order to help the evaluation of the residuals' significance, control charts, such as EWMA (exponentially weighted moving average), CuSum (cumulative sum) or Shewhart charts are used [27,28,30].

Other important applications developed due to the availability of real time data are the prediction of energy consumption, especially critical for building management [34–41] and electrical load management [42–54], and the forecast of energy production by renewable energy systems (mostly used for wind turbines and photovoltaic panels) [55–59].

Process control and performance monitoring are other fields interested in the innovation brought by data analytics tools. For instance, some data-driven models are used to give online estimation of key variables that, due to technical or economic limitations, would be too difficult to measure otherwise [60,61], while other models are used to perform process monitoring in order to detect anomalies [62].

In regards to maintenance, instead, depending on the specific application, different goals can be achieved: condition-monitoring, fault detection, diagnosis and prognostics. The more complex the objective, the more complex the model and the data used to achieve it. Therefore, while condition monitoring [63–65], fault detection [66,67] and diagnosis [12–14,68] may be achieved even with statistical models, prognostics models are usually developed through machine learning techniques [17,18,69–75]. Both supervised and unsupervised approaches are used in fault detection, diagnosis and prognosis (e.g., artificial neural networks, decision trees, ARMA (autoregressive moving average) or logistic regression models, etc.) [73].

Finally, the scientific literature related to energy accounting is still limited. In fact, while energy efficiency is finally regarded with attention in the industrial context, the cost of energy is still not widely analyzed. Some papers propose methodologies to predict the energy budget using variables of influence to estimate the energy consumption through statistical modelling [76,77]. Moreover, the differences between budget and actual costs can be analyzed through indicators that distinguish the effect of different specific causes (e.g., different prices, production volumes, on-site generation system efficiency, etc.) [76,78,79].

3. Methodology

The methodology described herein defines a series of steps for a real time energy performance control organized in a replicable process in order to facilitate their implementation in different industrial contexts.

The methodology was developed to be as widely applicable as possible (at least to a wide range of different industrial auxiliary systems) but is here particularly referred to CAS. It is fundamental to highlight that the paper focuses on energy performance rather than energy consumption. This entails a combined measure and integrates analysis of CAS' energy consumption and compressed air production which has the potential to provide much more information to the final user of the control system.

Moreover, one main innovative aspect of the proposed methodology is its capability of impacting, in a jointly manner, three different non-energy-related activities (operations, maintenance and accounting): (i) the definition of the best operating conditions of a system from an energy efficiency perspective; (ii) the identification of changes to energy consumption patterns or degradation of energy performances linked to sporadic faults or events and (iii) the achievement of a deeper understanding of the energy behavior of the overall system aimed at an improved energy accounting.

The methodology is developed following an empirical approach, on the basis of needs and requirements identified through a close collaboration with industrial practitioners involved in the case study and afferent to the different functions, such as energy, operations and maintenance management. Some of the tools integrated in the methodology adopt features from similar non-energy-related tools already in use, such as quality and process control tools as well as project management tools [9,10,27,28,76,77], in order to overcome implementation barriers mainly related to usability and resistance to change. As a result, people involved in the case study are able to work with familiar tools, even though based on new data analysis and real time control concepts and applied to the energy management context.

The rest of this section is organized as follows: the first subsection illustrates the overall methodology for real time energy performance control, while following subsections will be dedicated to each application.

3.1. Methodology for Real Time Energy Performance Control

The methodology is built on statistical tools for energy measurement and verification and energy monitoring and targeting [6,19,28], which were then customized for real time energy performance control (and auxiliary industrial systems, with particular focus on CAS) in order to draw a complete picture of the system's energy behavior, and therefore, create a reliable baseline.

Figure 1 gives a schematic representation of the most common variables to be monitored for CAS and the different possible boundaries of the analysis. Compressed air generation and utilization are represented together with their main inputs/outputs and all the main variables that are capable to influence the energy performance of the system, also called "energy drivers" [19]. Energy drivers are divided into two categories, i.e., controllable and non-controllable. Controllable energy drivers can be modified by the companies and set by operators, while non-controllable energy drivers cannot be modified (either because out of company's control, such as weather conditions, or because the company is not willing to change them to improve energy efficiency, such as production rates and working hours).

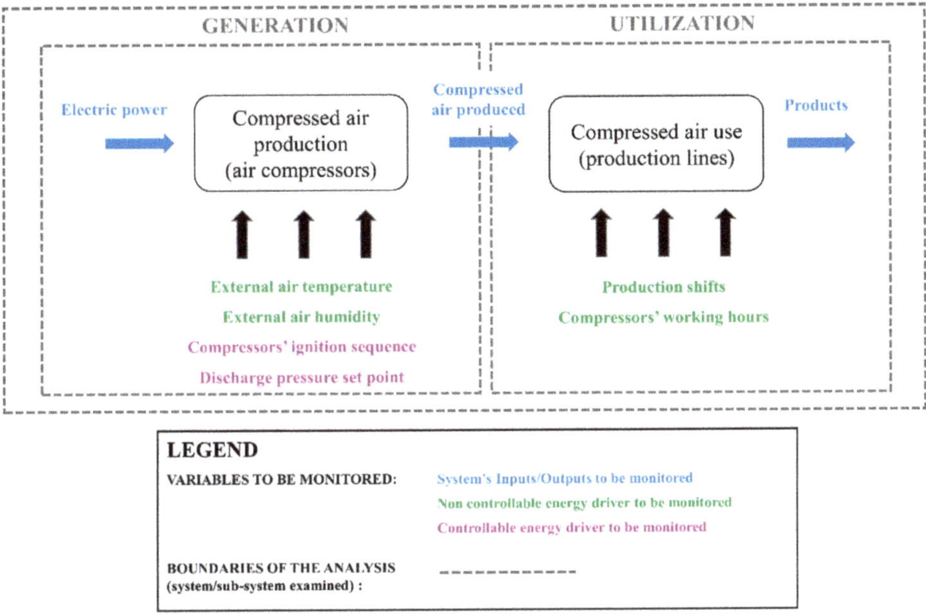

Figure 1. Schematic representation of the most common variables to be monitored for compressed air systems (CAS) and of the different boundaries of the analysis.

The methodology, represented in Figure 2, is made up of nine steps, briefly illustrated in the following.

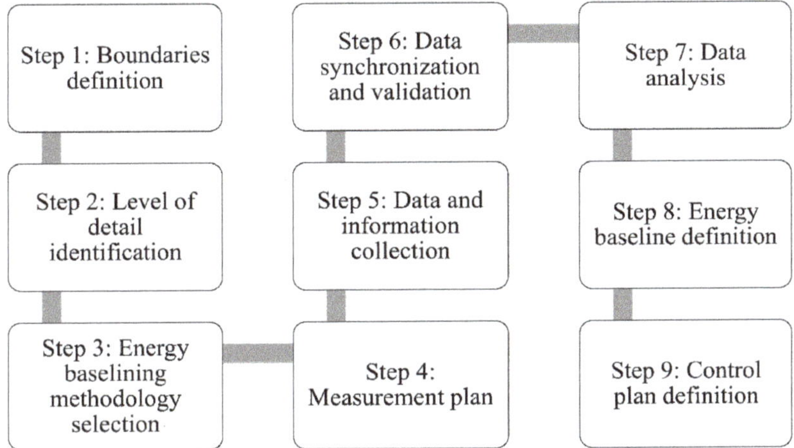

Figure 2. Conceptual map of the methodology developed for real time energy performance control. Such a process is iterative in nearly every step. Indeed, depending on the results of the specific step, it might be necessary to go back and change some of the assumptions or choices previously made.

Step 1 is the definition of the boundaries of the analysis. In this preliminary step, it is important to identify the main typologies of energy carriers (defined by the International Organization for Standardization as "substance or medium that can transport energy" [80]) and the transformations that they underwent. In particular, when examining a specific energy carrier, it is useful to distinguish between its generation and its utilization in order to be able to address the needs of different users in term of control (e.g., top management, operations managers, utilities managers, maintenance operators, etc.). For example, considering CAS, it is necessary to understand whether the analysis should focus on compressed air production (the generation of compressed air by air compressors using electricity), utilization (the use of compressed air by production equipment), or both.

Once boundaries are defined, it is necessary to identify the level of detail of the analysis (Step 2) in terms of space and time, determining, for example, whether it is more convenient to take into account the whole plant, a single department or just one production line or machine, and, moreover, defining the most suitable frequency at which analyze energy data. This second choice is made considering the thermodynamics of the process/system, the purpose of the control and the type of users of the implemented control system: a low frequency (e.g., from a few hours to a few days) is generally used for strategic control purposes whereas a higher frequency (e.g., from 10 to 30 min) is more suitable for operational control.

The third step is the energy baseline method definition. Among the several methods available to analyze data and to build the energy baseline, we chose statistical regression [25,27,81,82] as the most effective to organize the data collection accordingly. However, physical models, artificial neural networks or even machine learning techniques [15,28] are employable. First of all, it is important to highlight that the choice of the method mainly depends on the following factors: (i) available resources for the control system ramp-up; (ii) available resources for the control system maintenance; (iii) boundaries and chosen level of detail (i.e. the need to analyze several assets at the same time); and (iv) the need to keep a deep understanding of the thermodynamics controlling the process (i.e. highly automated tools generally allow a lower understanding, as they rely on "black box" methods).

As energy consumption is often dependent on several variables, the characterization of the energy behavior requires the collection of different types of information: consumption data, production data, environmental data, technical (users) and operational data [26]. Therefore, a measurement plan is implemented (Step 4) with the aim to perform an appropriate data collection for all the variables (inputs, outputs, energy drivers) that need to be monitored, paying special attention at their coherence (same period of time and frequency) and at identifying the appropriate measuring systems and the additional information that needs to be collected (e.g., maintenance schedules, production data). The amount of data and information collected and the consequent resources committed in this phase are defined accordingly to the entity of the potential saving achievable with real time control, while also establishing a duration of the data collection phase such as to guarantee a correct representation of all possible working conditions.

In the following phase (Step 5), the measurement plan illustrated above is implemented and data collection activities are performed.

Step 6 aims to check and solve any synchronization issues and to implement manual or automatic data handling, depending on the amount and quality of the collected data. Inaccuracies introduced by the measuring system are also corrected in this moment. Moreover, it is necessary to perform a first data validation in order to avoid poor/low quality data and meters' faults.

Afterwards, on the basis of the choice made in the third step, a model that characterizes the energy behavior of a system from a set of collected data is developed in Step 7 in order to enable the investigation of intrinsic and external causes of energy performance variability.

The first statistical tool is the correlation analysis of the previously collected energy drivers and energy consumption data to define the dependence of the energy behavior of the system on identified energy drivers, thus allowing preliminary qualitative considerations. Moreover, in order to get more quantitative information, it is possible to perform regression analyses, using the coefficient of determination (i.e., R^2) and the related probability (i.e., P_value) [81,83] to analyze the strength of the correlation and select the energy drivers on the basis of their actual influence.

Once the first model that included all the available data has been created, a validation and a first analysis of the overall performance of the system is carried out. Among the control charts used in quality control [83], two of them are particularly suitable for this kind of analysis.

First of all, let $E_{ac}(t)$ be the actual energy consumption of the system at time t and $E_{pd}(t)$ be the energy consumption predicted through the use of the baseline model for the same time t. The difference between these two values is the energy performance deviation, defined as $\Delta E(t)$ and calculated as follows:

$$\Delta E(t) = E_{ac}(t) - E_{pd}(t) \tag{1}$$

The evolution over time of this parameter ($\Delta E(t)$) can be observed in a first control chart—the control chart for energy performance deviations. Every point in the chart represented the energy performance deviation at a specific time (Figure 3a). Two lines, the upper and lower control limits, are added to the chart to support the analysis of the energy behavior of the system. These two control limits (UCL—upper control limit; LCL—lower control limit) are evaluated starting from σ, the standard deviation of the statistical distribution of $\Delta E(t)$, as follows:

$$UCL = +k \times \sigma \tag{2}$$

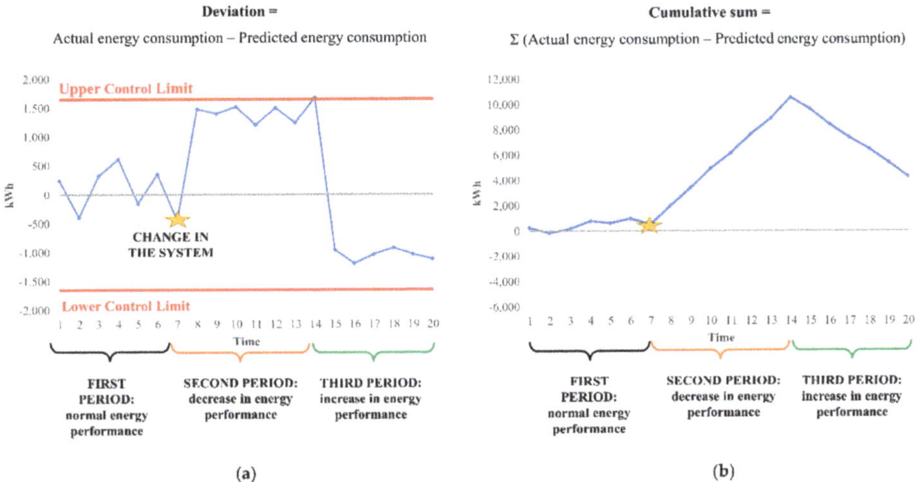

Figure 3. Qualitative representation of the two control charts applied for the joint analysis of energy behavior of a system: (**a**) control chart for energy performance deviations and (**b**) the cumulative sums chart (CuSum) chart. In the first period of the analysis (time interval from 1 to 7) the energy behavior of the system is normal, whereas in the second period (time interval from 7 to 14) the charts show a decrease in energy performance (positive shift of the average in (**a**) and upward trend in (**b**)). In the final period (time interval from 14 to 20), the system shows an improvement of the performance (negative shift of the average in (**a**) and downward trend in (**b**)).

$$LCL = -k \times \sigma \tag{3}$$

The coefficient k used in the equations is conventionally set to 3, but as the control chart's sensitivity is established by the width of its control limits, the coefficient k may also be set to a lower value in order to enact a stricter control. When the system examined shows an energy consumption behavior compliant with the baseline model, the energy performance deviations on the chart displays a normal statistical distribution with mean equal to zero, on the contrary, the presence of nonrandom patterns (e.g., points outside the control limits, mixtures or shits of the average) are signals of non-conformity with the baseline model and therefore of anomalous behavior.

Moreover, a second control chart, the cumulative sums chart, or CuSum, is used to deepen the analysis of the energy behavior of the system. Every point in the CuSum chart shows the value over time of the cumulative sum of energy performance deviations, defined as $\Delta EC(t)$ and is calculated as follows:

$$\Delta EC(t) = \sum_{t=0}^{t} \Delta E(t) \tag{4}$$

This type of control chart is useful because it highlights trends in energy performance behavior (Figure 3b). Indeed, a change in the slope of the CuSum represents a variation in the energy behavior of the system.

The joint examination of these two control charts supports the identification of different energy behaviors within the observed period or sub-periods and the identification of an accurate baseline model (i.e., the energy behavior that is taken as a reference to compare future performances).

The baseline model can be identified (Step 8) in the following three ways [19], depending on the choice of the set of data:

- considering the most recent available one; this choice is the best one when changes to technical, technological or structural configurations of the analyzed center occurred;
- considering the data set that showed the best energy performance; this choice is usually considered with a view to continuous improvement because it implies a more challenging outcome in terms of energy objectives;
- considering the data set that showed the most constant and stable energy behavior; this choice is generally made when none of the two previous options is applicable.

It is worth noting that, in any case, the sub-period chosen to create the baseline model should allow an accurate statistical data analysis, and therefore, it should be sufficiently long to include enough data. In some cases, where the cyclic nature of the energy consumption is evident (e.g., when the external temperature has a strong influence on the energy efficiency of the system), it might be appropriate to build different energy baselines for different time periods or conditions, defining when it is necessary to switch from one baseline model to another. In this phase, it is also possible to validate the baseline model by using all the available additional information, mainly related (but not limited to), operations and maintenance. This allows matching most of the variations from the baseline energy behavior occurring during the observed time period with actual events happened to the system (e.g., periodic maintenance interventions, shutdowns, etc.) and therefore, verifies the effectiveness of the model itself and of the control charts. This specific activity is usually time and resource consuming, as it is necessary to collect more information and to interview operators and maintenance personnel.

In the final step (Step 9), it is possible to use the baseline model to implement a continuous control over time. Real time data for both energy consumption and energy drivers are acquired by metering systems and their differences (i.e., $\Delta E(t)$) are evaluated at the chosen frequency (or at different/multiple frequencies, depending on users' needs) and represented on the control charts to highlight the presence of possible anomalies, enabling a real time control by the users. Moreover, the setting of control limits and other alarm conditions can automatize the generation of alerts to foster a timely response.

3.2. Definition of the Best Operating Conditions of a System From an Energy Efficiency Perspective

The methodology previously illustrated (Steps 1–7) is implemented in order to evaluate different operating conditions for the system and identify the best option from an energy efficiency perspective. In fact, it is sufficient to include in the data collection period, different sub-periods where operating conditions are different (e.g., different time sets, different pressure sets, different starting sequence, etc.). In many cases this happens automatically by considering a long data collection period or is achieved by switching such conditions on purpose during the data collection process.

Since different energy behaviors can be distinguished through the analysis of control charts (Step 7) and energy performance variations due to energy drivers' variability are already within the model, only the energy consumption variability due to the different operating conditions is made visible on the control charts. The evaluation of different operating conditions is therefore, made more accurate and effective. Moreover, the whole of Step 7 can also be iterated in different moments if a new evaluation is required during the control phase (e.g., when a new operating condition is introduced).

3.3. Identification of Changes to Energy Consumption Patterns or Degradation of Energy Performances Often Linked to Sporadic Faults or Events

The methodology previously illustrated (Steps 1–9) can also be implemented in order to identify a system's malfunctioning. The control charts can, in fact, be used over time to identify variations in energy behavior which can in turn be related to fault events through further investigations (in a similar way as the one described in Step 8).

In addition, it is possible to set up a registry (similar to fault registries already in use in industry for maintenance management purposes [84]) where all anomalies identified in the charts are collected together with relevant information such as time and entity of the deviation, values of all energy drivers, operating conditions, and causes of the anomaly. This kind of registry enables the analysis and classification of anomalies and allows identifying recurrences, patterns and similarities in the control charts, in order to promote the association of the most likely causes for each new anomaly in a quick and automatic way on the basis of past events (e.g., progressive obstruction of air compressors' filters resulted in a typical slope of the CuSum chart).

The creation of the registry and the analysis of past anomalies also increases the efficiency of the control system, as it is possible to identify anomalies related to inaccuracies of the control system itself rather than to actual issues of the observed physical system. Therefore it prevents false alarms (such as a meter's fault, missing data or a specific condition where the statistical model loses accuracy), thus enabling the user of the system to immediately activate the functions/units in the company that are most likely to be able to solve the issue. This registry can be compiled manually or automatically, implementing pattern recognition techniques [85], depending on system's characteristics and available resources.

3.4. Improved Energy Accounting

Energy accounting in industrial companies can be ineffective, as discerning between energy consumption variations to be expected (e.g., due to production volume variations) and actual anomalies (e.g., faults) is difficult without using an appropriate energy baseline. With the proposed methodology (Steps 1–8), once a proper energy baseline is created, it is possible to setup the earned value, a tool widely diffused for accounting in project management [86] whose application to energy management has already proved to be effective [76,87].

Such a tool requires a distinction between the actual values of energy consumption (measured, indicated in the followings as C_A) and the predicted ones (calculated using the baseline model, and indicated in the followings as C_P), calculated for budgeting purposes. In this case, the C_P values are calculated months before the energy is actually consumed by inputting the predicted values of the energy drivers into the baseline model. This means that the difference between C_A and C_P includes both energy efficiency and energy drivers' variations, and therefore it is not possible to distinguish actual inefficiencies from physiological variability. As a consequence, it is necessary to introduce another type of calculated energy consumption, named "flexible consumption" (C_F), which is calculated through the baseline model, using the actual values of the energy drivers as inputs. At this point, the difference between C_F and C_P represents the physiological variability of energy consumption (due to energy drivers' variations and here indicated as ΔP), while the difference between C_A and C_F represents the energy loss related to inefficiencies of the system (here indicated as ΔI).

$$\Delta P = C_F - C_P \tag{5}$$

$$\Delta I = C_A - C_F \tag{6}$$

Since this application aimed at facilitating energy accounting, it is necessary to shift from energy consumption to energy cost, multiplying C_A by a standard energy unit cost. Such a unit cost comprises energy purchasing and on-site energy production costs and allocates them to a single energy unit (e.g., €/kWh). Again, since the budget is usually calculated months before the energy is actually consumed, in order to take into account the possibility of differences in the energy price, it is necessary to distinguish between the predicted standard energy unit cost (indicated as c_P), which is calculated in advance, and the actual standard energy unit cost (indicated as c_A in the following), resulting from electricity and fuels bills. The energy budget variance due to differences in energy prices (Δp) is, therefore, calculated as follows:

$$\Delta p = (C_A \times c_A) - (C_A \times c_P) \tag{7}$$

This analysis is usually performed at time steps at which the energy budget is calculated, at least on a monthly basis. However, it could be in principle applied to any time step at which the energy baseline is calculated.

4. Case Study

The proposed methodology was applied to monitor and control the energy performance of the compressed air system of a pharmaceutical production plant located in central Italy, in order to get a first validation and to verify its applicability to the three non-energy-related activities illustrated in the previous section.

For what concerns Step 1, the company decided to focus on the compressed air production phase in order to identify additional improvement opportunities and to verify its correct operation and energy behavior (data were already available but had never been actually analyzed before). The plant was equipped with five screw compressors divided into two groups installed in separate sections of the industrial plant. We referred to these sections as room "A", hosting two of the five compressors, and room "B", with the other three compressors.

The maximum compressed air production was higher than the maximum compressed air demand, and usually only two or three compressors work simultaneously. Only Compressor 5, located in room "B" had a variable speed drive installed, and it served as "master" (continuously functioning), while the others worked as load/unload and served as "slaves". A central control system regulated the compressors in accordance with compressed air demand. Some general information on the compressors (nominal and stand-by nominal power) is reported in Figure 4. Data related to energy consumption and compressed air production, as well as compressed air pressure, external air temperature and humidity were available with a 15 min frequency (i.e., already collected, recorded and stored in the company's servers), while the number of hours worked by each compressor was available weekly (cumulated).

The level of detail of the analysis (Step 2) was defined by the company on the basis of three main observations: (i) the energy manager wanted to properly exploit existing data, which was considered to be relevant but had never been used to retrieve information on the system's performances, (ii) there was very little budget at that time to buy new meters, and the energy manager wanted to use results from a first system's analysis to justify additional budget allocation for new meters and (iii) for a first analysis of the system, the energy manager aimed at having general information on the whole system's performances rather than on single compressors.

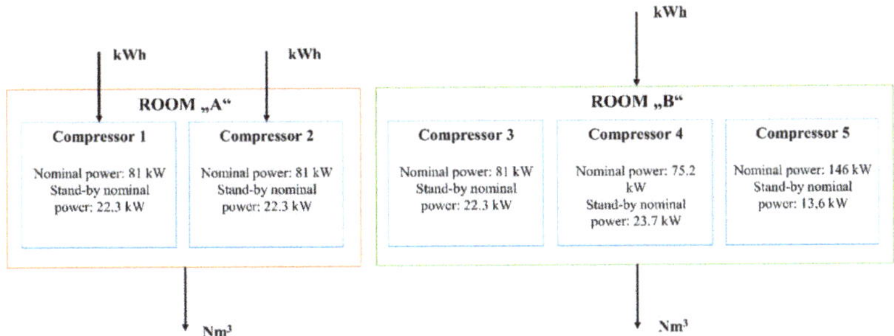

Figure 4. Scheme of compressors' groups and main data available.

All measured parameters (i.e., compressed air production, compressed air pressure, external air temperature and humidity, hours worked by each compressor) were considered as potential energy drivers and the first data analysis was conducted taking all of them into account, apart from the amount of worked hours, as it was available at a lower frequency. In fact, for this first analysis, aimed at defining the energy drivers as well as the energy behavior of the system, a daily frequency was considered to be the most appropriate, and a data collection period of one year was adopted.

Considering the amount of data to be processed and the company's needs, the baselining method selected was the statistical analysis (Step 3). It is important to point out that, although in this case a statistical regression was able to provide an appropriate baseline model, this method might not always represent the best choice. Indeed, for every specific case, the right baseline method should be identified taking into account the complexity and the dynamic of the system as well as the amount of data available.

Since there were constraints in terms of budget, and the data available were more than sufficient to conduct the analysis at the defined level of detail, the measurement plan and data collection (Steps 4–5) were performed considering the already available data. A synthesis of the measurement plan is given in the Table 1.

Table 1. Measurement plan.

		MEASURED DATA		
	Label and Unit	**Frequency**	**Collection Period**	**Meter/Collection Method**
Inputs/Outputs	Electric absorption compressor 1 [kW]	15 min	Last solar year (April to April)	Schneider PM9C energy meter; data on company's online server
	Electric absorption compressor 2 [kW]	15 min	Last solar year (April to April)	Schneider PM9C energy meter; data on company's online server
	Electric absorption room B [kW]	15 min	Last solar year (April to April)	Schneider PM9C energy meter; data on company's online server
	Compressed air flow room A [Nm3/h]	15 min	Last solar year (April to April)	Emerson Rosemount™ 485 Annubar™ (averaging Pitot tube); data on company's online server
	Compressed air flow room B [Nm3/h]	15 min	Last solar year (April to April)	Rosemount™ 485 Annubar™ (averaging Pitot tube); data on company's online server
Energy Drivers	External temperature [°C]	15 min	Last solar year (April to April)	Data transmitted by closest weather station
	External humidity [%]	15 min	Last solar year (April to April)	Data transmitted by closest weather station
	Pressure set point [bar]	15 min	Last solar year (April to April)	Juno Midas pressure transmitter; data on company's online server
	Compressors' working hours	Weekly	Last solar year (April to April)	Measure collected by operators on each compressor's screen and recorded on a shared spreadsheet

An accurate data synchronization and validation (Step 6) was performed in order to correctly aggregate data to the chosen frequency (from 15 min to one day), the level of detail (summing up data related to different compressors) and to avoid meters' inaccuracies in affecting the data analysis. First of all, some flaws were identified and corrected by observing and comparing measured data: a 15 minute delay was observed between electricity consumption and compressed air production (as shown in Figure 5) and bad compressed air flow meters' calibration (meters measuring compressed air flow while compressors were off, as shown in Figure 6).

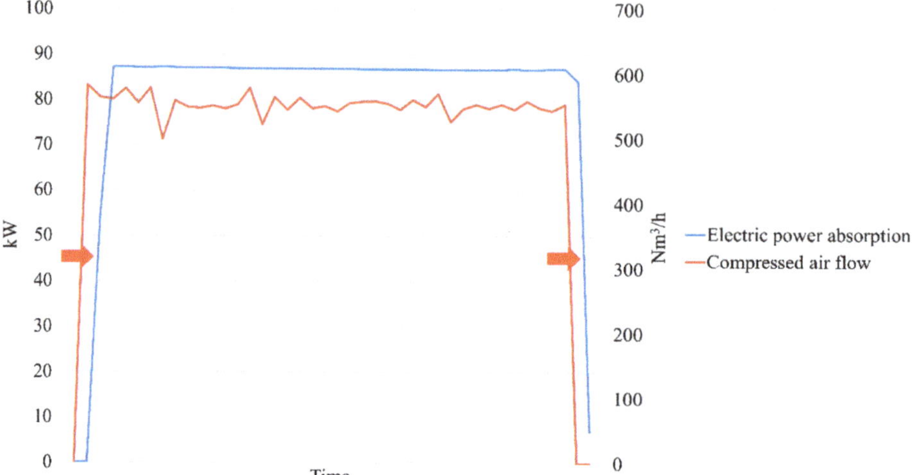

Figure 5. Electric power absorption and compressed air flow rate over time. The graph shows a 15 minutes delay between electric power absorption and compressed air production (data at a 15 minute frequency).

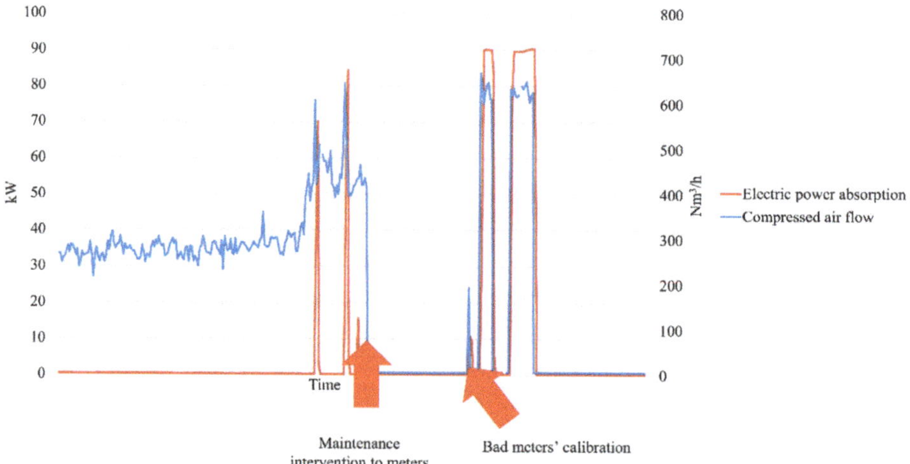

Figure 6. Maintenance issues to compressed air meters (data at a 15 minute frequency).

In addition, the data collection period was reduced from one year to seven months due to frequent failures and maintenance interventions occurred to the compressed air meters during the remaining five months (see Figure 6).

The first data analysis (Step 7) was conducted through correlation and regression, considering the system's energy consumption and the identified energy drivers in the data collection period. As a result, the only parameter showing a satisfying correlation with the energy consumption of the compressors and no multicollinearity with the other parameters, was the amount of compressed air produced, as reported in Table 2. The results of the multicollinearity analysis are represented in the form of a correlation matrix with a color scale in Figure 7.

Table 2. Results of the regression analysis.

Regression Analysis Results Considering All Parameters		Regression Analysis Results Considering Only Compressed Air Production	
Coefficient of determination (R^2)	0.96	Coefficient of determination R^2	0.94
P_value	9.45×10^{-137}	P_value	1.69×10^{-125}
P_value intercept	0.76	P_value intercept	0.34
P_value compressed air production	8.32×10^{-119}		
P_value external temperature	2.06×10^{-17}		
P_value external humidity	0.01		
P_value pressure	0.84		

	Compressed air production	External temperature	External humidity	Pressure	Energy consumption
Compressed air production	1				
External temperature	0.156	1			
External humidity	0.114	0.131	1		
Pressure	0.595	0.179	0.056	1	
Energy consumption	0.968	0.283	0.088	0.591	1

Figure 7. Correlation matrix. The color scale represents the value of the correlation between two variables: green highlights strong correlations while red highlights weak correlations.

The pressure referred to in the correlation analysis was the discharge pressure of the compressed air generation system and since it was considered a set point, the system was equipped with an advanced control in order to minimize its fluctuations. The pressure at which the compressed air was produced was around 6.5 bar, normally oscillating between 6.4 and 6.6 bar, during weekdays, while it was reduced to 6.2 bar (with a variation range between 6.1 and 6.3 bar) on weekends. These limited variations explained the limited effect on the regression model of this variable, usually quite critical for this kind of analysis. Moreover, the ambient conditions, were also examined in terms of external temperature and external humidity for a whole year in order to guarantee the analysis of the complete range of conditions possible, affecting poorly the regression model.

The resulting mathematical model of the system's energy consumption was as follows:

$$\text{Energy consumption [kWh]} = 0.1575 \times \text{Compressed air produced} \left[\text{Nm}^3\right] - 80.02 \qquad (8)$$

This is also represented in the scatterplot in Figure 8.

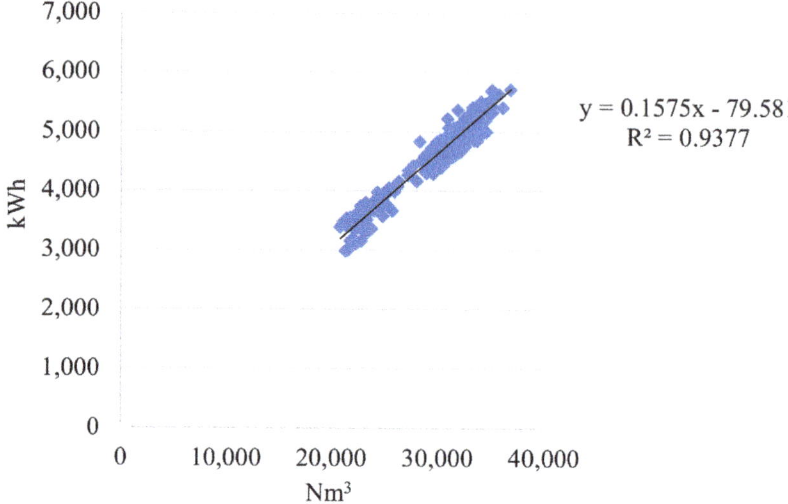

Figure 8. Scatterplot representing the relation between energy consumption and compressed air production.

The CuSum and the control chart for the energy performance deviations for the data collection period (Step 8) are given in Figure 9 and their joint observation shows the presence of four energy behaviors, highlighted using different colors. Indeed, the CuSum in Figure 9a shows four different trends clearly distinguishable and the control chart for the energy performance deviations in Figure 9b displays four different distributions with a different mean.

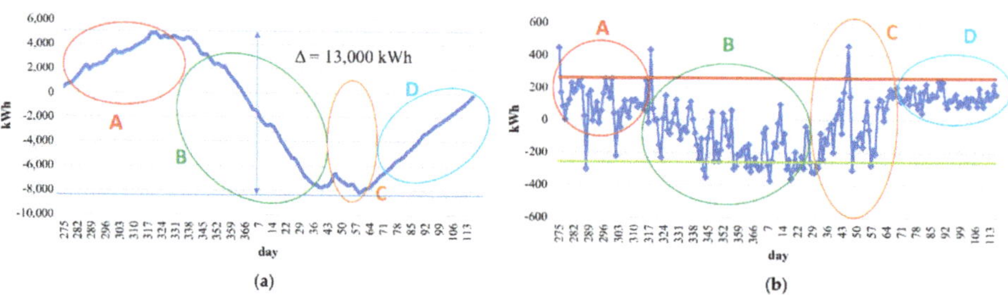

Figure 9. CuSum chart and control chart for the energy performance deviations over a seven month period (daily data): (**a**) CuSum chart; (**b**) control chart for the energy performance deviations.

The causes of these different behaviors were further investigated, mainly through the observation of daily consumption data over time and interviews to operators, and they were finally identified to be the following:

- Red period ("A"): Compressor 2 presented a higher specific consumption until around day 325 when a maintenance intervention solved the problem;
- Green period ("B"): a change to the activation sequence caused Compressor 3 to work mainly with Compressor 5, whereas the others were usually kept turned off;
- Orange period ("C"): Compressor 1 presented an evident malfunctioning, remaining stuck in stand-by for three whole consecutive days;

- Blue period ("D"): another change to the activation sequence caused Compressor 4 to work mainly with Compressor 5, whereas the others were usually kept turned off.

Thus, through the analyses, two different maintenance-related events were highlighted and it was revealed that the energy behavior of the system was strongly related to the set of operating compressors and therefore to the compressors' activation sequence (also confirmed by further analyses on the amount of hours worked by each compressor).

The baseline was then created using data from the green period ("B") which was the one showing the best energy performance. The data analysis (Step 7) was then repeated in order to develop the mathematical model for the baseline. Results are not given in full here for sake of brevity.

4.1. Definition of the Best Operating Conditions of a System from an Energy Efficiency Perspective

An optimal starting sequence was, therefore, identified (the one corresponding to the green period B in Figure 9) and uploaded into the central control system. Some trials were performed involving the operators working on the plant's technical systems and production lines and maintenance personnel, in order to make sure that the new starting sequence proposed suited the company's needs best.

Control charts, built using the selected baseline as a reference (Step 9), were then used for continuous monitoring and control. In this case, the company decided to build a control system with a slightly higher level of detail compared to the previous analysis. In particular, three CuSums were employed: one for the whole compressed air generation system, one for Room "A" and one for Room "B", in order to allow an easier and faster troubleshooting. Accordingly, two control charts for the energy performance deviations were also created: one for Room "A" and one for Room "B". The frequency used for the analysis also increased, aggregating data every four hours. This allowed management to use relevant information coming from control charts observation for defining operations and maintenance short-term strategies. It is worthy to note that some iterations of Steps 7 and 8 were needed in order to create different baselines for the two rooms (again, full details are not given here for sake of brevity). The energy performance control results (control charts plots) for about two months are given in Figure 10, which clearly show a stable system's behavior apart from some isolated issues mainly related to metering system's faults.

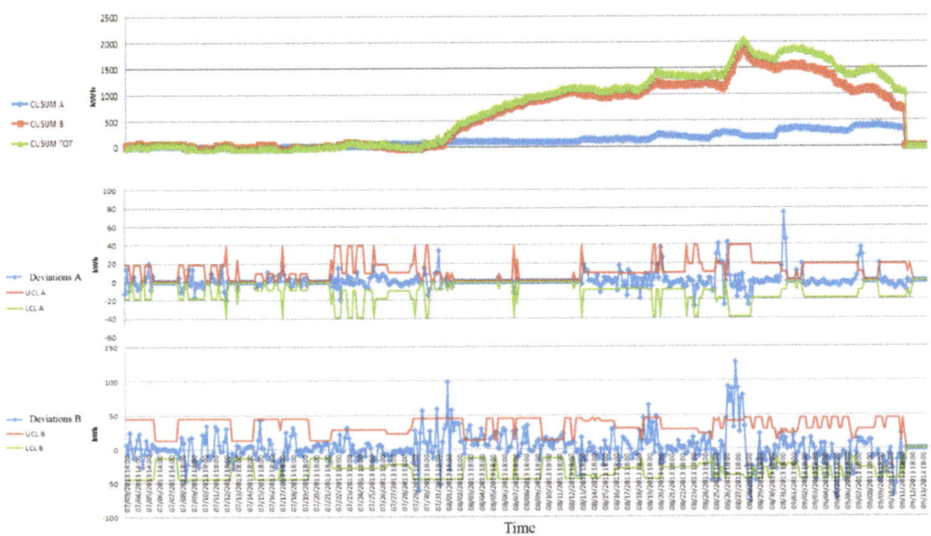

Figure 10. CuSum chart and control charts for the energy performance deviations for the two-month control period (data aggregated every four hours).

4.2. Identification of Changes to Energy Consumption Patterns or Degradation of Energy Performances Often Linked to Sporadic Faults or Events

Once the control system was finally up and running, the operations and maintenance team gathered in order to discuss each single event highlighted by the control charts and to start setting the anomalies registry up. Taking all occurred events into account, an initial classification was created by defining a severity scale (from 1 to 4) defined according to the deviation's entity. Then, most likely causes for each category were listed in order to facilitate troubleshooting activities. This initial version of the anomalies' classification is given in Figure 11. The troubleshooting system has not been automated yet, as data from a longer control period were needed in order to perform more relevant analyses.

Severity	Corresponding deviations' entity	Potential (registered) causes
1	<= 10 kWh	Missing data for 15 minutes.
2	<= 20 kWh	Missing data for 15 minutes; change in the ignition sequence due to compressors' rotation; uncontrolled variation in the pressure set point.
3	<= 30 kWh	Missing data for more than 15 minutes; change in the ignition sequence due to compressors' rotation.
4	> 30 kWh	Missing data for more than 15 minutes; change in the ignition sequence due to maintenance issues.

Figure 11. Classification of anomalies created on the basis of the anomalies' registry.

4.3. Improved Energy Accounting

The proposed methodology was also applied to energy accounting, in order to obtain more accurate predictions of the energy budget related to the use of compressed air in the production plant (intended as the total cost of electricity necessary to the generation of compressed air). Budgeting activities were performed for both compressed air generation and use, thus additional baselines models for the two production departments using compressed air were created following the same methodology. A new baseline was also defined for the compressed air generation system, as in this case, a monthly frequency was chosen for data analysis. Figure 12 gives a schematic representation of the energy accounting methodology proposed to predict the energy budget related to the compressed air used in the production plant, highlighting the boundaries of cost/responsibility centers (Dept. 1, Dept. 2, generation system, energy purchasing office).

From the monthly volume of production established for the following months, through the use of two different statistical models, the expected compressed air consumption to be used by the two departments in the plant was calculated. Then, using the sum of the expected compressed air consumption values for the two production departments as a driver for another statistical model, the electricity consumption of the compressed air generation system was also determined. The energy budget was then calculated as the product between the estimated electricity consumption and the predicted standard unit cost of electricity. Finally, the predicted standard unit cost of compressed air produced was calculated by dividing the budget by the total amount of compressed air consumption predicted. Table 3 gives all details related to the estimated energy budget of the compressed air system for three months.

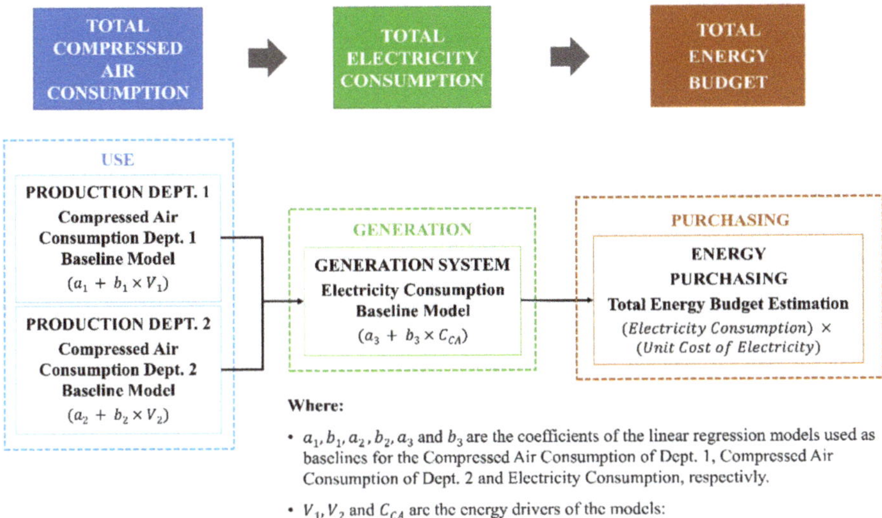

Figure 12. Schematic representation of the energy accounting methodology proposed to predict the energy budget related to the use of compressed air in the production plant.

Table 3. Estimated compressed air budget for the considered trimester.

ESTIMATED BUDGET				
Month		1	2	3
Production Dept. 1	Units	4,047,800	3,757,201	3,294,456
Production Dept. 2	Units	5,475,148	4,427,156	5,889,159
Compressed air consumption Dept. 1	Nm^3	12,805,432	11,219,241	8,693,420
Compressed air consumption Dept. 2	Nm^3	15,916,589	12,785,528	17,153,518
Electricity consumption of the generation phase	kWh	4,448,464	3,664,935	3,970,917
Standard unit cost of electricity	€/kWh	0.158	0.158	0.158
Standard unit cost of compressed air	€/m^3	0.024	0.024	0.024
Estimated budget Dept. 1	€	313,362	270,638	211,023
Estimated budget Dept. 2	€	389,495	308,421	416,382
Estimated budget for the generation phase	€	702,857	579,060	627,405

At the end of the trimester, once the actual flows of the energy consumed and actual standard unit costs were known, it was possible to analyze the variations occurred between predicted values and the actual ones to economically assess the performance of the plant in regards to production and use of compressed air (see Table 4). Multiplying the actual standard unit cost of compressed air and the consumption of the individual departments, individual values of budgets were calculated for the two departments.

Table 5 reports the flexible consumption of compressed air and electricity for the considered trimester, evaluated as described in Section 3.4. The flexible consumption of compressed air for the two departments was estimated through the two previous statistical models using the actual volumes of production of the two departments. Likewise, the flexible electricity consumption of the generation system was estimated through the third statistical model, inputting the actual amount of compressed air produced by the system.

Table 4. Actual compressed air budget for the considered trimester.

ACTUAL BUDGET				
Month		1	2	3
Production Dept. 1	Units	3,679,818	3,415,637	3,361,690
Production Dept. 2	Units	6,016,646	4,865,007	6,009,346
Compressed air consumption Dept. 1	Nm^3	10,531,292	8,339,483	9,268,500
Compressed air consumption Dept. 2	Nm^3	17,317,964	14,076,891	17,130,719
Electricity consumption of the generation phase	kWh	4,114,206	342,4346	4,220,230
Standard unit cost of electricity	€/kWh	0.159	0.159	0.159
Standard unit cost of compressed air	$€/m^3$	0.023	0.024	0.025
Estimated budget Dept. 1	€	247,372	202,558	235,587
Estimated budget Dept. 2	€	406,786	341,913	435,429
Estimated budget for the generation phase	€	654,159	544,471	671,017

Table 5. Flexible consumption of compressed air and electricity for the considered trimester.

FLEXIBLE CONSUMPTION				
Month		1	2	3
Compressed air consumption Dept. 1	Nm^3	10,796,859	9,354,867	9,060,406
Compressed air consumption Dept. 2	Nm^3	17,534,410	14,093,684	17,512,598
Electricity consumption of the generation phase	kWh	4,303,499	3,401,105	4,062,650

Table 6 reports the analysis of the budget variance, distinguishing among cost/responsibility centers and main causes. The total actual budget was lower than the predicted one in the first two months and higher in the third month, as reported in the column ΔTOT of the group generation system. Such a variation was mainly attributed to the lower production volumes of Department 1 (see row ΔP (units)—Dept. 1) while Department 2 showed an increased volume of production (see row ΔP (units)—Dept. 2). In the trimester there was also a constant positive variation attributed to the price of energy (see row Δp—generation system). The cause was indeed an electricity price higher than the estimated one because of a change in the procurement contract. Moreover, the performance of the compressed air generation system was better than the predicted one in the first month whereas in months two and three the efficiency worsened (see row ΔI—generation system). Lastly, the performance of the two departments in the use of compressed air generally improved in the trimester except for the performance of Department 1 in month three (see rows ΔI—Dept. 1 and Dept. 2).

Table 6. Budget variance analysis for the considered trimester.

Month			1	2	3
Budget performance—generation system	Δ TOT	€	−48,699	−34,589	43,612
	ΔI	€	−29,908	3672	24,898
	ΔP (Nm3)	€	−22,904	−41,685	14,494
	Δp	€	4114	3424	4220
Budget performance—Dept. 1	Δ TOT	€	−65,990	−68,081	24,564
	ΔI	€	−6499	−24,494	5051
	ΔP (units)	€	−49,152	−44,974	8908
	Δp	€	−10,339	1387	10,605
Budget performance—Dept. 2	Δ TOT	€	17,291	33,492	19,047
	ΔI	€	−5297	−405	−9270
	ΔP (units)	€	39,590	31,556	8716
	Δp	€	−17,002	2341	19,601

5. Conclusions

We developed a methodology to monitor and control energy performances of significant energy uses in industrial plants, through statistical analysis and baselining. Three different applications of the methodology with potential impact on non-energy-related activities such as operations, maintenance and accounting were presented, as well as their implementation in a real industrial case. The first step of the methodology was constructing a baseline of the system energy behavior, for example through statistical regression that correlated energy consumption to the main energy drivers. Then control charts allowed us to validate the baseline and identify optimal operating conditions and to control the energy performance over time and identify changes to energy consumption patterns or degradation of the energy performance related to sporadic faults or events.

Moreover, the knowledge of the energy behavior of the CAS enabled a reliable energy accounting system and allowed us to discern among the possible causes of budget variance.

The applicability of the proposed methodology to a real industrial environment was demonstrated for a pharmaceutical plant. In particular, we highlighted two different maintenance-related events as well as a quite strong dependency of the energy behavior on the set of the operating compressors, and therefore, on the compressors' activation sequence. Moreover, it was possible to define a first association between the anomalies identified by the control charts and their most likely causes in order to facilitate troubleshooting activities. Finally, the application of the proposed methodology to energy accounting gave more accurate predictions of the energy budget related to the use of compressed air in the departments of the production plant and helped top management in the analysis of the budget variance, distinguishing among reasons behind the occurrences, and thus helping in the allocation of the responsibility to different cost/responsibility centers.

Author Contributions: All authors contributed equally to the idea and the design of the methodology proposed; M.B. and A.S. were responsible for the case study application; M.B. and A.S. prepared the original draft; F.B., V.I. and S.U. contributed to the review and editing.

Funding: This research was funded by the Electrical System Research national funds, implemented under programme agreements between the Italian Ministry for Economic Development and ENEA (Energy, New Technology and Environment Agency), CNR (National Research Council), and RSE S.p.A. (a public Italian company that develops applied research in the electricity/energy sector).

Conflicts of Interest: The authors declare no conflict of interest.

Appendix A

This appendix is dedicated to the description of the relevant contributions examined in the scientific literature review phase of this research. Its aim is to summarize the main innovative applications and methods implemented through the analysis of data in the domains of energy, operations, maintenance and accounting. The research process started by using some main keywords such as "energy control", "energy performance control", "energy consumption control", "failure/fault detection", "failure/fault diagnosis", "condition monitoring/control", "accounting", "process control". The first results were examined also in terms of their references in order to expand the research.

Therefore, Table A1 reports the resulting list of relevant scientific publications about control tools in energy, process control, maintenance and accounting examined, reporting for each domain, specific application addressed, asset on which it was applied, type of contribution, methodological approach and presence of a specific case study.

Moreover, Figure A1 represents the evolution of publications that refer to the use of modelling techniques for different specific applications, highlighting their increase in the recent years.

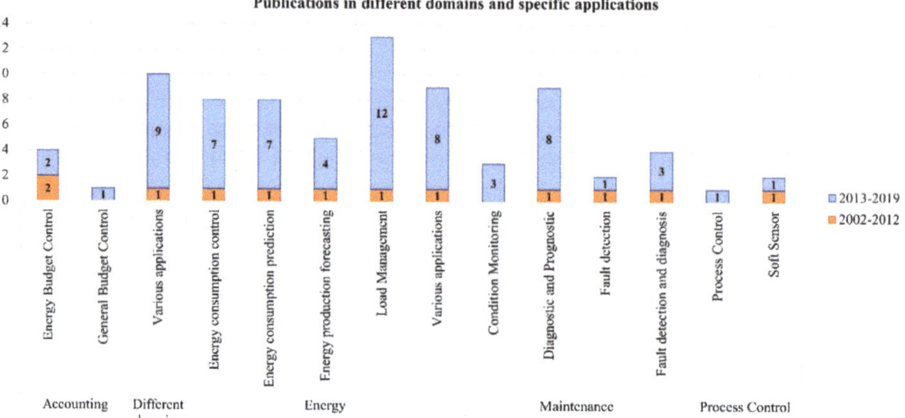

Figure A1. Publications that refer to the use of modelling techniques for energy control, forecasting, process control, accounting or maintenance, divided by domain and specific application and differentiated in two temporal categories by publication year: 2002–2012 and 2013–2019 (elaboration from data in Table A1).

Table A1. Summary of relevant publications about control tools in operations, maintenance and energy accounting.

	Ref.	Domain	Asset	Specific Application	Type of Contribution	Case Study	Approach Described
[79]	Capehart et al., 2002	Accounting	Industrial plant	Energy Budget Control	Methodological Contribution	Yes	Mathematical model
[76]	Cesarotti et al., 2009	Accounting	Industrial plant	Energy Budget Control	Methodological Contribution	Yes	Statistical model
[78]	Santolamazza et al., 2017	Accounting	Industrial plant	Energy Budget Control	Methodological Contribution	Yes	Mathematical model
[88]	Torregrossa et al., 2018	Accounting	Industrial plant	Energy Budget Control	Methodological Contribution	Yes	Machine Learning model
[77]	Pérez-Rave et al., 2017	Accounting	Industrial plant	General Budget Control	Methodological Contribution	Yes	Statistical/Machine Learning Model and Control Charts
[28]	Benedetti et al., 2016	Energy	Building	Energy consumption control	Methodological Contribution	Yes	Statistical/Machine Learning Model and Control Charts
[29]	Fan et al., 2018	Energy	Building	Energy consumption control	Literature Review/Conceptual Contribution	No	Different data modelling techniques
[8]	Hu et al., 2012	Energy	Industrial machine	Energy consumption control	Methodological Contribution	Yes	Statistical model
[27]	Nikula et al., 2016	Energy	Boiler	Energy consumption control	Methodological Contribution	Yes	Statistical/Machine Learning Model and Control Charts
[30]	Santolamazza et al., 2018	Energy	Compressor	Energy consumption control	Methodological Contribution	Yes	Statistical/Machine Learning Model and Control Charts
[31]	Shrouf and Miragliotta, 2015	Energy	Industrial plant	Energy consumption control	Literature Review/Conceptual Contribution	No	Not described
[32]	Sunthornnapha, 2017	Energy	Industrial plant	Energy consumption control	Methodological Contribution	Yes	Different data modelling techniques

Table A1. *Cont.*

	Ref.	Domain	Asset	Specific Application	Type of Contribution	Case Study	Approach Described
[33]	Torregrossa et al., 2017	Energy	Pump	Energy consumption control	Methodological Contribution	Yes	Machine Learning model
[34]	Amasyali and El-Gohary, 2018	Energy	Building	Energy consumption prediction	Literature Review/Conceptual Contribution	Yes	Different data modelling techniques
[35]	Christensen and Himme, 2017	Energy	Industrial machine	Energy consumption prediction	Methodological Contribution	Yes	Statistical model
[36]	Deb et al., 2017	Energy	Building	Energy consumption prediction	Literature Review/Conceptual Contribution	No	Different data modelling techniques
[37]	Foucquier et al., 2013	Energy	Building	Energy consumption prediction	Literature Review/Conceptual Contribution	No	Different data modelling techniques
[38]	Ngo, 2019	Energy	Building	Energy consumption prediction	Methodological Contribution	Yes	Different data modelling techniques
[39]	Pino-Mejías et al., 2017	Energy	Building	Energy consumption prediction	Literature Review/Conceptual Contribution	Yes	Different data modelling techniques
[40]	Wang et al., 2018	Energy	Industrial plant	Energy consumption prediction	Methodological Contribution	Yes	Machine Learning model
[55]	Das et al., 2018	Energy	Photovoltaic systems	Energy production forecasting	Literature Review/Conceptual Contribution	No	Different data modelling techniques
[56]	Foley et al., 2012	Energy	Wind turbines	Energy production forecasting	Literature Review/Conceptual Contribution	No	Different data modelling techniques
[57]	Sharma and Kakkar, 2018	Energy	Photovoltaic systems	Energy production forecasting	Methodological Contribution	Yes	Different data modelling techniques
[58]	Voyant et al., 2017	Energy	Photovoltaic systems	Energy production forecasting	Literature Review/Conceptual Contribution	No	Different data modelling techniques
[59]	Wang et al., 2016	Energy	Wind turbines	Energy production forecasting	Methodological Contribution	Yes	Different data modelling techniques
[42]	Ahmad et al., 2018	Energy	Building	Load Management	Literature Review/Conceptual Contribution	No	Different data modelling techniques
[43]	Ahmad et al., 2018	Energy	Electrical Grid	Load Management	Literature Review/Conceptual Contribution	Yes	Different data modelling techniques
[44]	Chou and Tran, 2018	Energy	Building	Load Management	Methodological Contribution	Yes	Different data modelling techniques
[45]	Diamantoulakis et al., 2015	Energy	Electrical Grid	Load Management	Literature Review/Conceptual Contribution	No	Different data modelling techniques
[46]	Ferreira et al., 2013	Energy	Electrical Grid	Load Management	Methodological Contribution	Yes	Machine Learning model
[47]	Grolinger et al., 2016	Energy	Building	Load Management	Methodological Contribution	Yes	Different data modelling techniques
[48]	Stoyanova et al., 2013	Energy	Building	Load Management	Methodological Contribution	Yes	Statistical/Machine Learning Model and Control Charts
[49]	Tsekouras et al., 2008	Energy	Electrical Grid	Load Management	Methodological Contribution	Yes	Machine Learning model
[50]	Tu et al., 2017	Energy	Electrical Grid	Load Management	Literature Review/Conceptual Contribution	No	Not described
[51]	Vázquez-Canteli and Nagy, 2019	Energy	Building	Load Management	Literature Review/Conceptual Contribution	No	Different data modelling techniques

Table A1. *Cont.*

	Ref.	Domain	Asset	Specific Application	Type of Contribution	Case Study	Approach Described
[52]	Wei et al., 2018	Energy	Building	Load Management	Literature Review/Conceptual Contribution	No	Different data modelling techniques
[53]	Yildiz et al., 2017	Energy	Building	Load Management	Literature Review/Conceptual Contribution	Yes	Different data modelling techniques
[54]	Zhou et al., 2013	Energy	Electrical Grid	Load Management	Literature Review/Conceptual Contribution	Yes	Different data modelling techniques
[89]	Debnath and Mourshed, 2018	Energy	Different Systems	Various applications	Literature Review/Conceptual Contribution	No	Different data modelling techniques
[90]	Jha et al., 2017	Energy	Different Renewable Energy Systems	Various applications	Literature Review/Conceptual Contribution	No	Different data modelling techniques
[91]	Koseleva and Ropaite, 2017	Energy	Building	Various applications	Literature Review/Conceptual Contribution	No	Not described
[92]	Lund et al., 2017	Energy	Different Systems	Various applications	Literature Review/Conceptual Contribution	No	Not described
[93]	Molina-Solana et al., 2017	Energy	Building	Various applications	Literature Review/Conceptual Contribution	No	Different data modelling techniques
[94]	Shrouf et al., 2014	Energy	Industrial plant	Various applications	Literature Review/Conceptual Contribution	No	Not described
[26]	Capobianchi et al., 2011	Energy	Industrial plant	Various applications	Methodological contribution	Yes	Statistical/Machine Learning Model and Control Charts
[95]	Yu et al., 2016	Energy	Building	Various applications	Literature Review/Conceptual Contribution	No	Different data modelling techniques
[96]	Zhou et al., 2013	Energy	Different Systems	Various applications	Literature Review/Conceptual Contribution	No	Different data modelling techniques
[41]	Kim and Kim, 2007	Energy	Chiller	Energy consumption prediction	Methodological Contribution	Yes	Statistical model
[63]	Engelberth et al., 2018	Maintenance	Compressor	Condition Monitoring	Methodological Contribution	Yes	Statistical model
[64]	Santolamazza et al., 2018	Maintenance	Compressor	Condition Monitoring	Methodological Contribution	Yes	Statistical/Machine Learning Model and Control Charts
[65]	Stetco et al., 2019	Maintenance	Wind turbines	Condition Monitoring	Literature Review/Conceptual Contribution	No	Different data modelling techniques
[69]	Diez-Olivan et al., 2019	Maintenance	Different Systems	Diagnostic and Prognostic	Literature Review/Conceptual Contribution	No	Different data modelling techniques
[17]	Guillén López et al., 2018	Maintenance	Different Systems	Diagnostic and Prognostic	Literature Review/Conceptual Contribution	Yes	Different data modelling techniques
[70]	Karim et al., 2016	Maintenance	Different Systems	Diagnostic and Prognostic	Literature Review/Conceptual Contribution	Yes	Not described
[71]	Kim et al., 2014	Maintenance	Industrial plant	Diagnostic and Prognostic	Methodological Contribution	Yes	Different data modelling techniques
[72]	Lee et al., 2015	Maintenance	Different Systems	Diagnostic and Prognostic	Literature Review/Conceptual Contribution	Yes	Different data modelling techniques
[16]	Lee et al., 20006	Maintenance	Different Systems	Diagnostic and Prognostic	Literature Review/Conceptual Contribution	Yes	Different data modelling techniques

Table A1. *Cont.*

	Ref.	Domain	Asset	Specific Application	Type of Contribution	Case Study	Approach Described
[73]	Lee et al., 2014	Maintenance	Different Systems	Diagnostic and Prognostic	Literature Review/Conceptual Contribution	Yes	Different data modelling techniques
[74]	Roy et al., 2016	Maintenance	Industrial machine	Diagnostic and Prognostic	Literature Review/Conceptual Contribution	No	Different data modelling techniques
[75]	Vogl et al., 2016	Maintenance	Industrial plant	Diagnostic and Prognostic	Literature Review/Conceptual Contribution	No	Different data modelling techniques
[66]	Romeo and Gareta, 2009	Maintenance	Boiler	Fault detection	Methodological Contribution	Yes	Different data modelling techniques
[67]	Xiao, 2016	Maintenance	Industrial plant	Fault detection	Methodological Contribution	Yes	Machine Learning model
[68]	Liu et al., 2018	Maintenance	Industrial machine	Fault detection and diagnosis	Literature Review/Conceptual Contribution	No	Different data modelling techniques
[14]	Qi et al., 2018	Maintenance	Compressor	Fault detection and diagnosis	methodological Contribution	Yes	Machine Learning model
[12]	Tran et al., 2015	Maintenance	Chiller	Fault detection and diagnosis	Methodological Contribution	Yes	Statistical/Machine Learning Model and Control Charts
[13]	Xiao et al., 2011	Maintenance	Chiller	Fault detection and diagnosis	Methodological Contribution	Yes	Statistical model
[62]	Y. Zhang et al., 2018	Process Control	Industrial machine	Process Control	Methodological Contribution	Yes	Statistical model
[61]	Kadlec et al., 2009	Process Control	Industrial plant	Soft Sensor	Literature Review/Conceptual Contribution	No	Different data modelling techniques
[60]	Shang et al., 2014	Process Control	Industrial plant	Soft Sensor	Literature Review/Conceptual Contribution	Yes	Different data modelling techniques
[87]	Cesarotti et al., 2007	Different domains	Industrial plant	Various applications	Literature Review/Conceptual Contribution	No	Not described
[9]	Ge et al., 2017	Different domains	Industrial plant	Various applications	Literature Review/Conceptual Contribution	No	Different data modelling techniques
[10]	Ge, 2017	Different domains	Industrial plant	Various applications	Literature Review/Conceptual Contribution	No	Different data modelling techniques
[97]	Kusiak et al., 2013	Different domains	Wind turbines	Various applications	Literature Review/Conceptual Contribution	No	Different data modelling techniques
[98]	Lee et al., 2013	Different domains	Different Systems	Various applications	Literature Review/Conceptual Contribution	No	Not described
[99]	Lee et al., 2013	Different domains	Industrial plant	Various applications	Literature Review/Conceptual Contribution	No	Different data modelling techniques
[100]	Tao et al., 2018	Different domains	Industrial plant	Various applications	Literature Review/Conceptual Contribution	Yes	Different data modelling techniques
[101]	J. Wang et al., 2018	Different domains	Industrial plant	Various applications	Literature Review/Conceptual Contribution	No	Different data modelling techniques
[102]	K. Zhang et al., 2018	Different domains	Industrial plant	Various applications	Literature Review/Conceptual Contribution	Yes	Different data modelling techniques
[103]	Zhang et al., 2017	Different domains	Industrial plant	Various applications	Literature Review/Conceptual Contribution	No	Different data modelling techniques

References

1. Benedetti, M.; Bertini, I.; Bonfà, F.; Ferrari, S.; Introna, V.; Santino, D.; Ubertini, S. Assessing and Improving Compressed Air Systems' Energy Efficiency in Production and Use: Findings from an Explorative Study in Large and Energy-Intensive Industrial Firms. *Energy Procedia* **2017**, *105*, 3112–3117. [CrossRef]

2.   Benedetti, M.; Bonfa, F.; Bertini, I.; Introna, V.; Ubertini, S. Explorative study on Compressed Air Systems' energy efficiency in production and use: First steps towards the creation of a benchmarking system for large and energy-intensive industrial firms. *Appl. Energy* **2018**, *227*, 436–448. [CrossRef]

3.   Salvatori, S.; Benedetti, M.; Bonfà, F.; Introna, V.; Ubertini, S. Inter-sectorial benchmarking of compressed air generation energy performance: Methodology based on real data gathering in large and energy-intensive industrial firms. *Appl. Energy* **2018**, *217*, 266–280. [CrossRef]

4.   European Parliament; The Council of The European Union. Directive 2012/27/EU of the European Parliament and of the Council of 25 October 2012 on energy efficiency, amending Directives 2009/125/EC and 2010/30/EU and repealing Directives 2004/8/EC and 2006/32. *OJ* **2012**, *315*, 1–56.

5.   ISO International Standard Organization. *ISO 50001 Energy Management Systems–Requirements with Guidance for Use 2018*; ISO International Standard Organization: Geneva, Switzerland, 2018.

6.   Efficiency Valuation Organization. *IPMVP Volume I–Concepts and Options for Determining Energy and Water Savings 2012*; Efficiency Valuation Organization: Washington, DC, USA, 2012.

7.   Tan, Y.S. Internet-of-Things Enabled Real-time Monitoring of Energy Efficiency on Manufacturing Shop Floors. *Procedia CIRP* **2017**, *6*, 376–381. [CrossRef]

8.   Hu, S.; Liu, F.; He, Y.; Hu, T. An on-line approach for energy efficiency monitoring of machine tools. *J. Clean. Prod.* **2012**, *27*, 133–140. [CrossRef]

9.   Ge, Z.; Song, Z.; Ding, S.X.; Huang, B. Data Mining and Analytics in the Process Industry: The Role of Machine Learning. *IEEE Access* **2017**, *5*, 20590–20616. [CrossRef]

10.   Ge, Z. Review on data-driven modeling and monitoring for plant-wide industrial processes. *Chemometr. Intell. Lab. Syst.* **2017**, *171*, 16–25. [CrossRef]

11.   Xenos, D.P.; Cicciotti, M.; Kopanos, G.M.; Bouaswaig, A.E.F.; Kahrs, O.; Martinez-Botas, R.; Thornhill, N.F. Optimization of a network of compressors in parallel: Real Time Optimization (RTO) of compressors in chemical plants–An industrial case study. *Appl. Energy* **2015**, *144*, 51–63. [CrossRef]

12.   Tran, D.A.T.; Chen, Y.; Chau, M.Q.; Ning, B. A robust online fault detection and diagnosis strategy of centrifugal chiller systems for building energy efficiency. *Energy Build.* **2015**, *108*, 441–453. [CrossRef]

13.   Xiao, F.; Zheng, C.; Wang, S. A fault detection and diagnosis strategy with enhanced sensitivity for centrifugal chillers. *Appl. Therm. Eng.* **2011**, *31*, 3963–3970. [CrossRef]

14.   Qi, G.; Zhu, Z.; Erqinhu, K.; Chen, Y.; Chai, Y.; Sun, J. Fault-diagnosis for reciprocating compressors using big data and machine learning. *Simul. Model. Pract. Theory* **2018**, *80*, 104–127. [CrossRef]

15.   Magoulès, F.; Zhao, H.-X. *Data Mining and Machine Learning in Building Energy Analysis*; John Wiley & Sons: Hoboken, NJ, USA, 2016; p. 187.

16.   Lee, J.; Ni, J.; Djurdjanovic, D.; Qiu, H.; Liao, H. Intelligent prognostics tools and e-maintenance. *Comput. Ind.* **2006**, *57*, 476–489. [CrossRef]

17.   Guillén López, A.J.; Crespo Márquez, A.; Macchi, M.; Gómez Fernández, J.F. Prognostics and Health Management in Advanced Maintenance Systems. In *Advanced Maintenance Modelling for Asset Management*; Crespo Márquez, A., González-Prida Díaz, V., Gómez Fernández, J.F., Eds.; Springer International Publishing: Cham, Germany, 2018; pp. 79–106. ISBN 978-3-319-58044-9.

18.   Lee, J.; Kao, H.-A.; Yang, S. Service Innovation and Smart Analytics for Industry 4.0 and Big Data Environment. *Procedia CIRP* **2014**, *16*, 3–8. [CrossRef]

19.   Benedetti, M.; Cesarotti, V.; Introna, V. From energy targets setting to energy-aware operations control and back: An advanced methodology for energy efficient manufacturing. *J. Clean. Prod.* **2018**, *167*, 1518–1533. [CrossRef]

20.   Roblek, V.; Meško, M.; Krapež, A. A Complex View of Industry 4.0. *SAGE Open* **2016**, *6*, 215824401665398. [CrossRef]

21.   Lu, Y. Industry 4.0: A survey on technologies, applications and open research issues. *J. Ind. Inf. Integr.* **2017**, *6*, 1–10. [CrossRef]

22.   Sharif Ullah, A. Modeling and simulation of complex manufacturing phenomena using sensor signals from the perspective of Industry 4.0. *Adv. Eng. Inf.* **2019**, *39*, 1–13. [CrossRef]

23.   Ghosh, A.K.; Ullah, A.S.; Kubo, A. Hidden Markov model-based digital twin construction for futuristic manufacturing systems. *AIEDAM* **2019**, *33*, 317–331. [CrossRef]

24.   Zhong, R.Y.; Xu, X.; Klotz, E.; Newman, S.T. Intelligent Manufacturing in the Context of Industry 4.0: A Review. *Engineering* **2017**, *3*, 616–630. [CrossRef]

25. Cesarotti, V.; Deli Orazi, S.; Introna, V. Improve Energy Efficiency in Manufacturing Plants through Consumption Forecasting and Real Time Control: Case Study from Pharmaceutical Sector. In Proceedings of the International Conference on Advances in Production Management Systems (APMS 2010); Cernobbio, Como, Italy, 11–13 October 2010.

26. Capobianchi, S.; Andreassi, L.; Introna, V.; Martini, F.; Ubertini, S. Methodology Development for a Comprehensive and Cost-Effective Energy Management in Industrial Plants. In *Energy Management Systems*; Kini, G., Ed.; InTech: London, UK, 2011; ISBN 978-953-307-579-2.

27. Nikula, R.-P.; Ruusunen, M.; Leiviskä, K. Data-driven framework for boiler performance monitoring. *Appl. Energy* **2016**, *183*, 1374–1388. [CrossRef]

28. Benedetti, M.; Cesarotti, V.; Introna, V.; Serranti, J. Energy consumption control automation using Artificial Neural Networks and adaptive algorithms: Proposal of a new methodology and case study. *Appl. Energy* **2016**, *165*, 60–71. [CrossRef]

29. Fan, C.; Xiao, F.; Li, Z.; Wang, J. Unsupervised data analytics in mining big building operational data for energy efficiency enhancement: A review. *Energy Build.* **2018**, *159*, 296–308. [CrossRef]

30. Santolamazza, A.; Cesarotti, V.; Introna, V. Evaluation of machine learning techniques to enact energy consumption control of compressed air generation in production plants. In Proceedings of the Summer School of Francesco Turco; AIDI–Italian Association of Industrial Operations Professors, Palermo, Italy, 12–14 September 2018; pp. 79–86.

31. Shrouf, F.; Miragliotta, G. Energy management based on Internet of Things: Practices and framework for adoption in production management. *J. Clean. Prod.* **2015**, *100*, 235–246. [CrossRef]

32. Sunthornnapha, T. Utilization of MLP and Linear Regression Methods to Build a Reliable Energy Baseline for Self-benchmarking Evaluation. *Energy Procedia* **2017**, *141*, 189–193. [CrossRef]

33. Torregrossa, D.; Hansen, J.; Hernández-Sancho, F.; Cornelissen, A.; Schutz, G.; Leopold, U. A data-driven methodology to support pump performance analysis and energy efficiency optimization in waste water treatment plants. *Appl. Energy* **2017**, *208*, 1430–1440. [CrossRef]

34. Amasyali, K.; El-Gohary, N.M. A review of data-driven building energy consumption prediction studies. *Renew. Sustain. Energy Rev.* **2018**, *81*, 1192–1205. [CrossRef]

35. Christensen, B.; Himme, A. Improving environmental management accounting: How to use statistics to better determine energy consumption. *J. Manag. Control* **2017**, *28*, 227–243. [CrossRef]

36. Deb, C.; Zhang, F.; Yang, J.; Lee, S.E.; Shah, K.W. A review on time series forecasting techniques for building energy consumption. *Renew. Sustain. Energy Rev.* **2017**, *74*, 902–924. [CrossRef]

37. Foucquier, A.; Robert, S.; Suard, F.; Stéphan, L.; Jay, A. State of the art in building modelling and energy performances prediction: A review. *Renew. Sustain. Energy Rev.* **2013**, *23*, 272–288. [CrossRef]

38. Ngo, N.-T. Early predicting cooling loads for energy-efficient design in office buildings by machine learning. *Energy Build.* **2019**, *182*, 264–273. [CrossRef]

39. Pino-Mejías, R.; Pérez-Fargallo, A.; Rubio-Bellido, C.; Pulido-Arcas, J.A. Comparison of linear regression and artificial neural networks models to predict heating and cooling energy demand, energy consumption and CO_2 emissions. *Energy* **2017**, *118*, 24–36. [CrossRef]

40. Wang, S.; Liang, Y.C.; Li, W.D.; Cai, X.T. Big Data enabled Intelligent Immune System for energy efficient manufacturing management. *J. Clean. Prod.* **2018**, *195*, 507–520. [CrossRef]

41. Kim, Y.-S.; Kim, K.-S. Simplified energy prediction method accounting for part-load performance of chiller. *Build. Environ.* **2007**, *42*, 507–515. [CrossRef]

42. Ahmad, T.; Chen, H.; Guo, Y.; Wang, J. A comprehensive overview on the data driven and large scale based approaches for forecasting of building energy demand: A review. *Energy Build.* **2018**, *165*, 301–320. [CrossRef]

43. Ahmad, T.; Chen, H.; Wang, J.; Guo, Y. Review of various modeling techniques for the detection of electricity theft in smart grid environment. *Renew. Sustain. Energy Rev.* **2018**, *82*, 2916–2933. [CrossRef]

44. Chou, J.-S.; Tran, D.-S. Forecasting energy consumption time series using machine learning techniques based on usage patterns of residential householders. *Energy* **2018**, *165*, 709–726. [CrossRef]

45. Diamantoulakis, P.D.; Kapinas, V.M.; Karagiannidis, G.K. Big Data Analytics for Dynamic Energy Management in Smart Grids. *Big Data Res.* **2015**, *2*, 94–101. [CrossRef]

46. Ferreira, A.M.S.; Cavalcante, C.A.M.T.; Fontes, C.H.O.; Marambio, J.E.S. A new method for pattern recognition in load profiles to support decision-making in the management of the electric sector. *Int. J. Electr. Power Energy Syst.* **2013**, *53*, 824–831. [CrossRef]

47. Grolinger, K.; L'Heureux, A.; Capretz, M.A.M.; Seewald, L. Energy Forecasting for Event Venues: Big Data and Prediction Accuracy. *Energy Build.* **2016**, *112*, 222–233. [CrossRef]

48. Stoyanova, I.; Marin, M.; Monti, A. Characterization of load profile deviations for residential buildings. In Proceedings of the IEEE PES ISGT Europe 2013, Lyngby, Denmark, 6–9 October 2013; pp. 1–5.

49. Tsekouras, G.J.; Kotoulas, P.B.; Tsirekis, C.D.; Dialynas, E.N.; Hatziargyriou, N.D. A pattern recognition methodology for evaluation of load profiles and typical days of large electricity customers. *Electr. Power Syst. Res.* **2008**, *78*, 1494–1510. [CrossRef]

50. Tu, C.; He, X.; Shuai, Z.; Jiang, F. Big data issues in smart grid–A review. *Renew. Sustain. Energy Rev.* **2017**, *79*, 1099–1107. [CrossRef]

51. Vázquez-Canteli, J.R.; Nagy, Z. Reinforcement learning for demand response: A review of algorithms and modeling techniques. *Appl. Energy* **2019**, *235*, 1072–1089. [CrossRef]

52. Wei, Y.; Zhang, X.; Shi, Y.; Xia, L.; Pan, S.; Wu, J.; Han, M.; Zhao, X. A review of data-driven approaches for prediction and classification of building energy consumption. *Renew. Sustain. Energy Rev.* **2018**, *82*, 1027–1047. [CrossRef]

53. Yildiz, B.; Bilbao, J.I.; Sproul, A.B. A review and analysis of regression and machine learning models on commercial building electricity load forecasting. *Renew. Sustain. Energy Rev.* **2017**, *73*, 1104–1122. [CrossRef]

54. Zhou, K.; Yang, S.; Shen, C. A review of electric load classification in smart grid environment. *Renew. Sustain. Energy Rev.* **2013**, *24*, 103–110. [CrossRef]

55. Das, U.K.; Tey, K.S.; Seyedmahmoudian, M.; Mekhilef, S.; Idris, M.Y.I.; Van Deventer, W.; Horan, B.; Stojcevski, A. Forecasting of photovoltaic power generation and model optimization: A review. *Renew. Sustain. Energy Rev.* **2018**, *81*, 912–928. [CrossRef]

56. Foley, A.M.; Leahy, P.G.; Marvuglia, A.; McKeogh, E.J. Current methods and advances in forecasting of wind power generation. *Renew. Energy* **2012**, *37*, 1–8. [CrossRef]

57. Sharma, A.; Kakkar, A. Forecasting daily global solar irradiance generation using machine learning. *Renew. Sustain. Energy Rev.* **2018**, *82*, 2254–2269. [CrossRef]

58. Voyant, C.; Notton, G.; Kalogirou, S.; Nivet, M.-L.; Paoli, C.; Motte, F.; Fouilloy, A. Machine learning methods for solar radiation forecasting: A review. *Renew. Energy* **2017**, *105*, 569–582. [CrossRef]

59. Wang, J.; Song, Y.; Liu, F.; Hou, R. Analysis and application of forecasting models in wind power integration: A review of multi-step-ahead wind speed forecasting models. *Renew. Sustain. Energy Rev.* **2016**, *60*, 960–981. [CrossRef]

60. Shang, C.; Yang, F.; Huang, D.; Lyu, W. Data-driven soft sensor development based on deep learning technique. *J. Process Control* **2014**, *24*, 223–233. [CrossRef]

61. Kadlec, P.; Gabrys, B.; Strandt, S. Data-driven Soft Sensors in the process industry. *Comput. Chem. Eng.* **2009**, *33*, 795–814. [CrossRef]

62. Zhang, K.; Peng, K.; Chu, R.; Dong, J. Implementing multivariate statistics-based process monitoring: A comparison of basic data modeling approaches. *Neurocomputing* **2018**, *290*, 172–184. [CrossRef]

63. Engelberth, T.; Krawczyk, D.; Verl, A. Model-based method for condition monitoring and diagnosis of compressors. *Procedia CIRP* **2018**, *72*, 1321–1326. [CrossRef]

64. Santolamazza, A.; Cesarotti, V.; Introna, V. Anomaly detection in energy consumption for Condition-Based maintenance of Compressed Air Generation systems: An approach based on artificial neural networks. *IFAC PapersOnLine* **2018**, *51*, 1131–1136. [CrossRef]

65. Stetco, A.; Dinmohammadi, F.; Zhao, X.; Robu, V.; Flynn, D.; Barnes, M.; Keane, J.; Nenadic, G. Machine learning methods for wind turbine condition monitoring: A review. *Renew. Energy* **2019**, *133*, 620–635. [CrossRef]

66. Romeo, L.M.; Gareta, R. Fouling control in biomass boilers. *Biomass Bioenergy* **2009**, *33*, 854–861. [CrossRef]

67. Xiao, W. A probabilistic machine learning approach to detect industrial plant faults. *Int. J. Progn. Health Manag.* Available online: https://www.scopus.com/inward/record.uri?eid=2-s2.0-84966359883&partnerID=40&md5=536a49cd2e7d4809b45c73215dfecba7 (accessed on 2 July 2019).

68. Liu, R.; Yang, B.; Zio, E.; Chen, X. Artificial intelligence for fault diagnosis of rotating machinery: A review. *Mech. Syst. Signal Proc.* **2018**, *108*, 33–47. [CrossRef]

69. Diez-Olivan, A.; Del Ser, J.; Galar, D.; Sierra, B. Data fusion and machine learning for industrial prognosis: Trends and perspectives towards Industry 4.0. *Inf. Fusion* **2019**, *50*, 92–111. [CrossRef]

70. Karim, R.; Westerberg, J.; Galar, D.; Kumar, U. Maintenance Analytics–The New Know in Maintenance. *IFAC PapersOnLine* **2016**, *49*, 214–219. [CrossRef]

71. Kim, H.; Na, M.G.; Heo, G. Application of monitoring, diagnosis, and prognosis in thermal performance analysis for nuclear power plants. *Nucl. Eng. Technol.* **2014**, *46*, 737–752. [CrossRef]

72. Lee, J.; Ardakani, H.D.; Yang, S.; Bagheri, B. Industrial Big Data Analytics and Cyber-physical Systems for Future Maintenance & Service Innovation. *Procedia CIRP* **2015**, *38*, 3–7.

73. Lee, J.; Wu, F.; Zhao, W.; Ghaffari, M.; Liao, L.; Siegel, D. Prognostics and health management design for rotary machinery systems—Reviews, methodology and applications. *Mech. Syst. Signal Proc.* **2014**, *42*, 314–334. [CrossRef]

74. Roy, R.; Stark, R.; Tracht, K.; Takata, S.; Mori, M. Continuous maintenance and the future–Foundations and technological challenges. *CIRP Ann.* **2016**, *65*, 667–688. [CrossRef]

75. Vogl, G.W.; Weiss, B.A.; Helu, M. A review of diagnostic and prognostic capabilities and best practices for manufacturing. *J. Intell. Manuf.* **2016**, *30*, 79–95. [CrossRef]

76. Cesarotti, V.; Di Silvio, B.; Introna, V. Energy budgeting and control: A new approach for an industrial plant. *Int. J. Energy Sector Manag.* **2009**, *3*, 131–156. [CrossRef]

77. Pérez-Rave, J.; Muñoz-Giraldo, L.; Correa-Morales, J.C. Use of control charts with regression analysis for autocorrelated data in the context of logistic financial budgeting. *Comput. Ind. Eng.* **2017**, *112*, 71–83. [CrossRef]

78. Santolamazza, A.; Introna, V.; Cesarotti, V. Energy budget control in manufacturing systems with on-site energy generation: An advanced methodology for analyzing specific cost variations. In Proceedings of the Summer School Francesco Turco, AIDI–Italian Association of Industrial Operations Professors, Palermo, Italy, 13–15 September 2017; pp. 404–410.

79. Capehart, L.; Turner, C.; Kennedy, J. *Guide to Energy Management*, 4th ed.; Fairmont Press: Lilburn, GA, USA, 2002; ISBN 978-0-8247-4120-4.

80. ISO International Standard Organization. *ISO/IEC 13273-1:2015 Energy Efficiency and Renewable Energy Sources–Common International Terminology Energy Efficiency 2015*; ISO International Standard Organization: Geneva, Switzerland, 2015.

81. Morvay, Z.K.; Gvozdenac, D.D. *Applied Industrial Energy and Environmental Management*; John Wiley & Sons: Hoboken, NJ, USA, 2008; ISBN 978-0-470-71437-9.

82. Song, B.; Ao, Y.; Xiang, L.; Lionel, K.Y.N. Data-driven Approach for Discovery of Energy Saving Potentials in Manufacturing Factory. *Procedia CIRP* **2018**, *69*, 330–335. [CrossRef]

83. Montgomery, D.C. *Introduction to Statistical Quality Control*; John Wiley & Sons: Hoboken, NJ, USA, 2007.

84. Braaksma, A.J.J.; Klingenberg, W.; Veldman, J. Failure mode and effect analysis in asset maintenance: A multiple case study in the process industry. *Int. J. Prod. Res.* **2013**, *51*, 1055–1071. [CrossRef]

85. Bishop, C. *Pattern Recognition and Machine Learning*; Springer-Verlag: New York, NY, USA, 2006.

86. Kerzner, H. *Project Management, a System Approach to Planning, Scheduling, and Controlling*, 7th ed.; John Wiley & Sons: Hoboken, NJ, USA, 2000.

87. Cesarotti, V.; Ciminelli, V.; Di Silvio, B.; Fedele, T.; Introna, V. Energy budgeting and control for industrial plant through consumption analysis and monitoring. In Proceedings of the IASTED International Conference on Power and Energy Systems, Clearwater, FL, USA, 3–5 January 2007; pp. 389–394.

88. Torregrossa, D.; Leopold, U.; Hernández-Sancho, F.; Hansen, J. Machine learning for energy cost modelling in wastewater treatment plants. *J. Environ. Manag.* **2018**, *223*, 1061–1067. [CrossRef]

89. Debnath, K.B.; Mourshed, M. Forecasting methods in energy planning models. *Renew. Sustain. Energy Rev.* **2018**, *88*, 297–325. [CrossRef]

90. Jha, S.K.; Bilalovic, J.; Jha, A.; Patel, N.; Zhang, H. Renewable energy: Present research and future scope of Artificial Intelligence. *Renew. Sustain. Energy Rev.* **2017**, *77*, 297–317. [CrossRef]

91. Koseleva, N.; Ropaite, G. Big Data in Building Energy Efficiency: Understanding of Big Data and Main Challenges. *Procedia Eng.* **2017**, *172*, 544–549. [CrossRef]

92. Lund, H.; Østergaard, P.A.; Connolly, D.; Mathiesen, B.V. Smart energy and smart energy systems. *Energy* **2017**, *137*, 556–565. [CrossRef]

93. Molina-Solana, M.; Ros, M.; Ruiz, M.D.; Gómez-Romero, J.; Martin-Bautista, M.J. Data science for building energy management: A review. *Renew. Sustain. Energy Rev.* **2017**, *70*, 598–609. [CrossRef]

94. Shrouf, F.; Ordieres, J.; Miragliotta, G. Smart factories in Industry 4.0: A review of the concept and of energy management approached in production based on the Internet of Things paradigm. In Proceedings of the 2014 IEEE International Conference on Industrial Engineering and Engineering Management, Selangor Darul Ehsan, Malaysia, 9–12 December 2014; pp. 697–701.

95. Yu, Z.; Haghighat, F.; Fung, B.C.M. Advances and challenges in building engineering and data mining applications for energy-efficient communities. *Sustain. Cities Soc.* **2016**, *25*, 33–38. [CrossRef]

96. Zhou, K.; Fu, C.; Yang, S. Big data driven smart energy management: From big data to big insights. *Renew. Sustain. Energy Rev.* **2016**, *56*, 215–225. [CrossRef]

97. Kusiak, A.; Zhang, Z.; Verma, A. Prediction, operations, and condition monitoring in wind energy. *Energy* **2013**, *60*, 1–12. [CrossRef]

98. Lee, J.; Lapira, E.; Bagheri, B.; Kao, H. Recent advances and trends in predictive manufacturing systems in big data environment. *Manuf. Lett.* **2013**, *1*, 38–41. [CrossRef]

99. Lee, J.; Lapira, E.; Yang, S.; Kao, A. Predictive Manufacturing System–Trends of Next-Generation Production Systems. *IFAC Proc. Vol.* **2013**, *46*, 150–156. [CrossRef]

100. Tao, F.; Qi, Q.; Liu, A.; Kusiak, A. Data-driven smart manufacturing. *J. Manuf. Syst.* **2018**, *48*, 157–169. [CrossRef]

101. Wang, J.; Ma, Y.; Zhang, L.; Gao, R.X.; Wu, D. Deep learning for smart manufacturing: Methods and applications. *J. Manuf. Syst.* **2018**, *48*, 144–156. [CrossRef]

102. Zhang, Y.; Ma, S.; Yang, H.; Lv, J.; Liu, Y. A big data driven analytical framework for energy-intensive manufacturing industries. *J. Clean. Prod.* **2018**, *197*, 57–72. [CrossRef]

103. Zhang, Y.; Ren, S.; Liu, Y.; Si, S. A big data analytics architecture for cleaner manufacturing and maintenance processes of complex products. *J. Clean. Prod.* **2017**, *142*, 626–641. [CrossRef]

Article

Experimental Investigation of Productivity, Specific Energy Consumption, and Hole Quality in Single-Pulse, Percussion, and Trepanning Drilling of IN 718 Superalloy

Shoaib Sarfraz [1],*, Essam Shehab [1,2], Konstantinos Salonitis [1] and Wojciech Suder [1]

[1] Manufacturing Department, School of Aerospace, Transport and Manufacturing, Cranfield University, Cranfield, Bedfordshire MK43 0AL, UK; e.shehab@cranfield.ac.uk (E.S.); k.salonitis@cranfield.ac.uk (K.S.); w.j.suder@cranfield.ac.uk (W.S.)

[2] School of Engineering and Digital Sciences, Nazarbayev University, Nur-Sultan 010000, Kazakhstan

* Correspondence: shoaib.sarfraz@cranfield.ac.uk; Tel.: +44-(0)-7729-475536

Received: 7 October 2019; Accepted: 3 December 2019; Published: 4 December 2019

Abstract: Laser drilling is a high-speed process that is used to produce high aspect ratio holes of various sizes for critical applications, such as cooling holes in aero-engine and gas turbine components. Hole quality is always a major concern during the laser drilling process. Apart from hole quality, cost and productivity are also the key considerations for high-value manufacturing industries. Taking into account the significance of improving material removal quantity, energy efficiency, and product quality, this study is performed in the form of an experimental investigation and multi-objective optimisation for three different laser drilling processes (single-pulse, percussion, and trepanning). A Quasi-CW fibre laser was used to produce holes in a 1 mm thick IN 718 superalloy. The impacts of significant process parameters on the material removal rate (MRR), specific energy consumption (SEC), and hole taper have been discussed based on the results collected through an experimental matrix that was designed using the Taguchi method. The novelty of this work focuses on evaluating and comparing the performance of laser drilling methods in relation to MRR, SEC, and hole quality altogether. Comparative analysis revealed single-pulse drilling as the best option for MRR and SEC as the MRR value reduces with percussion and trepanning by 99.70% and 99.87% respectively; similarly, percussion resulted in 14.20% higher SEC value while trepanning yielded a six-folds increase in SEC as compared to single-pulse drilling. Trepanning, on the other hand, outperformed the rest of the drilling processes with 71.96% better hole quality. Moreover, optimum values of parameters simultaneously minimising SEC and hole taper and maximising MRR are determined using multi-objective optimisation.

Keywords: laser drilling; percussion; trepanning; productivity; cost; material removal rate (MRR); specific energy consumption (SEC); Taguchi; hole taper; IN 718

---

1. Introduction

Machining is a fundamental method to transform raw material into a finished product. Machining processes of various types are involved in crafting the solid structure into intricate parts of desired geometry. Despite the usage of advanced conventional machining technologies, manufacturing of complex parts with high accuracy has remained a challenge for the manufacturing industry. For instance, certain complex parts, such as gas turbine or aero-engine components need highly accurate and miniature-sized machining, which can be of microsize, such as holes in nozzle guide vane, turbine blade, fuel injector, and combustion chamber, are mainly in milli to microsize;

therefore, the accomplishment of these complex holes warrants the selection of a highly accurate drilling process.

Inconel 718 is extensively used in the aerospace industry, particularly for manufacturing of aero-engine components that operate under high-temperatures. Conventional machining is difficult for this material because of its high strength and work hardening properties [1]. The machinability of superalloys can be improved using different machining methods, such as ultrasonic machining, electrochemical machining, water jet machining, and laser-assisted machining [2–4]. Laser drilling is a high power, high speed, and non-contact machining process, which is specified for the drilling of holes of various shapes and sizes in almost any material, such as composites, metals and non-metals [5]. During recent years, the laser drilling method has been proven as an important industrial process for producing cooling holes for aero-engine components where the size of hole ranges between 0.25 and 1.0 mm [6]. During this method, a laser beam is focused on the workpiece surface, where the thermal energy transforms the substrate material into a molten metal that can be removed easily using the pressurised assist gas, as shown in Figure 1. In addition to that, the laser beam can heat the material instantly to its vaporisation temperature, and the vaporised material exits out of the hole. At this stage, vapour pressure may also be produced, which contributes to the expulsion of molten metal out of the hole cavity [7]; at the same time, the holes produced by the laser reveal some defects, such as recast layer, heat affected zone (HAZ), and hole taper, which may limit the utilisation of the laser drilling process in the industry. From the manufacturing perspective, product quality is always important. In the laser drilling process, the drilled hole quality is assessed by examining its geometrical and metallurgical features, such as circularity, hole taper, microcracks, HAZ, and recast layer thickness [8]. Different drilling methods can be used to produce a particular hole geometry. Depending upon the application requirements, a distinctive method will be selected, as shown in Figure 2.

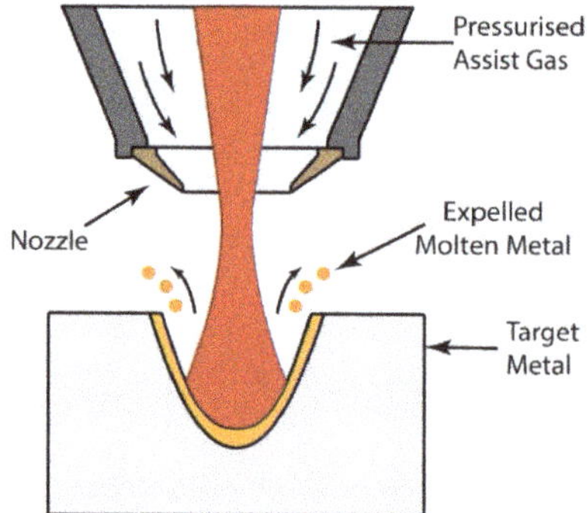

Figure 1. Laser drilling process—schematic diagram [9].

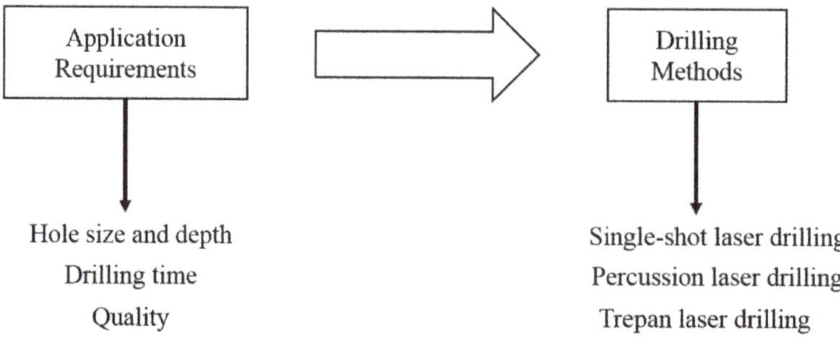

Figure 2. Laser drilling methods and the application requirements.

Methods that are commonly used for laser drilling include single-shot laser drilling, percussion, and trepan laser drilling. Single-shot laser drilling, also known as single-pulse drilling, is the most basic method in which a single high-energy pulse from the laser produces a hole throughout the material thickness. High productivity can be achieved with this simple drilling method. Single-shot drilling is preferable when production throughput has priority over quality. The percussion laser drilling method is quite similar to single-shot drilling and is directed by delivering consecutive laser pulses to a particular spot of the material. Using percussion drilling, high-quality holes are achieved as compared to single-pulse drilling. The fact is that less energy is applied to the material every time the pulse is fired; hence, avoiding the thermal defects, such as HAZ. Higher dimensional accuracy can be achieved with percussion drilling; however, this process is slower in contrast with single-pulse drilling. Trepan laser drilling or trepanning is used when the required shape has a size of large diameter. In this process, the hole is initially pierced into the substrate in the same way as percussion drilling followed by spiral configuration to cut a circular disc or cylindrical core from the material by rotating the laser beam around the circumference of the hole. The cylindrical core falls out after the required hole size is created. The drilling time is relatively longer as compared to other methods [10]. The relationship between drilling time and hole quality for different laser drilling processes is presented in Figure 3.

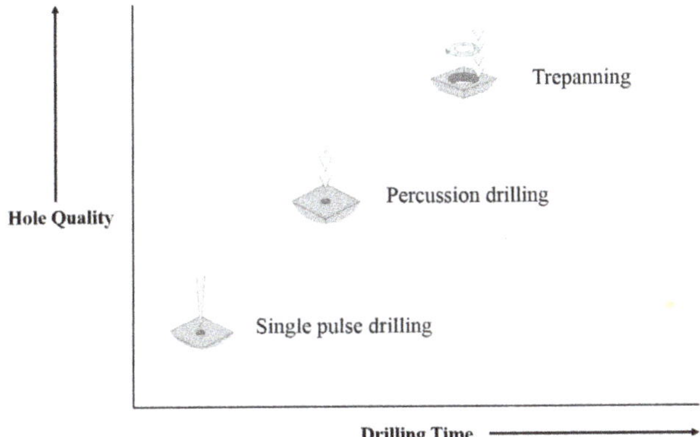

Figure 3. Drilling time and hole quality relationship using different laser drilling methods. Source: [8].

The laser drilling process is complex, and there are several parameters that affect the quality of manufactured holes. For improved drilling performance, researchers have been experimenting with various approaches, including different laser drilling methods and with process parameters of various levels. Panda et al. [11] investigated the influence of laser drilling process parameters on hole quality during percussion drilling of high carbon steel and found laser pulse width/duration as a critical parameter that increases the heat affected zone at higher values. Yilbas [12] employed a parametric study to observe the effects of different laser machining parameters on the drilled hole quality. This study revealed that pulse energy, pulse duration, pulse frequency, and laser focus position control the hole quality. In another study, Yilbas and Aleem [13] found that pulse energy, assist gas pressure, and focal position are the important parameters that influence the overall quality of the laser drilled hole. Ng and Li [7] found that high peak power and short pulse width combination improve the repeatability of holes. The Taguchi method was used by Chien and Hou [14] to analyse the impacts of different laser drilling process parameters on hole quality during trepanning. It was observed that improved hole quality could be obtained when higher pulse energy and lower trepan speed is used. An experimental investigation was performed by Morar et al. [15] to investigate the hole quality during laser trepanning of nickel-based superalloy; pulse width, pulse energy, and trepan speed were observed as the most influencing parameters affecting the quality of the drilled hole. Rajesh et al. [16] examined the effects of several laser drilling parameters on drilled hole quality and reported that pulse duration significantly influences the hole taper. Dhaker and Pandey [17] investigated the parameters influencing hole quality during laser trepanning. They concluded that the hole quality could be significantly improved by the proper control of laser drilling parameters.

Furthermore, productivity is an important attribute of the laser drilling process that is defined by the material removal rate (MRR). In the laser drilling process, MRR is influenced by the applied laser drilling parameters, i.e., pulse width, pulse frequency, pulse energy, and assist gas [11,18–21]. Higher productivity is always desirable for manufacturing industries as it reduces the cost of manufacturing of a component [2,22].

Energy consumption, needed for the manufacturing of products, is also the major concern of the manufacturing community because of the constant increase in energy cost and due to ecological effect linked with the production of energy and its use [23]. Reducing energy consumption is one of the top priorities of both national and international policies. The hefty CO_2 emissions are the result of extensive use of energy in various manufacturing processes and are responsible for climatic changes. It is found that a large proportion (20–40%) of energy is wasted when performing industrial operations [24]. The International Energy Agency (IEA) underlined the necessity of energy efficiency evaluation in the direction of two-thirds energy intensity reduction of the world economy before 2050 [25]. Consequently, there is a need to evaluate the energy consumed during the manufacturing processes.

The energy efficiency of a laser-based process is low, but on the other hand, the material can be removed more precisely. Dahmen et al. [26] revealed that lasers could impart to sustainable manufacturing because of the minimal use of consumables, confined heat input even at low energy, saving of cost and energy for heat treatment, and with the aid of hybrid methods. Utilising more economical laser sources, for instance, disc or fibre lasers can also be examined as a possible energy efficient method. Similar findings were reported by Kaierle et al. [27]. An investigation was performed by Apostolos et al. [28] and Franco et al. [29] to evaluate laser drilling process energy efficiency by examining different process parameters using CO2 and femtosecond-pulsed fibre laser respectively. The results revealed that optimising the process parameters could lead to reducing the energy consumption of the process. Reduction in energy consumption will provide a great advantage to the industries by alleviating the cost of energy and at the same time reducing the energy crisis and air pollution problems.

Manufacturing industries are continuously striving to enhance their competitive position through improved productivity and quality at a minimum possible cost that shows the importance of these factors for the industrial sector. From the literature, it has been found that little or no research is reported that characterise the laser drilling methods in terms of MRR, SEC, and hole quality altogether. Therefore, the objective of the presented study is to deliver a clear understanding of the impacts of different laser drilling methods and process parameters on productivity (material removal rate), cost (specific energy consumption), and hole quality (hole taper) in laser drilling of IN 718 superalloy. Three different laser drilling processes have been investigated, i.e., single-pulse, percussion, and trepanning. Further analysis has been performed using multi-objective optimisation to achieve the optimum levels of process parameters for maximum MRR, with minimum SEC and hole taper.

2. Materials and Methods

2.1. Experimental Setup

Laser drilling of nickel superalloy was performed at 90° to the material surface using three different methods, i.e., single-pulse, percussion, and trepanning. Inconel® alloy 718 (Goodfellow, UK) was used as a base material in this study. The size of the specimen was 100 × 100 × 1 mm (Figure 4a). Energy dispersive X-ray (EDX) analysis was performed to verify the chemical composition of the material and is provided in Table 1.

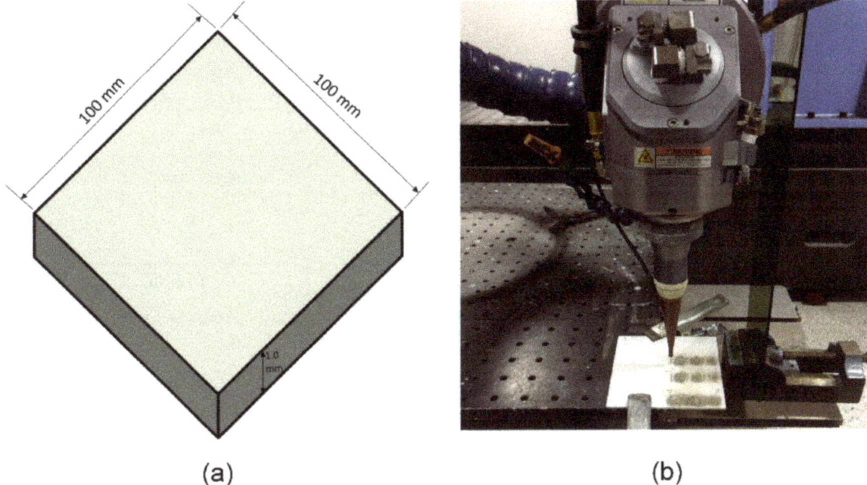

(a)   (b)

Figure 4. (**a**) Dimensions of work material used in the experiment. (**b**) Laser drilling experimental setup.

Table 1. Chemical composition (wt%) of IN 718 superalloy.

Cr	Fe	Nb	Mo	Ti	Al	Mn	Si	Ni
19	18.5	5.13	3.05	0.9	0.5	0.18	0.18	Rest

A Quasi-CW fibre laser (Model YLS-2000/20000-QCW) from IPG Photonics, UK, was used for this study. The laser drilling setup prepared for the experiments is presented in Figure 4b. The specification of laser system includes wavelength: 1070 nm, maximum average power: 2000 W, peak power: 20,000 W, maximum pulse energy: 200 J, pulse duration: 0.2–10 ms, and maximum pulse frequency: 2 kHz. The hole pitch was set at 5 mm to prevent the potential effects from adjacent holes. The laser beam was directed at the workpiece material using an optical lens with 200 mm focal length. The diameter of fibre used and laser beam spot size was 200 μm and 285 μm, respectively. The lens was equipped with a gas nozzle co-axially to deliver and assist gas and get protection from the flushing material.

2.2. Experimental Design

Three different laser drilling methods were performed in this study, namely, single-pulse drilling, percussion, and trepanning. Therefore, different input parameters were selected for each method. For assessing the performance of single-pulse drilling, pulse energy and pulse duration with selected ranges were used as input parameters. Three process variables namely pulse energy, pulse width and number of pulses (NOP) per hole were used for percussion drilling. Moreover, for trepanning, the process parameters used were pulse energy, pulse width, pulse frequency, and trepan speed. Some of the parameters were held constant during the entire experimentation and are presented in Table 2. In this study, the input variables were chosen because of their significant impact on hole quality, material removal rate, and specific energy consumption [29–36]. The ranges of input parameters were selected after the trial experimentation so that drilling of holes gives better hole quality and material removal rate with minimum energy consumption. For each method, nine experiments in total were designed using the Taguchi L9 orthogonal array. The process parameters with levels for the employed drilling methods are provided in Table 3.

Table 2. Constant parameters during experiment.

S. No.	Parameters	Values
1	Frequency (percussion)	10 Hz
2	Programmed radius (trepanning)	0.125 mm
3	Gas pressure	100 psi
4	Assist gas	Air
5	Focal plane position	On top surface

Table 3. Variable parameters.

Drilling Method(s)	Process Variables	Unit	Levels		
			Low Level	Medium Level	High Level
Single-pulse drilling	Pulse energy	J	20	30	40
	Pulse width	ms	2	3	4
Percussion	Pulse energy	J	5	6	7
	Pulse width	ms	0.5	1	1.5
	NOP/hole		5	10	15
Trepanning	Pulse energy	J	5	6	7
	Pulse width	ms	0.5	1	1.5
	Pulse frequency	Hz	20	30	40
	Trepan speed	mm/min	30	40	50

2.3. Response Measurements

The productivity, cost and quality of each laser drilling method were measured using material removal rate (MRR), specific energy consumption (SEC), and hole taper (HT), respectively. Each experimental run was performed four times, and the average value was considered to minimise the error defects during experimentation and measurement.

2.3.1. Productivity

The productivity of the laser drilling process was determined by the material removal rate, which specifies the amount of material removed per unit time. For the employed drilling techniques, MRR was determined using Equation (1).

$$\text{MRR} = \frac{V}{t} \tag{1}$$

where MRR denotes the material removal rate in mm^3/s, V represents the volume of material removed in mm^3, and t is the drilling time measured in seconds during the process. The final geometry of drilled holes was assumed as a frustum of the cone because of hole taper. Therefore, the volume of material removed (V) was computed employing Equation (2) [18].

$$V = \frac{1}{3}\pi T\left(R_{ent}^2 + R_{ent}^2 R_{ex}^2 + R_{ex}^2\right) \tag{2}$$

where V expresses the volume of material removed in mm^3, R_{ent} and R_{ex} are the entry and exit side radii of the drilled hole, respectively, in millimetres, and T is the workpiece thickness in mm.

For each hole, a total of seven measurements were recorded for both entry and exit diameters ensuring coverage of minimum, maximum, and average values (Figure 5a). The arithmetic mean of these measurements was calculated to get the average value of hole diameter for both entry and exit sides. These measurements were taken using an optical microscope (LEICA CTR6000, Leica, Germany), as presented in Figure 5b. Before the measurements, all samples were cleaned using a series of 240, 1200, and 2500 grade silicon carbide papers to make sure that the debris from the surface of the specimen had been eliminated.

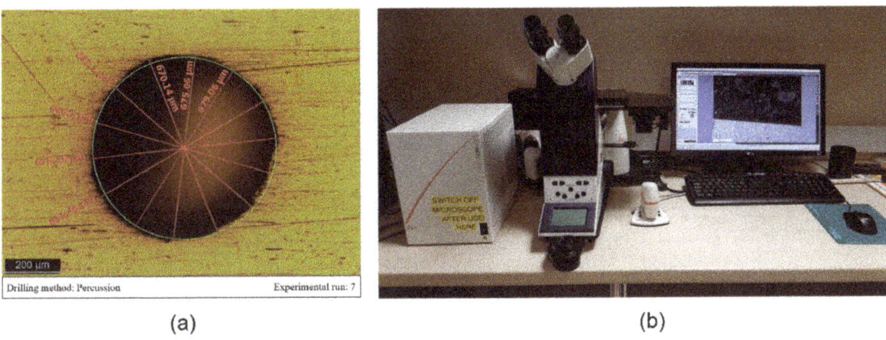

(a)   (b)

Figure 5. (a) Strategy for measuring hole diameter. (b) Leica optical microscope.

2.3.2. Cost

Specific energy consumption determines the energy consumed to remove a unit volume of material. SEC shows how efficiently the material is removed in terms of energy utilization, and it affects the cost of energy. For the single-pulse drilling method, Equation (3) was used for the calculation of SEC, while Equations (4) and (5) were used for percussion and trepanning methods, respectively [18,29].

$$SEC_{single\ pulse\ drilling} = \frac{E}{V}, \tag{3}$$

$$SEC_{percussion} = \frac{NOP \times E}{V}, \tag{4}$$

$$SEC_{trepanning} = \frac{P_{avg}}{MRR}, \tag{5}$$

$$P_{avg} = Applied\ pulse\ energy \times pulse\ frequency, \tag{6}$$

where SEC denotes the specific energy consumption in J/mm^3, E is the applied pulse energy in joules (J), V is the volume of material removed in mm^3, NOP is the number of pulses, P_{avg} is the average laser power in watts (W), and MRR is the material removal rate in mm^3/s. The expression used for P_{avg} calculation is given in Equation (6) where the applied pulse energy is measured in joules (J) and pulse frequency in hertz (Hz).

2.3.3. Quality

The quality of the produced hole was defined by the hole taper, which is the ratio of the difference between the entry and exit hole diameter and the plate thickness. The taper angle was measured in degrees. The following relation (Equation (7)) was used to determine the hole taper angle.

$$HT(°) = \tan^{-1}\left(\frac{D_{ent} - D_{ex}}{2 \times T}\right) \tag{7}$$

where HT represents the hole taper in degrees, D_{ent} and D_{ex} are the entry and exit side diameters of the drilled hole, respectively, both measured in millimetres, and T denotes the workpiece thickness in mm.

3. Results and Discussion

3.1. Development of Mathematical Models

For the mathematical modelling of response variables, a regression analysis was conducted using statistical software (Design-Expert®version10, Stat-Ease, USA). Analysis of variance (ANOVA) was applied to examine the significance level of process parameters concerning the output responses and to verify the accuracy of developed models.

3.1.1. Single-Pulse Drilling

For single-pulse drilling, the fit summary for MRR suggested a quadratic model as the best fit model. For SEC and hole taper, a 2 Factorial Interaction (FI) model was suggested for both responses. The ANOVA results indicate that both input variables pulse energy and pulse width contributed significantly in all responses. The ANOVA table, including significant terms along with adequacy measures (R^2, adjusted R^2, and predicted R^2), are listed in Table 4. It is clearly evident that all the models are significant, with p values less than 0.05. The adequacy measures for all developed models are approximately 1, which affirms the adequacy of the mathematical models. Moreover, the low values of coefficient of variation (CoV) 3.11%, 2.04%, and 4.91% (for MRR, SEC, and hole taper, respectively) specifies the reliability and improved precision. The concluding empirical models for responses MRR, SEC, and hole taper are provided in Equations (8)–(10).

$$MRR = +208.30078 + (3.59646 \times pulse\ energy) - (105.05632 \times pulse\ width) - (0.73563 \times pulse\ energy \times pulse\ width) + \left(15.01390 \times Pulse\ width^2\right) \tag{8}$$

$$SEC = +59.55444 + (0.99583 \times pulse\ energy) + (2.00667 \times pulse\ width) + (0.46150 \times pulse\ energy \times pulse\ width) \tag{9}$$

$$HT = +2.23639 + (0.34358 \times pulse\ energy) + (0.81917 \times pulse\ width) - (0.077250 \times pulse\ energy \times pulse\ width) \tag{10}$$

Table 4. Analysis of variance for material removal rate (MRR), specific energy consumption (SEC), and hole taper (single-pulse drilling).

Source	SS	df	MS	F Value	P Value	
			For MRR			
Model	10,058.37	4	2514.59	406.50	<0.0001	significant
A-pulse energy	1158.57	1	1158.57	187.29	0.0002	
B-pulse width	8232.51	1	8232.51	1330.83	<0.0001	
AB	216.46	1	216.46	34.99	0.0041	
B^2	450.83	1	450.83	72.88	0.0010	
Residual	24.74	4	6.19			
Cor Total	10,083.11	8				
Mean	79.95		R^2		0.9975	
Std. Dev.	2.49		Pred R^2		0.9849	
CoV%	3.11		Adj R^2		0.9951	
PRESS	151.92		Adeq Precision		54.954	
			For SEC			
Model	4992.44	3	1664.15	212.55	<0.0001	significant
A-pulse energy	3399.59	1	3399.59	434.21	<0.0001	
B-pulse width	1507.65	1	1507.65	192.56	<0.0001	
AB	85.19	1	85.19	10.88	0.0215	
Residual	39.15	5	7.83			
Cor Total	5031.58	8				
Mean	136.98		R^2		0.9922	
Std. Dev.	2.80		Pred R^2		0.9626	
CoV%	2.04		Adj R^2		0.9876	
PRESS	188.17		Adeq Precision		42.516	

Table 4. *Cont.*

Source	SS	df	MS	F Value	P Value	
			For HT			
Model	23.36	3	7.79	49.93	0.0004	significant
A-pulse energy	7.50	1	7.50	48.11	0.0010	
B-pulse width	13.47	1	13.47	86.37	0.0002	
AB	2.39	1	2.39	15.30	0.0113	
Residual	0.78	5	0.16			
Cor Total	24.14	8				
Mean	8.05		R^2		0.9677	
Std. Dev.	0.39		Pred R^2		0.8698	
CoV%	4.91		Adj R^2		0.9483	
PRESS	3.14		Adeq Precision		19.877	

SS: Sum of squares, MS: Mean square.

3.1.2. Percussion

Fit summary for percussion indicated 2FI relationship as the best fit model for all responses. All the process parameters contributed significantly in MRR, SEC, and hole taper. The p values (<0.05) shows that all models for the percussion are significant. The ANOVA results (Table 5) revealed the adequacy of developed models with all adequacy measure values close to unity. The developed empirical models for MRR, SEC, and hole taper are presented in Equations (11)–(13) respectively.

$$\text{MRR} = +0.52756 + (0.022800 \times pulse\ energy) - (0.13907 \times pulse\ width) - \left(0.033580 \times \tfrac{NOP}{hole}\right) + \left(0.00864 \times pulse\ width \times \tfrac{NOP}{hole}\right) \tag{11}$$

$$\text{SEC} = +204.12578 - (50.19433 \times pulse\ energy) - (417.90267 \times pulse\ width) + \left(33.48080 \times \tfrac{NOP}{hole}\right) + (81.91600 \times pulse\ energy \times pulse\ width) \tag{12}$$

$$\text{HT} = +9.53556 - (0.37667 \times pulse\ energy) - (0.7 \times pulse\ width) - \left(0.16767 \times \tfrac{NOP}{hole}\right) + \left(0.02 \times pulse\ energy \times \tfrac{NOP}{hole}\right) \tag{13}$$

Table 5. Analysis of variance for MRR, SEC, and hole taper (percussion).

Source	SS	df	MS	F Value	P Value	
			For MRR			
Model	0.099	4	0.025	282.52	<0.0001	significant
A-pulse energy	1.949×10^{-3}	1	1.949×10^{-3}	22.14	0.0093	
B-pulse width	4.161×10^{-3}	1	4.161×10^{-3}	47.26	0.0023	
C-NOP/hole	0.093	1	0.093	1059.75	<0.0001	
BC	1.166×10^{-3}	1	1.166×10^{-3}	13.25	0.0220	
Residual	3.522×10^{-4}	4	8.804×10^{-5}			
Cor Total	0.100	8				
Std. Dev.	9.383×10^{-3}		R^2		0.9965	
Mean	0.28		Pred R^2		0.9673	
CoV %	3.40		Adj R^2		0.9929	
PRESS	3.268×10^{-3}		Adeq Precision		43.191	

Table 5. *Cont.*

Source	SS	df	MS	F Value	P Value	
			For MRR			
			For SEC			
Model	1.479×10^5	4	36,969.30	531.68	<0.0001	significant
A-pulse energy	6037.58	1	6037.58	86.83	0.0007	
B-pulse width	8123.97	1	8123.97	116.84	0.0004	
C-NOP/hole	1.051×10^5	1	1.051×10^5	1511.37	<0.0001	
AB	4193.89	1	4193.89	60.32	0.0015	
Residual	278.13	4	69.53			
Cor Total	1.482×10^5	8				
Mean	311.36		R^2		0.9981	
Std. Dev.	8.34		Pred R^2		0.9857	
CoV%	2.68		Adj R^2		0.9962	
PRESS	2120.22		Adeq Precision		58.303	
			For HT			
Model	1.51	4	0.38	251.95	< 0.0001	significant
A-pulse energy	0.19	1	0.19	124.73	0.0004	
B-pulse width	0.46	1	0.46	305.97	< 0.0001	
C-NOP/hole	0.34	1	0.34	227.00	0.0001	
AC	0.025	1	0.025	16.65	0.0151	
Residual	6.006×10^{-3}	4	1.501×10^{-3}			
Cor Total	1.52	8				
Mean	6.10		R^2		0.9960	
Std. Dev.	0.039		Pred R^2		0.9745	
CoV%	0.64		Adj R^2		0.9921	
PRESS	0.039		Adeq Precision		48.186	

SS: Sum of squares, MS: Mean square.

3.1.3. Trepanning

Fit summary results suggested a linear relation as the best fit model for all response variables. The main effects of pulse energy (A), pulse width (B), pulse frequency (C), and trepan speed (D) are found as significant model terms. ANOVA results for all the output responses are provided in Table 6. The adequacy measure (~1) and adequate precision (>4) values specify that the models are adequate. The empirical models developed for MRR, SEC, and hole taper are given in Equations (14)–(16) respectively.

$$\text{MRR} = +0.036617 + (0.00135 \times pulse\ energy) - (0.025200 \times pulse\ width) + \\ (0.000315 \times pulse\ frequency) + (0.00295 \times trepan\ speed) \tag{14}$$

$$\text{SEC} = -627.29167 + (273.15833 \times pulse\ energy) + (222.11667 \times pulse\ width) + \\ (43.85617 \times pulse\ frequency) - (31.86150 \times trepan\ speed) \tag{15}$$

$$\text{HT} = +6.84056 - (0.53167 \times pulse\ energy) - (0.90333 \times pulse\ width) - \\ (0.038500 \times pulse\ frequency) + (0.036167 \times trepan\ speed) \tag{16}$$

Table 6. Analysis of variance for MRR, SEC, and hole taper (trepanning).

Source	SS	df	MS	F Value	P Value	
			For MRR			
Model	6.245×10^{-3}	4	1.561×10^{-3}	1638.98	<0.0001	significant
A-pulse energy	1.094×10^{-5}	1	1.094×10^{-3}	11.48	0.0276	
B-pulse width	9.526×10^{-4}	1	9.526×10^{-4}	1000.06	<0.0001	
C-pulse frequency	5.954×10^{-5}	1	5.954×10^{-5}	62.50	0.0014	
D-trepan speed	5.222×10^{-3}	1	5.222×10^{-3}	5481.89	<0.0001	
Residual	3.810×10^{-6}	4	9.525×10^{-7}			
Cor Total	6.248×10^{-3}	8				
Mean	0.15		R^2		0.9994	
Std. Dev.	9.760×10^{-4}		Pred R^2		0.9960	
CoV%	0.66		Adj R^2		0.9988	
PRESS	2.507×10^{-5}		Adeq Precision		113.893	
			For SEC			
Model	2.285×10^6	4	5.712×10^5	77.60	0.0005	significant
A-pulse energy	4.477×10^5	1	4.477×10^5	60.82	0.0015	
B-pulse width	74003.72	1	74003.72	10.05	0.0338	
C-pulse frequency	1.154×10^6	1	1.154×10^6	156.78	0.0002	
D-trepan speed	6.091×10^5	1	6.091×10^5	82.75	0.0008	
Residual	29443.40	4	7360.85			
Cor Total	2.314×10^6	8				
Mean	1275.00		R^2		0.9873	
Std. Dev.	85.80		Pred R^2		0.9229	
CoV%	6.73		Adj R^2		0.9746	
PRESS	1.784E+005		Adeq Precision		19.724	
			For HT			
Model	4.59	4	1.15	46.55	0.0013	significant
A-pulse energy	1.70	1	1.70	68.74	0.0012	
B-pulse width	1.22	1	1.22	49.61	0.0021	
C-pulse frequency	0.89	1	0.89	36.05	0.0039	
D-trepan speed	0.78	1	0.78	31.81	0.0049	
Residual	0.099	4	0.025			
Cor Total	4.69	8				
Mean	3.04		R^2		0.9790	
Std. Dev.	0.16		Pred R^2		0.7887	
CoV%	5.17		Adj R^2		0.9579	
PRESS	0.99		Adeq Precision		20.087	

SS: Sum of squares, MS: Mean square.

3.2. Validation of Developed Models

The developed empirical models have been validated by confirmation through validation experiments. For each method, three additional confirmatory tests were conducted with input parameter values selected randomly (other than used for model development) within the design space. The results obtained from the confirmatory tests have been presented in Table 7. The predicted and measured values of the confirmatory tests were used to calculate the percentage error (using Equation (17)). It can be observed from Table 7 that all the percentage error values lie between 1% and 5%, which establishes the accuracy and validity of developed models.

$$Percentage\ error = \left| \frac{measured\ value - predicted\ value}{predicted\ value} \right| \times 100 \tag{17}$$

Table 7. Confirmation test results.

Drilling Method (s)	Trial No.	Input Parameters					Output Responses			
		Pulse Energy (J)	Pulse Width (ms)	NOP/Hole	Pulse Frequency (Hz)	Trepan Speed (mm/min)		MRR (mm³/s)	SEC (J/mm³)	Taper (°)
Single-pulse drilling	1	22	3.5				Me	45.84	128.10	6.58
							Pr	47	124.02	6.71
							% error	2.47	3.29	2.0
	2	25	3.5				Me	47.64	134.09	6.62
							Pr	50.07	131.85	6.93
							% error	4.85	1.70	4.52
	3	34	2.5				Me	101.5	131.57	9.74
							Pr	99.25	137.66	9.40
							% error	2.27	4.42	3.62
Percussion	4	5.5	0.7	6			Me	0.409	157.42	6.32
							Pr	0.390	151.79	6.63
							% error	4.76	3.71	4.64
	5	5.5	0.8	8			Me	0.334	230.2	6.75
							Pr	0.328	222.01	6.44
							% error	1.72	3.69	4.77
	6	6.5	1.2	12			Me	0.228	427.10	5.62
							Pr	0.230	417.09	5.80
							% error	1.01	2.40	3.02
Trepanning	7	5.5	0.7		23	32	Me	0.134	1040	3.43
							Pr	0.128	1019.69	3.56
							% error	4.49	1.99	3.54
	8	6.5	1.2		34	47	Me	0.170	1424.36	2.81
							Pr	0.164	1408.40	2.69
							% error	3.64	1.13	4.40
	9	5.5	1.2		23	47	Me	0.1532	671.72	3.59
							Pr	0.160	652.82	3.65
							% error	4.07	2.90	1.56

Me: Measured value, Pr: Predicted value

3.3. Response Surface Plots

The effects of input variables (single-pulse drilling: pulse energy and pulse width; percussion: pulse energy, pulse width, and number of pulses per hole; trepanning: pulse energy, pulse width, pulse frequency, and trepan speed) on MRR, SEC, and HT for single-pulse, percussion, and trepanning have been analysed using 3D response surface graphs as provided in the sections below. It is important to mention that these graphs represent the simultaneous effects of two input variables while keeping other input variables at the centre level.

3.3.1. Single-Pulse Drilling

Figure 6a shows the material removal rate (MRR) achieved during single-pulse drilling for different pulse energies at the three different pulse widths used. It is evident that MRR increases slightly with the increase in pulse energy. On the other hand, a significant decrease in MRR is observed with an increase in pulse width because of the increase in drilling time, which is directly dependent on the applied pulse width. The combination of minimum pulse width and maximum pulse energy results in maximum MRR because of high power intensity availability, which promotes the melting rate of the material and produces less heat loss, and as a result, enhances the material removal phenomenon [30].

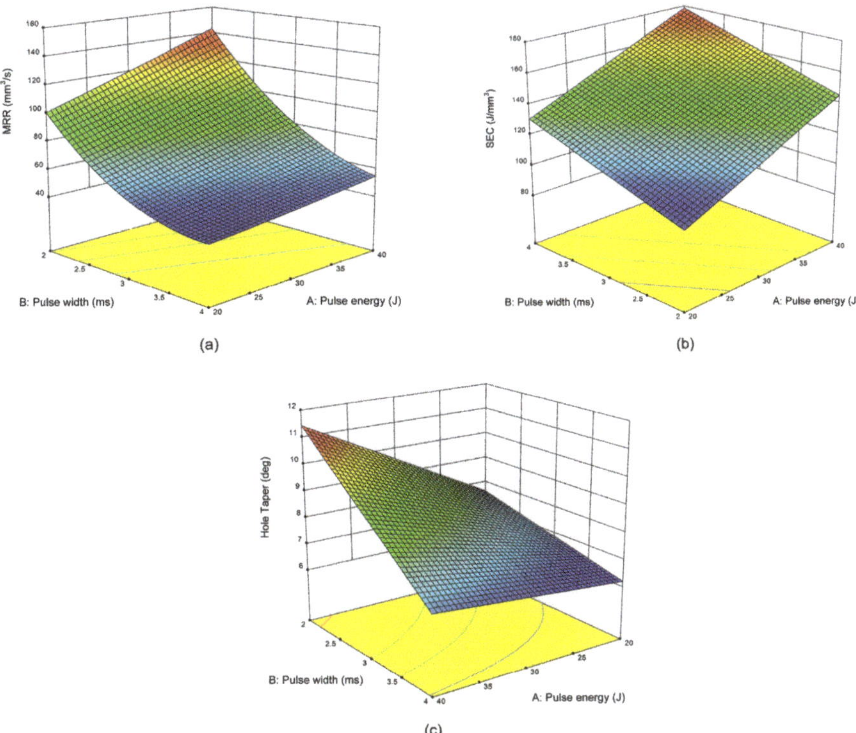

Figure 6. Surface plot showing the effects of pulse energy and pulse width on (**a**) MRR, (**b**) SEC, and (**c**) hole taper for single-pulse drilling.

The impacts of pulse energy and pulse width on the SEC are presented in Figure 6b. An increasing trend is observed with an increment in pulse width and pulse energy. The graph demonstrates that keeping the pulse width constant a significant increase in SEC value is observed with an increase in pulse energy because of the high energy consumed during the process [29]. It is also evident that keeping the pulse energy constant, SEC increases with the increase in pulse width because of longer pulse duration, which consumes more energy to transfer into the workpiece material.

Figure 6c depicts the effects of pulse energy and pulse width on hole taper. The graph demonstrates that there is a substantial decrease in the value of hole taper when the pulse width is increased from 2 ms to 4 ms because it permits enough interaction time between the workpiece and laser beam to allow the expulsion of molten material from the hole (bottom side) more effectively [35]. On the other hand, a small increase in hole taper value is observed when pulse energy is changed from 20 J to 40 J. When a laser beam with high pulse energy interacts with the top side of the workpiece, it melts and vaporizes the material instantly and increases the mean (entrance) hole diameter [12]; however, the intensity of the laser beam decreases as it passes through the thickness, which results in a small exit hole diameter, less material removal, and produces a high hole taper. This variation is consistent with the findings of Chatterjee et al. [37] and Yilbas [12].

3.3.2. Percussion

The effects of pulse energy and pulse width on MRR and SEC for percussion are presented in Figures 7a and 8a, respectively. Similar effects have been observed for pulse energy and pulse duration on MRR and SEC as in the case of single-pulse drilling; however, this process is a multi-pulse process.

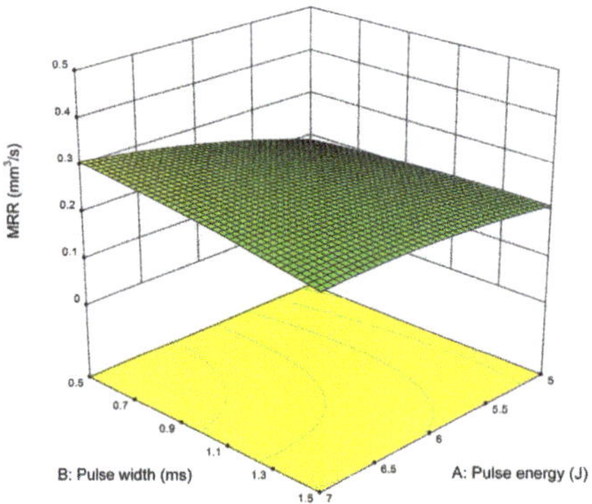
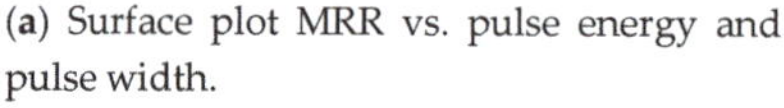

(**a**) Surface plot MRR vs. pulse energy and pulse width.

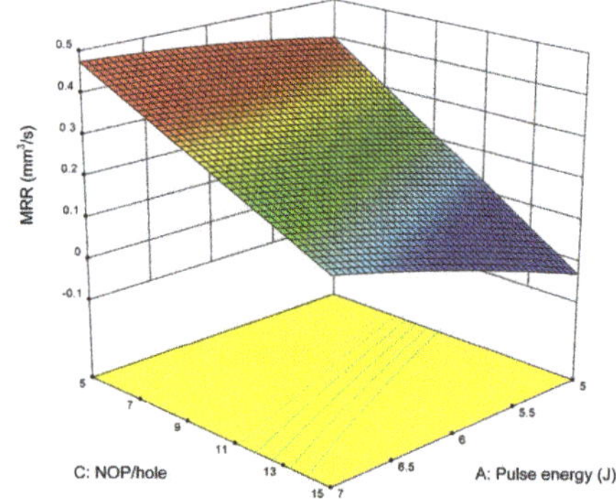

(**b**) Surface plot MRR vs. pulse energy and number of pulses (NOP)/hole.

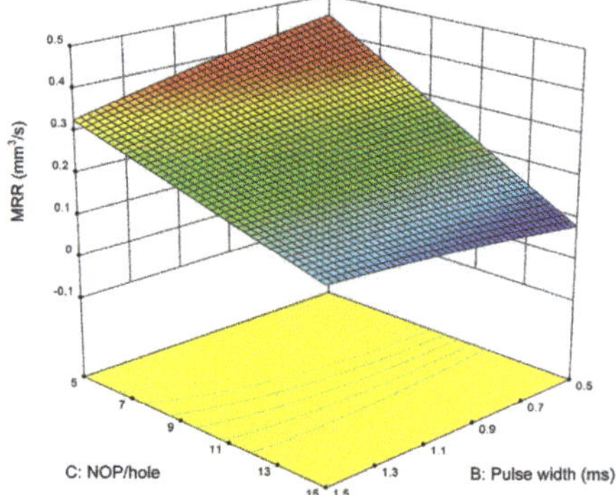

(**c**) Surface plot MRR vs. pulse width and NOP/hole.

Figure 7. Effects of parameters on MRR for percussion.

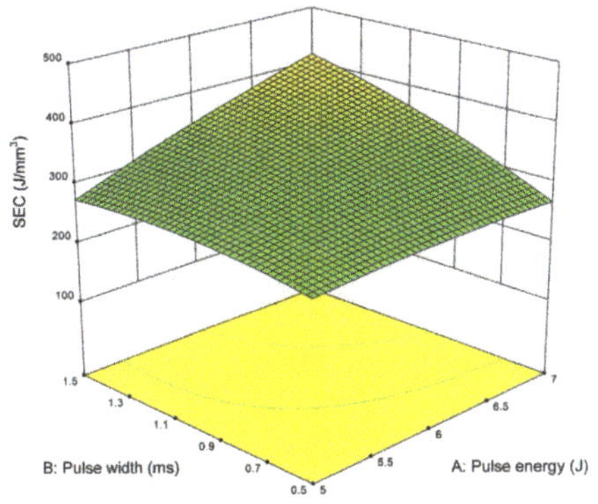

(**a**) Surface plot SEC vs. pulse energy and pulse width.

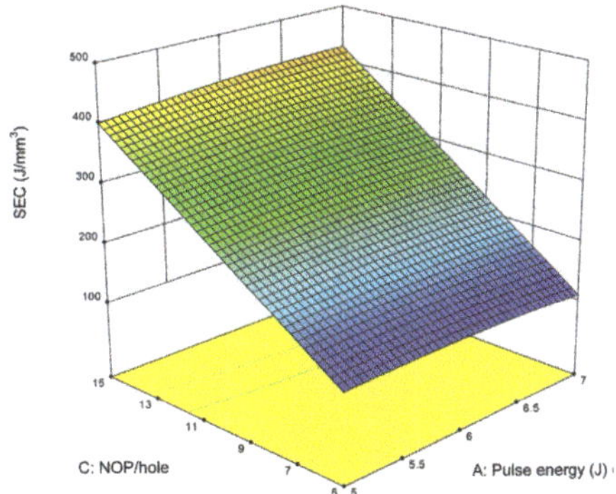

(**b**) Surface plot SEC vs. pulse energy and NOP/hole.

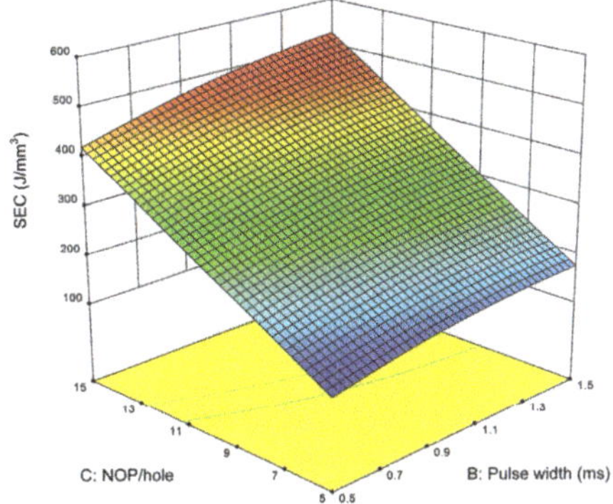

(**c**) Surface plot SEC vs. pulse width and NOP/hole.

Figure 8. Effects of parameters on SEC for percussion.

Figure 7b shows the impacts of pulse energy and NOP per hole on MRR. It is noted that MRR decreases with the increase in NOP per hole and increases with the increase in pulse energy. It is also revealed that the combination of minimum NOP and high pulse energy results in maximum MRR. This is due to the fact that higher NOP need more time for drilling, whereas, high pulse energy increases the transfer rate of heat energy into the substrate without affecting the drilling time, resulting in a rapid increase in melt volume and eventually results in higher MRR.

The surface plot (Figure 7c) presents the inverse effect of pulse width and NOP per hole on MRR. It can also be observed that MRR is affected more by NOP than the pulse width.

Figure 8b depicts the impacts of pulse energy and NOP per hole on the SEC. The figure indicates that SEC increases with the increment in pulse energy and NOP. It can also be noted that SEC is affected more by NOP than pulse energy. Both pulse energy and NOP has a direct relation with SEC and therefore results in higher SEC value. Similar findings have been reported by Bandyopadhay et al. [38].

The effects of pulse width and NOP per hole on SEC have been provided in Figure 8c. The SEC is maximum at higher values of pulse width and NOP per hole. It is also evident that the impact of NOP on SEC is higher as compared to pulse width. Pulse width is the duration during which energy is provided to the drilling zone. The increase in pulse width consumes more energy to supply at the drilling zone [39], resulting in a higher SEC value.

Figure 9a demonstrates the impacts of pulse energy and pulse width on hole taper for percussion drilling. It is clear that the hole taper is less sensitive to variation in pulse energy as compared to pulse width. Furthermore, hole taper decreases with the increase in values of both parameters. This reason is that an increase in pulse energy and pulse width results in high energy availability per pulse, which enhances the penetration capability of the laser beam into the workpiece. As a result, large hole size is produced at the exit side of the hole, and the difference between entry and exit side hole diameters decreases, thus reducing the hole taper [39].

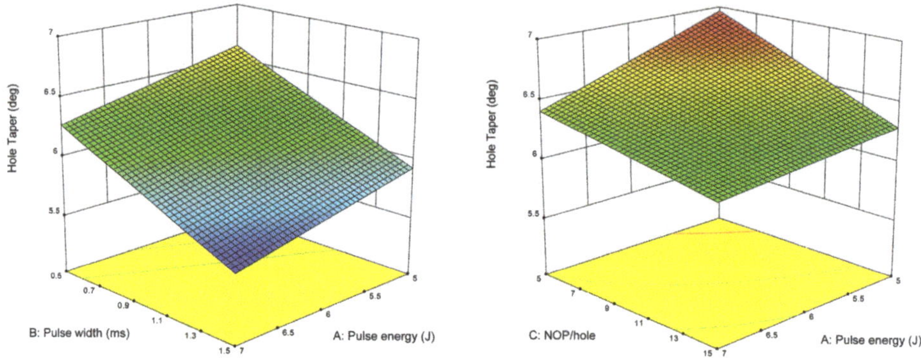

(**a**) Surface plot HT vs. pulse energy and pulse width.

(**b**) Surface plot HT vs. pulse energy and NOP/hole.

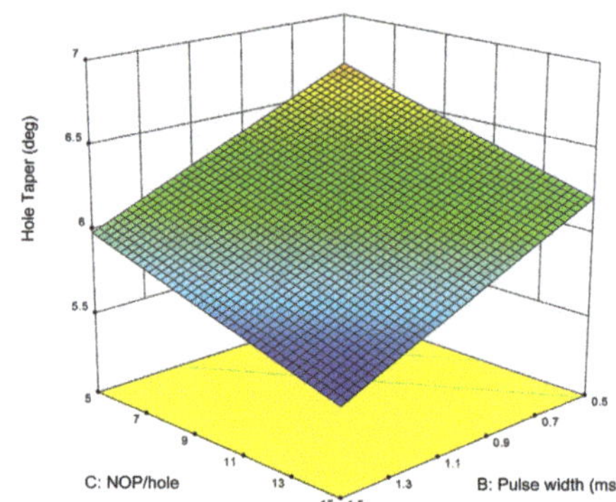

(**c**) Surface plot HT vs. pulse width and NOP/hole.

Figure 9. Effects of parameters on hole taper for percussion.

The impacts of pulse energy and NOP per hole on hole taper are presented in Figure 9b. It can be observed that the hole taper decreases with the increase in pulse energy and NOP per hole. The decrease in hole taper at higher NOP value is the result of additional laser pulses that assist in removing material from the hole on the bottom side after the formation of the through-hole, thereby enlarging the exit hole diameter, which eventually produces lower hole taper [40]. It is also evident that the effect of NOP on the hole taper is large as compared to pulse energy.

The 3D relationship of pulse width and NOP per hole on hole taper is illustrated in Figure 9c. It is noted that the minimum hole taper can be obtained at high levels of pulse width and NOP per hole. Moreover, hole taper decreases with the increase in pulse width. This behaviour is because of an increase in radiation time with the pulse width, which results in a longer interaction time between the workpiece and laser beam and provides sufficient heat at the exit hole side, and consequently increases the melted volume at the exit hole surface and produces lower hole taper [41].

3.3.3. Trepanning

Figures 10a and 11a illustrate the impacts of pulse energy and pulse width on MRR and SEC for trepanning. The trends are similar to singe pulse and percussion drilling.

Figure 10b shows the direct influence of pulse energy and pulse frequency on MRR. It can be observed that the combination of maximum pulse frequency and pulse energy results in high MRR value. This is because high pulse frequency and pulse energy values result in a short time gap between pulses and allow more energy to enter into the workpiece material. Consequently, more amount of material is removed. Similar findings have been reported by Mishra and Yadava [39].

The 3D response surface plot shown in Figure 10c presents the direct influence of pulse energy and trepan speed on MRR. It can also be observed that MRR is affected more by trepan speed than pulse energy. Pulse energy has a direct relation with heat flow. Increase in pulse energy allows a large amount of heat to enter into the material and consequently increases the melt front temperature to produce a large-melt volume. Furthermore, the increase in trepan speed removes the material faster, which eventually results in higher MRR.

The impacts of pulse width and pulse frequency on MRR exhibit that MRR decreases by increasing pulse width (Figure 10d). On the contrary, a positive trend is noticed with the increase in pulse frequency. It is also clear that MRR is more sensitive to pulse width in comparison with pulse frequency.

Figure 10e describes the influence of pulse width and trepan speed on MRR. It is evident from the graph that pulse width has less effect on MRR as compared to trepan speed. Moreover, maximum MRR is achieved at a lower level of pulse width and a higher level of trepan speed. This is because, at fast trepan speed, the laser beam overlap increases, which removes the material more effectively [10], and heat energy produced at low pulse width (high peak power) produces more melt volume, thus higher MRR.

The 3D relationship of pulse frequency and trepan speed on MRR is presented in Figure 10f. The combination of minimum pulse frequency and trepan speed results in a lower MRR value. MRR increases with the increase in pulse frequency and trepan speed because of high laser power availability and large beam overlap.

The impacts of pulse energy and pulse frequency on SEC exhibit that SEC increases by increasing pulse energy (Figure 11b). SEC also increases with the increment in pulse frequency. This is due to the fact that the average power of laser increases at higher values of pulse energy and pulse frequency and, therefore, consumes more energy [29].

Figure 11c depicts the effects of pulse energy and trepan speed on SEC. The surface plot shows a direct influence of pulse energy on SEC. On the contrary, a negative trend is observed with an increase in trepan speed. An increase in the trepan speed can decrease the drilling time, which eventually reduces the energy consumption value [29].

The 3D response surface plot shown in Figure 11d presents the effects of pulse width and pulse frequency on SEC. It can be identified that the SEC value increases with the increase in pulse width and pulse frequency. It is also clear that pulse frequency influences SEC more than the pulse width. The reason for this is that at higher pulse frequency, the laser consumes more power [28].

Figure 11e describes the influence of pulse width and trepan speed on SEC. It is clear from the surface plot that pulse width has less effect on SEC as compared to trepan speed. Moreover, minimum SEC is achieved at a lower level of pulse width. At higher pulse width, the heat energy transferred to the workpiece material is for a longer duration, which ultimately consumes more energy [39].

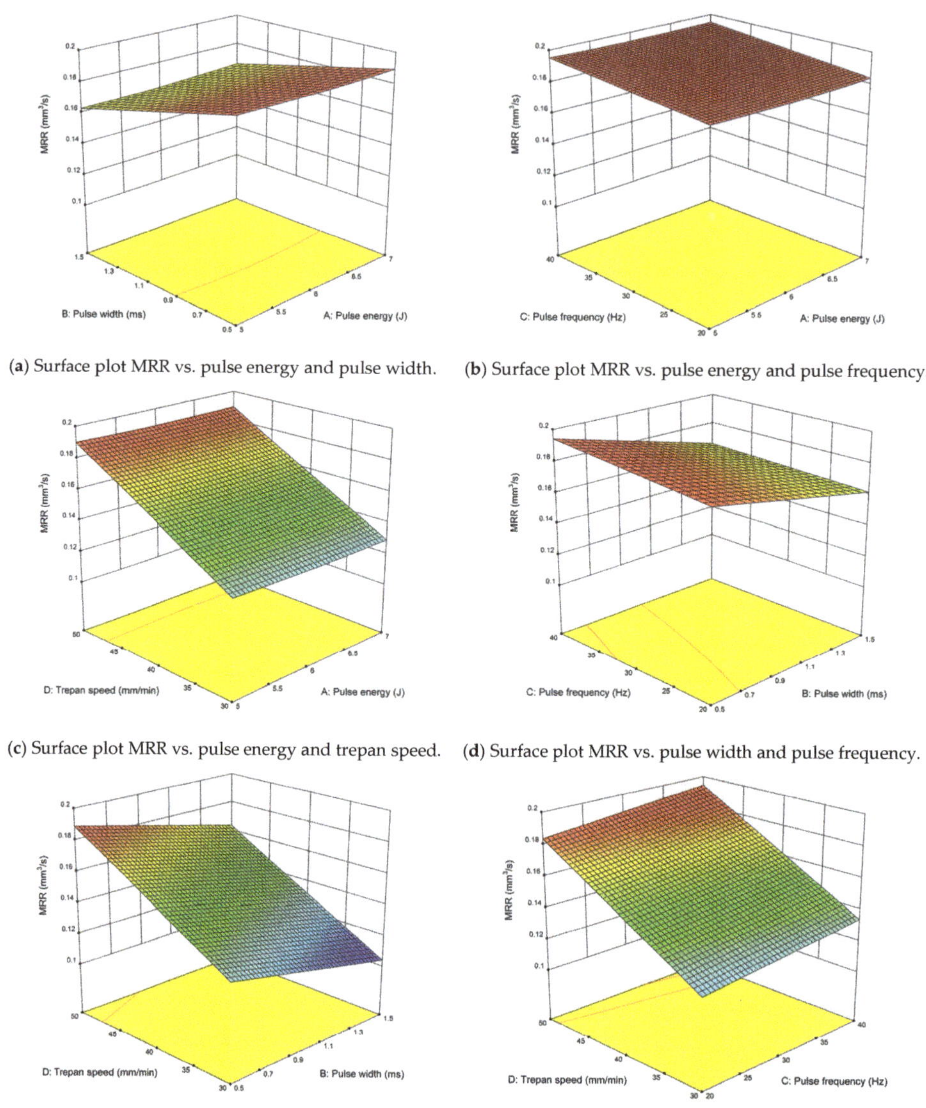

(**a**) Surface plot MRR vs. pulse energy and pulse width. (**b**) Surface plot MRR vs. pulse energy and pulse frequency.

(**c**) Surface plot MRR vs. pulse energy and trepan speed. (**d**) Surface plot MRR vs. pulse width and pulse frequency.

(**e**) Surface plot MRR vs. pulse width and trepan speed. (**f**) Surface plot MRR vs. pulse frequency and trepan speed.

Figure 10. Effects of parameters on MRR for trepanning.

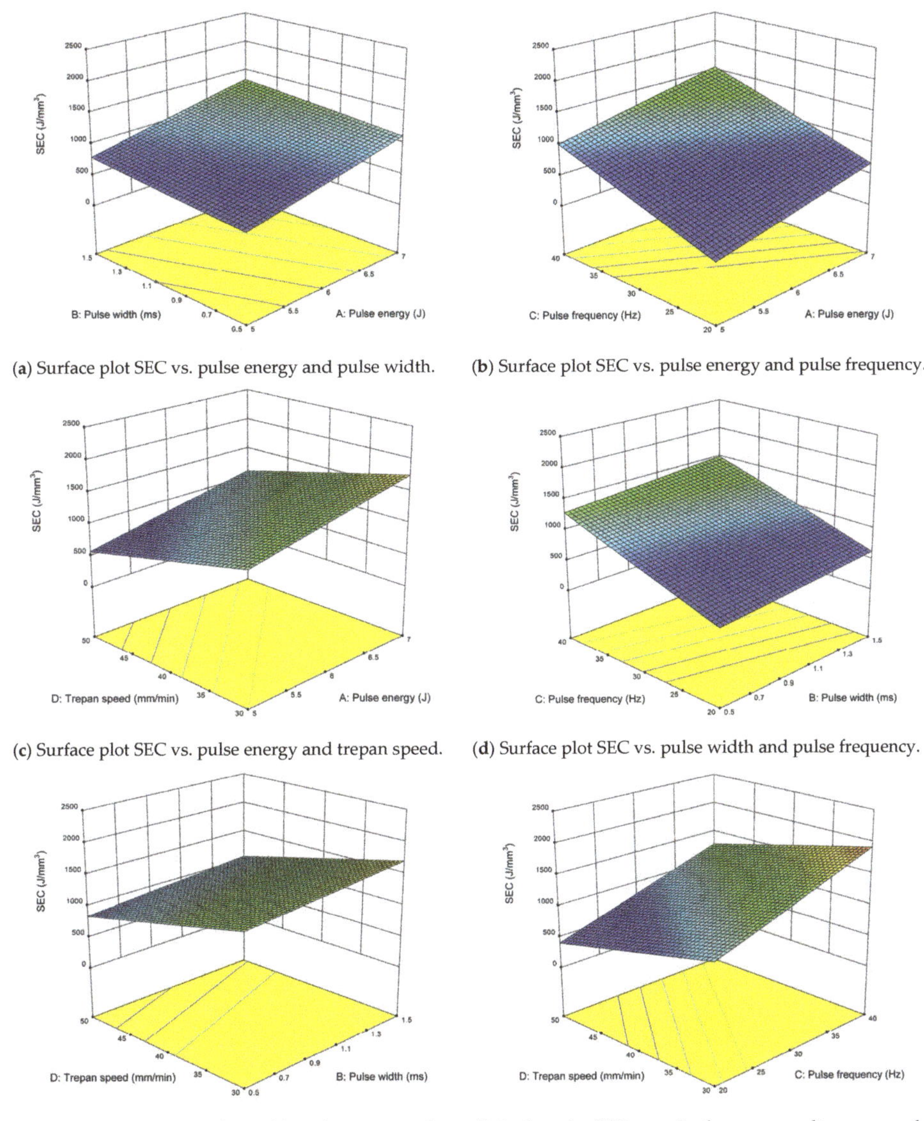

(**a**) Surface plot SEC vs. pulse energy and pulse width.   (**b**) Surface plot SEC vs. pulse energy and pulse frequency.

(**c**) Surface plot SEC vs. pulse energy and trepan speed.   (**d**) Surface plot SEC vs. pulse width and pulse frequency.

(**e**) Surface plot SEC vs. pulse width and trepan speed.   (**f**) Surface plot SEC vs. pulse frequency and trepan speed.

Figure 11. Effects of parameters on SEC for trepanning.

The response surface plot in Figure 11f describes the effects of pulse frequency and trepan speed on SEC. The graph demonstrates that SEC is minimum at low levels of pulse frequency and high levels of trepan speed and maximum at high levels of pulse frequency and low levels of trepan speed. Furthermore, SEC is found more sensitive to variation in pulse frequency as compared to the trepan speed.

The impacts of pulse energy and pulse width on hole taper for trepanning are presented in Figure 12a. Similar trends have been found as in the case of percussion drilling.

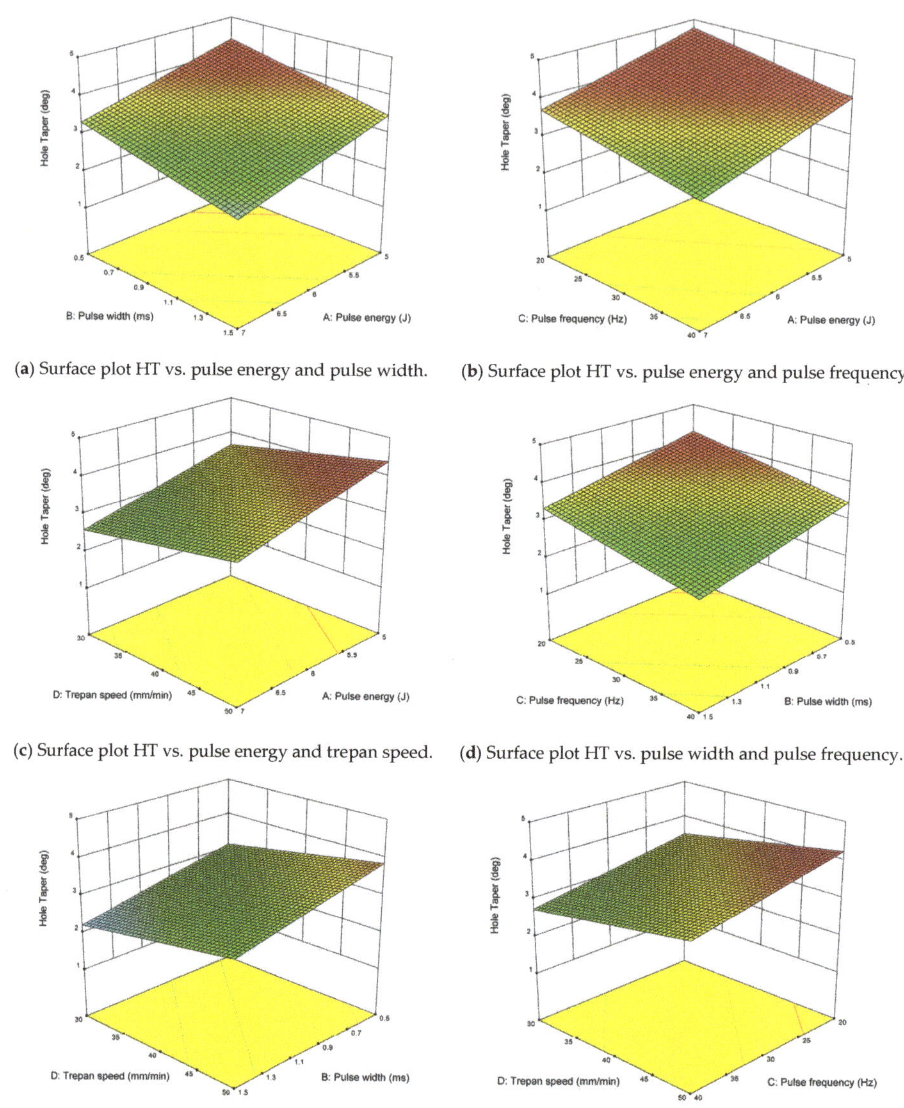

(**a**) Surface plot HT vs. pulse energy and pulse width.

(**b**) Surface plot HT vs. pulse energy and pulse frequency.

(**c**) Surface plot HT vs. pulse energy and trepan speed.

(**d**) Surface plot HT vs. pulse width and pulse frequency.

(**e**) Surface plot HT vs. pulse width and trepan speed.

(**f**) Surface plot HT vs. pulse frequency and trepan speed.

Figure 12. Effects of parameters on hole taper for trepanning.

Figure 12b represents the effects of pulse energy and pulse frequency on hole taper. A decreasing trend is observed with the increase in pulse energy and pulse frequency. The laser power increases at higher values of pulse frequency, which impart more heat into the substrate material and therefore results in efficient melting (removal) of material, particularly at the exit side of the hole. As a result, the difference between entry and exit hole diameters decreases and lower hole taper is produced [39].

The impacts of pulse energy and trepan speed on hole taper exhibit that hole taper decreases by increasing pulse energy (Figure 12c). On the contrary, an increase in the trepan speed results in increased hole taper. It is also evident that hole taper is less sensitive to trepan speed as compared to pulse energy. The reason for this behaviour is that an increase in trepan speed does not provide enough time to distribute the required heat into the work material and eventually results in higher hole taper.

The effects of pulse width and pulse frequency on hole taper have been described in Figure 12d. It can be identified that minimum hole taper is observed at the maximum level of pulse width and pulse frequency because of high laser power availability.

Figure 12e depicts the influence of pulse width and trepan speed on hole taper. It is clearly seen that the combination of maximum pulse width and minimum trepan speed results in smaller hole taper value.

Figure 12f shows the effects of pulse frequency and trepan speed on hole taper. At a low level of trepan speed, hole taper increases with an increase in pulse frequency. A similar effect is observed at high levels of trepan speed.

3.4. Performance Comparison of Single-Pulse, Percussion, and Trepanning Drilling

One of the objectives of this research was to compare the performance of single-pulse, percussion, and trepanning drilling; therefore, the effectiveness of each method in terms of maximum values of MRR and minimum values of SEC and hole taper has been summarized, as shown in Figure 13. Single-pulse drilling is taken as a reference to compare the corresponding values of different drilling methods. The increment and decrement in corresponding drilling method values from single-pulse drilling are presented with positive and negative percentages. It is evident from the figure that the performance of single-pulse drilling is better in case of MRR as the MRR reduces by 99.70% when using percussion drilling and 99.87% when trepanning was employed. SEC increases by 14.20% and 626.50% when using percussion and trepanning, respectively, indicating that single-pulse drilling outperformed the others with minimum SEC value. In the case of hole taper, trepanning yields better results by decreasing it by 72.92%, whereas percussion gives the second best value with 11.22% reduction.

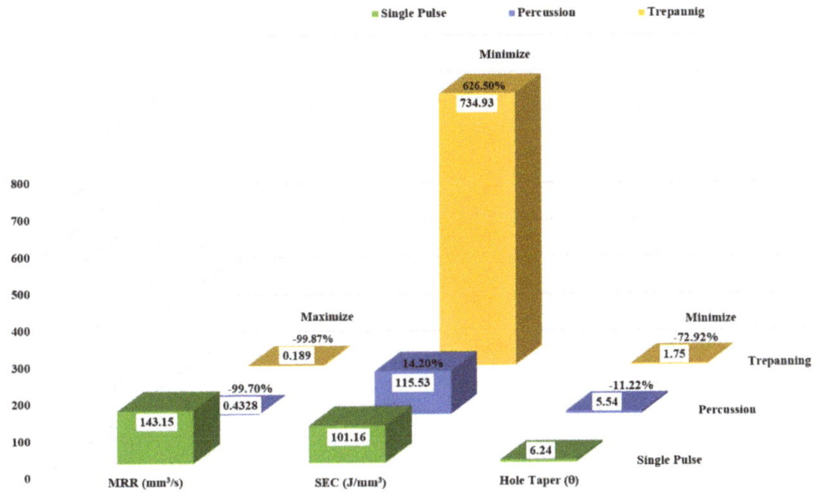

Figure 13. Comparison plot of single-pulse, percussion and trepanning for MRR, SEC, and hole taper.

4. Multi-Objective Optimisation

For the manufacturing industries, optimum levels of process parameters are very important aimed at maximising productivity and quality while minimising the energy cost. Maximising the MRR makes the process faster, which means a higher amount of material can be removed in minimum (drilling) time, and minimising the SEC results in higher efficiency of the process because, at this stage, a higher amount of material is removed with minimum energy consumption. Therefore, a multi-objective optimization based on desirability function was used for simultaneous optimisation of these conflicting responses, i.e., MRR, SEC, and hole taper. The purpose of the desirability function was to combine the effects of multiple responses into a single desirability value using mathematical transformation. The range of desirability lies between 0 and 1, where 0 indicates least desirable, and 1 depicts most desirable. The steps are discussed in detail in [42]. The following optimisation criteria were applied:

Pulse energy = in range; pulse width = in range; NOP/hole = in range; pulse frequency = in range; trepan speed = in range; MRR = maximize; SEC = minimize; hole taper = minimize.

The achieved desirability, along with optimum process parameters and predicted response values are summarised in Table 8. It is evident from the table that multi-objective optimisation provides maximum desirability of 74.8%, 79.9%, and 68.9% for single-pulse, percussion, and trepanning drilling, respectively, when the full range of process parameters is used and all responses possess equal weights. It can also be observed that single-pulse drilling is the best option if the productivity and cost are given the priority over quality, as it results in maximum MRR with a lower SEC value. On the other hand, the best hole quality is obtained with trepanning, but at the expense of higher energy consumption and lower MRR.

Table 8. Optimisation results.

Drilling Method (s)	Optimum Input Parameters					Predicted Responses			Desirability
	Pulse Energy (J)	Pulse Width (ms)	NOP/Hole	Pulse Frequency (Hz)	Trepan Speed (mm/min)	MRR (mm^3/s)	SEC (J/mm^3)	Hole Taper (°)	
Single-Pulse	20	2				100.75	101.94	7.66	0.748
Percussion	7	0.98	5			0.426	171.55	6.08	0.799
Trepanning	7	1.5		20	50	0.162	901.94	2.80	0.689

5. Conclusions

This research was aimed to investigate the productivity, cost, and quality during fibre laser drilling of IN 718 superalloy. Three different laser drilling processes, namely single-pulse, percussion, and trepanning drilling were employed to examine and model the impacts of laser drilling process parameters on material removal rate, specific energy consumption, and hole taper. Taguchi L9 orthogonal array was employed for the design of experiments, and empirical models were developed to predict the output responses. Finally, a multi-objective optimisation was performed to attain the optimum levels of process parameters for maximum MRR with minimum SEC and hole taper. The following concluding remarks are found from this investigation:

- In single-pulse drilling, pulse width is the main driver for MRR (productivity); for percussion and trepanning, NOP/hole and trepan speed are the most significant input parameters influencing the MRR.
- Pulse energy, NOP/hole, and pulse frequency are the most influencing parameters affecting the SEC (cost) in single-pulse, percussion, and trepanning, respectively.
- The process parameters significantly affecting the hole taper (quality) are pulse width during single-pulse drilling and percussion and pulse energy during trepanning.
- The developed mathematical models are reliable and adequate for predicting the response variables at a 95% confidence interval.
- Single-pulse drilling presents better MRR and SEC as compared to percussion and trepanning. Taking single-pulse drilling as a reference, 99.70% less MRR was attained using percussion drilling, and the value further reduced by 99.87% through trepanning. Similarly, percussion drilling yielded 14.20% more SEC while trepanning resulted in a six-fold increase in SEC (626.50%) as compared to single-pulse drilling.

- Concerning the hole taper, trepanning outperformed the rest of the drilling processes with 72.92% better hole quality, where percussion only resulted in 11.22% improvement.
- Multi-objective optimisation results in desirability values of 74.8%, 79.9%, and 68.9% for single-pulse, percussion, and trepanning drilling, respectively.

This research will serve as a guide for the practitioners to select a suitable laser drilling method with optimum levels of laser drilling process parameters for the required MRR, SEC, and hole taper values.

Further research is in progress to evaluate the response variables for pulsed Nd:YAG laser drilling. The surface integrity of generated holes will be analysed along with other response variables to improve the drilling performance.

Author Contributions: Conceptualization, S.S., E.S., K.S. and W.S.; data curation, S.S.; formal analysis, S.S.; funding acquisition, E.S.; investigation, S.S.; methodology, S.S.; project administration, E.S., K.S. and W.S.; resources, E.S and W.S.; software, S.S.; supervision, E.S., K.S. and W.S.; validation, S.S, E.S. and W.S.; visualization, S.S., E.S., K.S. and W.S.; writing—original draft, S.S.; writing—review, editing, layout formatting S.S., E.S., K.S. and W.S.

Funding: The work was supported by the Punjab Educational Endowment Fund (PEEF, Pakistan) and Cranfield University (United Kingdom).

Acknowledgments: The authors would like to thank Dr Jeferson Araujo de Oliveira (StressMap—The Open University, UK) for providing the microscope facility. Laser drilling facilities provided by IPG Photonics Corporation, coordinated by Mr Stan Wilford (Senior Sales Manager IPG Photonics, UK) are also gratefully acknowledged.

Conflicts of Interest: The authors declare no conflict of interest.

References

1. Donachie, M.J.; Donachie, S.J. *Superalloys: A Technical Guide*, 2nd ed.; ASM International: Materials Park, OH, USA, 2002.
2. Sarfraz, S.; Shehab, E.; Salonitis, K.; Suder, W.; Zahoor, S. Evaluation of productivity and operating cost of laser drilling process—a case study. In Proceedings of the 16th International Conference on Manufacturing Research, Skövde, Sweden, 11–13 September 2018; pp. 9–14.
3. Wojciechowski, S.; Przestacki, D.; Chwalczuk, T. The evaluation of surface integrity during machining of inconel 718 with various laser assistance strategies. *MATEC Web Conf.* **2017**, *136*, 01006. [CrossRef]
4. Chwalczuk, T.; Przestacki, D.; Szablewski, P.; Felusiak, A. Microstructure characterisation of Inconel 718 after laser assisted turning. *MATEC Web Conf.* **2018**, *188*, 02004. [CrossRef]
5. Sarfraz, S.; Shehab, E.; Salonitis, K. A review of technical challenges of laser drilling manufacturing process. In Proceedings of the 15th International Conference on Manufacturing Research, London, UK, 5–7 September 2017; pp. 51–56.
6. Yilbas, B.S. *Laser Drilling: Practical Applications*; SpringerBriefs in Applied Sciences and Technology; Springer Science & Business Media: Berlin/Heidelberg, Germany, 2013; ISBN 978-3-642-34981-2.
7. Ng, G.K.L.; Li, L. Repeatability characteristics of laser percussion drilling of stainless-steel sheets. *Opt. Lasers Eng.* **2003**, *39*, 25–33. [CrossRef]
8. Gautam, G.D.; Pandey, A.K. Pulsed Nd:YAG laser beam drilling: A review. *Opt. Laser Technol.* **2018**, *100*, 183–215. [CrossRef]
9. Abidou, D.; Yusoff, N.; Nazri, N.; Omar Awang, M.A.; Hassan, M.A.; Sarhan, A.A.D. Numerical simulation of metal removal in laser drilling using radial point interpolation method. *Eng. Anal. Bound. Elem.* **2017**, *77*, 89–96. [CrossRef]
10. Marimuthu, S.; Antar, M.; Dunleavey, J.; Hayward, P. Millisecond fibre laser trepanning drilling of angular holes. *Int. J. Adv. Manuf. Technol.* **2019**, *102*, 2833–2843. [CrossRef]
11. Panda, S.; Mishra, D.; Biswal, B.B. Determination of optimum parameters with multi-performance characteristics in laser drilling—A grey relational analysis approach. *Int. J. Adv. Manuf. Technol.* **2011**, *54*, 957–967. [CrossRef]
12. Yilbas, B.S. Parametric study to improve laser hole drilling process. *J. Mater. Process. Technol.* **1997**, *70*, 264–273. [CrossRef]

13. Yilbas, B.S.; Aleem, A. Laser hole drilling quality and efficiency assessment. *Proc. Inst. Mech. Eng. Part B J. Eng. Manuf.* **2004**, *218*, 225–233. [CrossRef]

14. Chien, W.T.; Hou, S.C. Investigating the recast layer formed during the laser trepan drilling of Inconel 718 using the Taguchi method. *Int. J. Adv. Manuf. Technol.* **2007**, *33*, 308–316. [CrossRef]

15. Morar, N.I.; Roy, R.; Mehnen, J.; Marithumu, S.; Gray, S.; Roberts, T.; Nicholls, J. Investigation of recast and crack formation in laser trepanning drilling of CMSX-4 angled holes. *Int. J. Adv. Manuf. Technol.* **2018**, *95*, 4059–4070. [CrossRef]

16. Rajesh, P.; Nagaraju, U.; Gowd, G.H.; Vardhan, T.V. Experimental and parametric studies of Nd:YAG laser drilling on austenitic stainless steel. *Int. J. Adv. Manuf. Technol.* **2015**, *93*, 65–71. [CrossRef]

17. Dhaker, K.L.; Pandey, A.K. Particle swarm optimisation of hole quality characteristics in laser trepan drilling of inconel 718. *Def. Sci. J.* **2019**, *69*, 37–45. [CrossRef]

18. Mishra, S.; Yadava, V. Prediction of hole characteristics and hole productivity during pulsed Nd:YAG laser beam percussion drilling. *Proc. Inst. Mech. Eng. Part B J. Eng. Manuf.* **2013**, *227*, 494–507. [CrossRef]

19. Ghoreishi, M.; Low, D.K.Y.; Li, L. Comparative statistical analysis of hole taper and circularity in laser percussion drilling. *Int. J. Mach. Tools Manuf.* **2002**, *42*, 985–995. [CrossRef]

20. Mishra, S.; Yadava, V. Prediction of material removal rate due to laser beam percussion drilling in aluminium sheet using the finite element method. *Int. J. Mach. Mach. Mater.* **2013**, *14*, 342. [CrossRef]

21. Dietrich, J.; Brajdic, M.; Engbers, S.; Klinkmueller, L.; Kelbassa, I. Enhancing the quality and productivity of laser drilling by changing process gas parameters while processing. In Proceedings of the 28th International Congress on Applications of Lasers & Electro-Optics, Orlando, FL, USA, 2–5 November 2009; pp. 729–732.

22. Sarfraz, S.; Shehab, E.; Salonitis, K.; Suder, W.; Muhammad, S. Towards cost modelling for laser drilling process. In Proceedings of the 25th ISPE Inc. International Conference on Transdisciplinary Engineering, Modena, Italy, 3–6 July 2018; pp. 611–618.

23. Owodunni, O. Awareness of energy consumption in manufacturing processes. *Procedia Manuf.* **2017**, *8*, 152–159. [CrossRef]

24. Allen, D.; Bauer, D.; Bras, B.; Gutowski, T.; Murphy, C.; Piwonka, T.; Sheng, P.; Sutherland, J.; Thurston, D.; Wolff, E. Environmentally benign manufacturing: Trends in Europe, Japan, and the USA. *J. Manuf. Sci. Eng.* **2002**, *124*, 908. [CrossRef]

25. IEA (International Energy Agency). *Energy Technology Perspectives 2012: Pathways to a Clean Energy System*; OECD/IEA: Paris, France, 2012.

26. Dahmen, M.; Güdükkurt, O.; Kaierle, S. The ecological footprint of laser beam welding. *Phys. Procedia* **2010**, *5*, 19–28. [CrossRef]

27. Kaierle, S.; Dahmen, M.; Güdükkurt, O. Eco-Efficiency of Laser Welding Applications. In Proceedings of the International Society for Optical Engineering (SPIE), SPIE Eco-Photonics, Strasbourg, France, 28–30 March 2011; p. 80650T.

28. Apostolos, F.; Panagiotis, S.; Konstantinos, S.; George, C. Energy efficiency assessment of laser drilling process. *Phys. Procedia* **2012**, *39*, 776–783. [CrossRef]

29. Franco, A.; Rashed, C.A.A.; Romoli, L. Analysis of energy consumption in micro-drilling processes. *J. Clean. Prod.* **2016**, *137*, 1260–1269. [CrossRef]

30. Tam, S.C.; Yeo, C.Y.; Jana, S.; Lau, M.W.S.; Lim, L.E.N.; Yang, L.J.; Noor, Y.M. Optimization of laser deep-hole drilling of Inconel 718 using the Taguchi method. *J. Mater. Process. Technol.* **1993**, *37*, 741–757. [CrossRef]

31. Ng, G.K.L.; Li, L. The effect of laser peak power and pulse width on the hole geometry repeatability in laser percussion drilling. *Opt. Laser Technol.* **2001**, *33*, 393–402. [CrossRef]

32. Goyal, R.; Dubey, A.K. Modeling and optimization of geometrical characteristics in laser trepan drilling of titanium alloy. *J. Mech. Sci. Technol.* **2016**, *30*, 1281–1293. [CrossRef]

33. Bahar, N.D.; Marimuthu, S.; Yahya, W.J. Pulsed Nd: YAG laser drilling of aerospace materials (Ti-6Al-4V). *IOP Conf. Ser. Mater. Sci. Eng.* **2016**, *152*, 012056. [CrossRef]

34. Marimuthu, S.; Antar, M.; Dunleavey, J.; Chantzis, D.; Darlington, W.; Hayward, P. An experimental study on quasi-CW fibre laser drilling of nickel superalloy. *Opt. Laser Technol.* **2017**, *94*, 119–127. [CrossRef]

35. Goyal, R.; Dubey, A.K. Quality improvement by parameter optimization in laser trepan drilling of superalloy sheet. *Mater. Manuf. Process.* **2014**, *29*, 1410–1416. [CrossRef]

36. Tagliaferri, F.; Genna, S.; Leone, C.; Palumbo, B.; De Chiara, G. Experimental study of fibre laser microdrilling of aerospace superalloy by trepanning technique. *Int. J. Adv. Manuf. Technol.* **2017**, *93*, 3203–3210. [CrossRef]

37. Chatterjee, S.; Mahapatra, S.S.; Bharadwaj, V.; Choubey, A.; Upadhyay, B.N.; Bindra, K.S. Drilling of micro-holes on titanium alloy using pulsed Nd:YAG laser: Parametric appraisal and prediction of performance characteristics. *Proc. Inst. Mech. Eng. Part B J. Eng. Manuf.* **2018**, *233*, 1872–1889. [CrossRef]
38. Bandyopadhyay, S.; Sarin Sundar, J.K.; Sundararajan, G.; Joshi, S.V. Geometrical features and metallurgical characteristics of Nd:YAG laser drilled holes in thick IN718 and Ti-6Al-4V sheets. *J. Mater. Process. Technol.* **2002**, *127*, 83–95. [CrossRef]
39. Mishra, S.; Yadava, V. Modeling and optimization of laser beam percussion drilling of nickel-based superalloy sheet using Nd: YAG laser. *Opt. Lasers Eng.* **2013**, *51*, 681–695. [CrossRef]
40. Ghoreishi, M.; Low, D.K.Y.; Li, L. Statistical modelling of laser percussion drilling for hole taper and circularity control. *Proc. Inst. Mech. Eng. Part B J. Eng. Manuf.* **2002**, *216*, 307–319. [CrossRef]
41. Mishra, S.; Yadava, V. Modeling and optimization of laser beam percussion drilling of thin aluminum sheet. *Opt. Laser Technol.* **2013**, *48*, 461–474. [CrossRef]
42. Derringer, G.; Suich, R. Simultaneous Optimization of Several Response Variables. *J. Qual. Technol.* **1980**, *12*, 214–219. [CrossRef]

Article

Parametric Modelling and Multi-Objective Optimization of Electro Discharge Machining Process Parameters for Sustainable Production

Misbah Niamat [1], Shoaib Sarfraz [2,*], Wasim Ahmad [3], Essam Shehab [4] and Konstantinos Salonitis [2]

[1]   Mechanical Engineering Department, Muhammad Nawaz Sharif University of Engineering and Technology, Multan 66000, Pakistan; misbahniamat@gmail.com
[2]   Manufacturing Department, School of Aerospace, Transport and Manufacturing, Cranfield University, Cranfield, Bedfordshire MK43 0AL, UK; k.salonitis@cranfield.ac.uk
[3]   Faculty of Industrial Engineering, University of Engineering and Technology, Taxila 47080, Pakistan; wasim.ahmad@uettaxila.edu.pk
[4]   School of Engineering and Digital Sciences, Nazarbayev University, Nur-Sultan 010000, Kazakhstan; essam.shehab@nu.edu.kz
*   Correspondence: shoaib.sarfraz@cranfield.ac.uk; Tel.: +44-(0)-7729-475536

Received: 19 November 2019; Accepted: 18 December 2019; Published: 19 December 2019

Abstract: Electro Discharge Machining (EDM) can be an element of a sustainable manufacturing system. In the present study, the sustainability implications of EDM of special-purpose steels are investigated. The machining quality (minimum surface roughness), productivity (material removal rate) improvement and cost (electrode wear rate) minimization are considered. The influence and correlation of the three most important machining parameters including pulse on time, current and pulse off time have been investigated on sustainable production. Empirical models have been established based on response surface methodology for material removal rate, electrode wear rate and surface roughness. The investigation, validation and deeper insights of developed models have been performed using ANOVA, validation experiments and microstructure analysis respectively. Pulse on time and current both appeared as the prominent process parameters having a significant influence on all three measured performance metrics. Multi-objective optimization has been performed in order to achieve sustainability by establishing a compromise between minimum quality, minimum cost and maximum productivity. Sustainability contour plots have been developed to select suitable desirability. The sustainability results indicated that a high level of 75.5% sustainable desirability can be achieved for AISI L3 tool steel. The developed models can be practiced on the shop floor practically to attain a certain desirability appropriate for particular machine limits.

Keywords: electric discharge machining; response surface methodology; sustainability; productivity; cost; surface quality; microstructure

---

1. Introduction

The enormously growing demand for tool steel during the last decades is due to its embedded properties including wear resistance, corrosion resistance, hardness and exceptional property of retaining cutting end at exalted temperatures. Moreover, its cost-effectiveness also makes it an ideal candidate over carbides, titanium and inconel materials. Therefore, tool steel is being extensively employed for tools and die manufacturing in automotive, nuclear and aerospace industries [1,2]. Tool steel is commercially available in a number of series such as D, A, H, L and M. L series is a special purpose low alloy steel and is available in a number of grades including L1, L2, L3, L6 and L7 [3]. Due to hard to cut nature of tool steels, their machining through conventional methods results in dimensional inaccuracies, residual stresses and higher surface roughness and tool wear. These limitations in conventional techniques of machining are being addressed by utilizing non-conventional or special purpose machining processes. Electric discharge machining (EDM) is a non-conventional machining process used for manufacturing complex profiles with accuracy [4]. Minimum chattering, residual stresses and mechanical vibration are eminent advantages obtained by EDM owing to the absence of direct interaction of tool with work part while machining.

With a growing competition of production rates among industries, there is a visible increase in resources utilization and emissions of toxic materials to the environment. This has initiated a sustainability study of manufacturing systems and technologies. Sustainability has three pillars—the environmental, economic and the social. Sustainability in manufacturing is concerned with the manufacturing of products having minimum adverse environmental effect, safety of employees, conservation of natural resources and energy and are economically viable for customers. Machining is a major constituent of manufacturing system and sustainability in machining is related to environmental friendliness (cutting fluids), minimum cost and energy consumption, higher production and quality, better waste management and safety of worker [5]. The productivity, cost and quality of electric discharge machined parts are measured through performance measures: material removal rate (MRR), electrode wear rate (EWR) and surface roughness (SR) respectively [6]. Machining performance measures are directly associated with process parameters including pulse on time (Pon), current, pulse off time (Poff), voltage, flushing pressure, polarity, servo speed, frequency, gap and jump distance as shown in Figure 1a. The illustration of selected process parameters (Pon, Poff and current) is presented in Figure 1b. The Pareto chart is based on a detailed literature review of more than fifty research papers published in the last 15 years for electric discharge machining of tool steel (only those papers are selected in which MRR, EWR and SR separately or in combination have been evaluated). For investigating the influence of process parameters on productivity, cost and quality of electric discharged machined part, it can be observed from Figure 1 that current, pulse on time (Pon) and pulse off time (Poff) are widely applicable process parameters as identified by researchers.

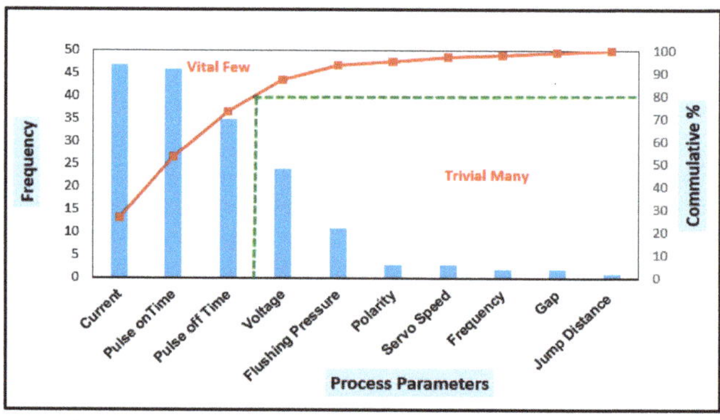

(a)

Figure 1. *Cont.*

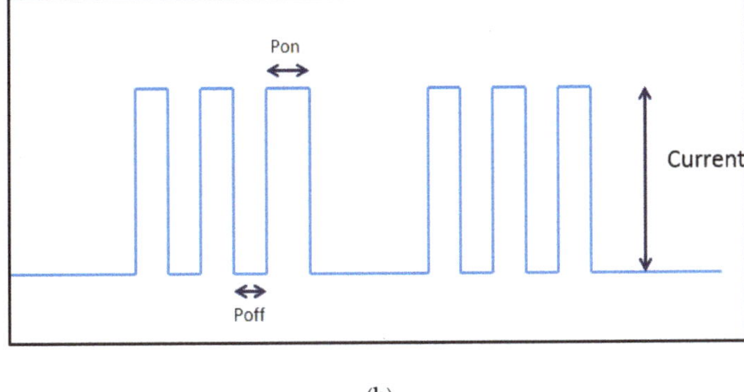

(**b**)

Figure 1. (**a**) Pareto chart for process parameters selection (**b**) Selected process parameters.

Literature also suggests that the current and Pon has a direct influence on MRR, EWR and SR; all three performance measures increase when current is increased [7–26]. Whereas, Pulse off time exhibits inverse effects that are, MRR, EWR and SR decrease at higher Poff [19,25]. Moreover, It had also been reported that Poff does not significantly influence MRR, EWR and SR [22].

Different mathematical and statistical approaches have been presented by various researchers for modelling and optimization of performance measures related to EDM of tool steel. These techniques include conventional methods, taguchi (orthogonal array), response surface methodology (RSM), genetic algorithm (GA), fuzzy logic and grey relational analysis (GRA) [7–10,16–18,23–26]. The application of these techniques for different tool steel materials is presented in Figure 2 [7–43]. The figure represents a literature summary of experimental works for different tool steels along with performance measures and experimental techniques. Each performance measure is indicated by a symbol as shown in Figure 2.

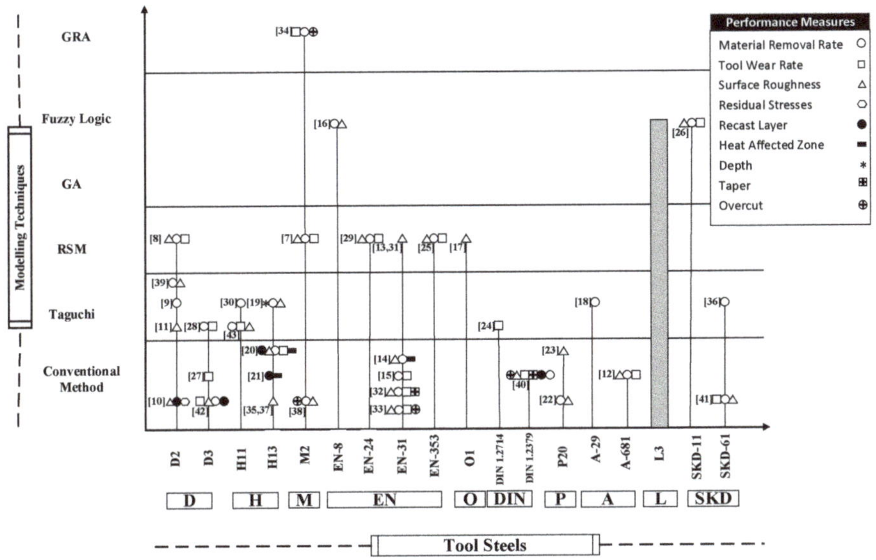

Figure 2. Literature review.

Although various researchers have attempted to investigate and model the impact of process parameters on EWR, SR and MRR, limited or no research has been reported on special-purpose low alloy tool steel (AISI L3) to predict MRR, EWR and SR. Furthermore, literature reveals that productivity and quality are inversely related [22,25] and current and pulse on time are the most decisive factors [44]. Productivity increases with high-energy consumption (higher values of current and pulse on time), whereas quality improves at low energy utilization (lower current and pulse on time) as observed by Mandaloi et al. [7], Payal et al. [14] and Singh et al. [15] for machining AISI M2 and EN-31 tool steels.

Sustainability in machining operation is related to different aspects such as tool life, surface quality of machined parts, production ratio, energy consumption, environment issues, usage of lubricants and coolants and safety and welfare of workers. All these elements are broadly classified as economic, environmental and social pillars of sustainability considering a machining operation [45]. This research aims to achieve sustainable production based on the economic aspect of sustainability while electric discharge machining of low alloying special-purpose AISI L3 tool steel. L3 alloy is selected as it possesses higher hardenability, because of higher percentages of Cr, V and C which make it suitable for making tools and dies. Empirical models have been derived adopting Response Surface Methodology (RSM) for MRR, EWR and SR and adequacy of models have been checked by analysis of the variance (ANOVA). Analysis of process parameters has been performed using surface plots. Moreover, sustainability has been achieved employing desirability based multi-objective optimization. In the end, micrographs have been discussed to reveal the machined surface in terms of voids, pits and micro-cracks.

2. Experimental Procedure

This segment illustrates the chemical properties of the material, preparation of material samples and tools, experimental setup and measurement of responses. Experiments have been performed on AISI L3 tool steel, the chemical composition is presented in Table 1. All samples of work material were prepared by extracting cylindrical samples having 20 mm length and 22 mm diameter, while copper rods in the form of cylindrical shape having dimensions 50.8 mm × 19 mm (length × diameter) were employed as electrodes as shown in Figure 3a,b respectively. The machining was performed on die-sinker electric discharge machine CM655C (75 N). Workpieces and electrodes were subjected to grinding and polishing before conducting experiments.

Table 1. Chemical properties of AISI L3 low alloy tool steel.

Composition	C	Cr	Mn	P	Si	S	V	Fe
Weightage (%)	1.02	1.3	0.59	0.01	0.32	0.01	0.18	Balance

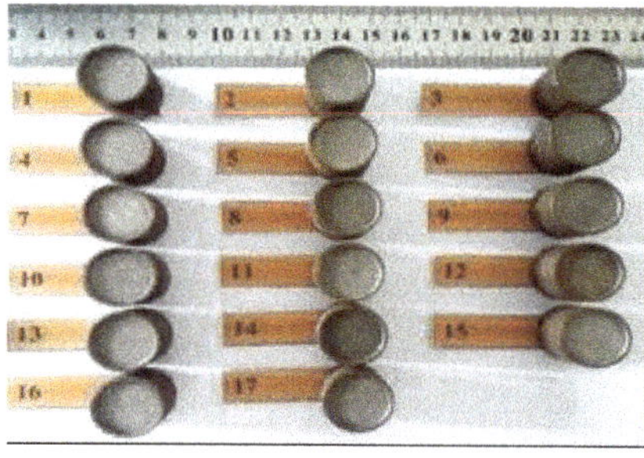

(a) L3 Material specimens

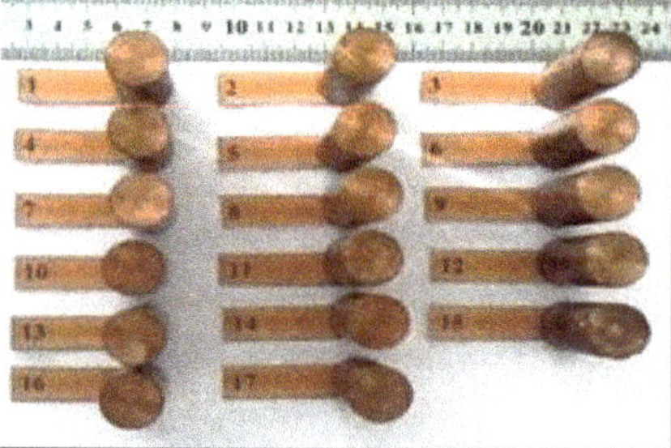

(b) Copper electrodes

Figure 3. Samples of work material.

Three process parameters including, (i) current, (ii) pulse on time (Pon) and (iii) pulse off time (Poff) were investigated for machining AISI L3 tool steel. During each experiment, the material was removed up-to a depth of 1.5 mm and to ensure the uniformity of machining a new electrode was used for each experimental run. Moreover, each experimental run was repeated thrice (three times) to accurately estimate the variation in the obtained values and to avoid uncertainty in the results.

The productivity, cost and quality of electric discharge machined parts were measured using MRR, EWR and SR respectively. Moreover, the analysis of the microstructure of the samples was performed for surface quality evaluation. Before taking micrographs, samples were treated firstly by dipping them into a well-prepared mixture of resin and hardener and then dried for stabilization. Micrographs have been taken using the Scanning Electron Microscope (SEM), TESCAN (MIRA 3 XMU type). All micrographs were taken in nanospace, keeping 101× magnification and 10 kV High Voltage (HV) value. Weight difference method was used for the measurement of MRR and EWR [46–48]. For this purpose, weight balance Mettler PE 1600 was used before and after the individual experiment. The following relations (Equations (1) and (2)) were used for the calculation of MRR and EWR.

$$\text{MRR} \left(\text{mm}^3 / \text{min} \right) = \frac{W_p\ (g) - W_a(g)}{\text{Machining Time (min)} \times \text{Density (g/mm}^3\text{)}} \tag{1}$$

$$\text{EWR} \left(\text{mm}^3 / \text{min} \right) = \frac{E_p\ (g) - E_a(g)}{\text{Machining Time (min)} \times \text{Density (g/mm}^3\text{)}} \tag{2}$$

where Wp and Wa are the weights of workpieces prior to machining and afterwards; whereas, Ep and Ea are the weights of electrodes prior to machining and afterwards. Compressed air followed by dipping in acetone was employed to remove debris and kerosene remains on machined specimens. Surface roughness (SR) was recorded by surface roughness meter, SJ-410-Surftest. Three observations were made at random locations and the average was taken as final reading for further analysis.

3. Experimental Design

The higher and lower levels for the three parameters are chosen based on the literature review and trial runs as long as the machined parts remained within the acceptable quality range. The selected parameters together with their chosen ranges are provided in Table 2. Modelling and analysis of performance measures (MRR, EWR and SR) have been carried out through RSM employing the Box Behnken Design (BBD). Overall, seventeen (17) experimental runs were performed with twelve (12) factorial and five (5) center points. The experimental runs along with process parameters and observed values of performance measures have been presented in Table 3.

Table 2. Process parameters with their levels.

Process Parameters	Levels		
	Low	**Middle**	**High**
Pulse On Time (μs)	200	400	600
Current (A)	10	13	16
Pulse Off Time (μs)	50	100	150

Table 3. Design matrix.

Exp Run	Process Parameters			Performance Measures		
	Pulse On Time	Current	Pulse Off Time	Material Removal Rate	Electrode Wear Rate	Surface Roughness
	(Pon)		(Poff)	(MRR)	(EWR)	(SR)
	μs	A	μs	mm^3/min	mm^3/min	μm
1	200	10	100	1.57	1.14	0.64
2	600	10	100	3.49	2.59	1.18
3	200	16	100	2.84	1.77	3.10
4	600	16	100	8.46	4.58	3.58
5	200	13	50	3.07	1.39	1.65
6	600	13	50	5.30	2.50	2.40
7	200	13	150	0.77	1.00	1.20
8	600	13	150	5.49	4.09	1.63
9	400	10	50	4.77	2.24	1.21
10	400	16	50	8.70	3.56	3.56
11	400	10	150	4.57	2.64	0.67
12	400	16	150	6.42	3.57	3.10
13	400	13	100	7.50	3.11	2.10
14	400	13	100	7.80	3.00	2.19
15	400	13	100	7.69	3.26	2.23
16	400	13	100	7.72	3.12	2.12
17	400	13	100	7.77	3.11	2.22

4. Results, Analysis and Discussions

This section is comprised of results discussion and statistical analysis using RSM. Moreover, mathematical model selection and adequacy confirmation through ANOVA are discussed. Influences of process parameters on performance measures have been evaluated using 3D graphs (surface plots).

4.1. Development of Empirical Models

Modelling of the performance measures MRR, EWR and SR have been performed through regression analysis using commercial software (Design Expert ®10.06). Analysis of Variance have been used to test the significance of factors and developed models.

4.1.1. Material Removal Rate (MRR)

After detailed experimentation, linear, quadratic and cubic models were tested to select the fitted model. The results revealed quadratic expression as the preferable model for material removal rate (MRR) (based on minimum p-value and R^2, adjusted R^2 and predicted R^2 (close to 1)). The results obtained through ANOVA are presented in Table 4. It can be observed that main effects Pon (A), current (B), poff (C), interaction effects Pon and current (AB), Pon and Poff (AC), current and Poff (BC) and quadratic effects Pon (A^2), current (B^2) and Poff (C^2) were the significant terms of MRR model. The statistical measures R^2, adjusted R^2 and predicted R^2 have also been provided. Form the results, it is evident that the fitted regression model is significant at 95% confidence interval with 'p' value under 0.05. Furthermore, resulted values of statistical terms (R^2, adjusted R^2 and predicted R^2) are nearly 1 which depicted that model is satisfactory to adopt. The resulted empirical model for MRR is provided in Equation (3).

$$\begin{aligned}
\text{MRR} = {}&-22.22697 + (0.043425 \times \text{Pon}) + (1.88978 \times \text{Current}) + (0.089232 \times \text{Poff}) \\
&+ \left(1.54167 \times 10^{-3} \times \text{Pon} \times \text{Current}\right) + \left(6.225 \times 10^{-5} \times \text{Pon} \times \text{Poff}\right) \\
&- \left(3.46667 \times 10^{-3} \times \text{Current} \times \text{Poff}\right) - \left(7.57938 \times 10^{-5} \times \text{Pon}^2\right) \\
&- \left(0.063806 \times \text{Current}^2\right) - \left(4.027 \times 10^{-4} \times \text{Poff}^2\right)
\end{aligned} \tag{3}$$

Table 4. ANOVA results for material removal rate, electrode wear rate and surface roughness.

Material Removal Rate					
Source	**Sum of Squares**	**df**	**Mean Square**	**F Value**	**p-Value Prob > F**
Model	100.14	9	11.13	546.97	<0.0001
A-Pon	26.25	1	26.25	1290.18	<0.0001
B-Current	18.06	1	18.06	887.81	<0.0001
C-Poff	2.63	1	2.63	129.46	<0.0001
AB	3.42	1	3.42	168.25	<0.0001
AC	1.55	1	1.55	76.20	<0.0001
BC	1.08	1	1.08	53.17	0.0002
A^2	38.70	1	38.70	1902.51	<0.0001
B^2	1.39	1	1.39	68.26	<0.0001
C^2	4.27	1	4.27	209.79	<0.0001
Residual	04	7	0.020		
Lack of Fit	0.087	3	0.029	2.10	0.2432
Pure Error	0.055	4	0.014		
Cor. Total	100.28	16			
Std. Dev.	0.14		R-Squared		0.9986
Mean	5.53		Adj. R-Squared		0.9968
C.V. %	2.58		Pred. R-Squared		0.9852
PRESS	1.48		Adeq Precision		73.693
Electrode wear rate					
Source	**Sum of Squares**	**df**	**Mean Square**	**F Value**	**p-value Prob > F**
Model	15.82	7	2.26	104.91	<0.0001
A-Pon	8.95	1	8.95	415.23	<0.0001
B-Current	2.96	1	2.96	137.60	<0.0001
C-Poff	0.32	1	0.32	15.04	0.0037
AB	0.46	1	0.46	21.46	0.0012
AC	0.98	1	0.98	45.49	<0.0001
A^2	1.92	1	1.92	89.18	<0.0001
C^2	0.16	1	0.16	7.23	0.0248
Residual	0.19	9	0.022		
Lack of Fit	0.16	5	0.032	3.74	0.1129
Pure Error	0.034	4	8.55×10^{-3}		
Cor. Total	16.02	16			
Std. Dev.	0.15		R-Squared		0.9879
Mean	2.75		Adj. R-Squared		0.9785
C.V. %	5.35		Pred. R-Squared		0.9398
PRESS	0.96		Adeq Precision		35.252
Surface roughness					
Source	**Sum of Squares**	**df**	**Mean Square**	**F Value**	**p-value Prob > F**
Model	13.42	7	1.92	653.02	<0.0001
A-Pon	0.61	1	0.61	206.02	<0.0001
B-Current	11.62	1	11.62	3955.57	<0.0001
C-Poff	0.62	1	0.62	209.78	<0.0001
AC	0.026	1	0.026	8.72	0.0162
A^2	0.22	1	0.22	76.51	<0.0001
B^2	0.14	1	0.14	48.54	<0.0001
C^2	0.21	1	0.21	70.03	<0.0001
Residual	0.026	9	2.937×10^{-3}		
Lack of Fit	0.013	5	2.51×10^{-3}	0.72	0.6406
Pure Error	0.014	4	3.47×10^{-3}		
Cor. Total	13.45	16			
Std. Dev.	0.054		R-Squared		0.9980
Mean	2.05		Adj. R-Squared		0.9965
C.V. %	2.65		Pred. R-Squared		0.9905
PRESS	0.13		Adeq Precision		79.960

4.1.2. Electrode Wear Rate (EWR)

The fit model details for electrode wear rate (EWR) also confirms that the quadratic expression is the most suitable relationship (based on minimum p-value and R^2, adjusted R^2 and predicted R^2 (close to 1)). The regression terms (main effect, interaction and quadratic) which are significant for EWR include Pon (A), current (B) and Poff (C), Pon and current (AB) and Pon and off (AC), Pon (A^2) and Poff (C^2), respectively. The ANOVA results along with adequacy measures have been provided in Table 4. The resulted empirical model with less than 0.5 'p' value is shown in Equation (4) that can be used successfully for prediction.

$$
\begin{aligned}
\text{EWR} = {} & -0.54243 + \left(6.46294 \times 10^{-3} \times \text{Pon}\right) - (0.023750 \times \text{Current}) \\
& -\left(4.06579 \times 10^{-4} \times \text{Poff}\right) + \left(5.66667 \times 10^{-4} \times \text{Pon} \times \text{Current}\right) \\
& +\left(4.95 \times 10^{-5} \times \text{Pon} \times \text{Poff}\right) - \left(1.68651 \times 10^{-5} \times \text{Pon}^2\right) \\
& -\left(7.68421 \times 10^{-5} \times \text{Poff}^2\right)
\end{aligned}
\tag{4}
$$

4.1.3. Surface Roughness (SR)

The details of the fit model for surface roughness also recommended the quadratic model as the most appropriate model (based on minimum p-value and R^2, adjusted R^2 and predicted R^2 (close to 1)). Pon (A) current (B) and Poff (C) main terms, Pon and Poff (AC) interaction terms and Pon (A^2), current (B^2) and Poff (C^2) quadratic terms have significant effects on surface roughness model. The ANOVA results including regression terms have been demonstrated as Table 4. The developed model is valid because it illustrated a good relationship between parameters and performance measure as 'p' value is under 0.05 (Confidence interval= 95). Moreover, the obtained values of regression metrics are approximately 1 which establish the suitability of model. The empirical model of surface roughness is presented in Equation (5) which is effective for prediction.

$$
\begin{aligned}
\text{SR} = {} & -1.71756 + \left(6.795 \times 10^{-3} \times \text{Pon}\right) - (0.12989 \times \text{Current}) + (0.015330 \times \text{Poff}) \\
& -\left(8 \times 10^{-6} \times \text{Pon} \times \text{Poff}\right) - \left(5.775 \times 10^{-6} \times \text{Pon}^2\right) + \left(0.020444 \times \text{Current}^2\right) \\
& -\left(8.84 \times 10^{-5} \times \text{Poff}^2\right)
\end{aligned}
\tag{5}
$$

4.2. Validation of Model

Statistical analysis has been conducted to assure the fitness of regression models. Furthermore, for experimental validation of empirical models, additional confirmation experiments have been performed. In order to confirm either the developed model are the best representation of actual responses, normal probability plots have been plotted as shown in Figures 4a–6a for MRR, EWR and SR respectively. It is observed that the residuals generally fall on a straight line implying that the errors are normally distributed. Moreover, the comparison plots of predicted against actual values of performance measures MRR, EWR and SR are established and shown in Figures 4b–6b respectively. These plots confirm the normal distribution of error for all responses since all the theoretically assumed (predicted) and actual response values lie on the recommended straight-line or in its close approximation. Consequently, the established models are appropriate and are less likely to violate assumptions.

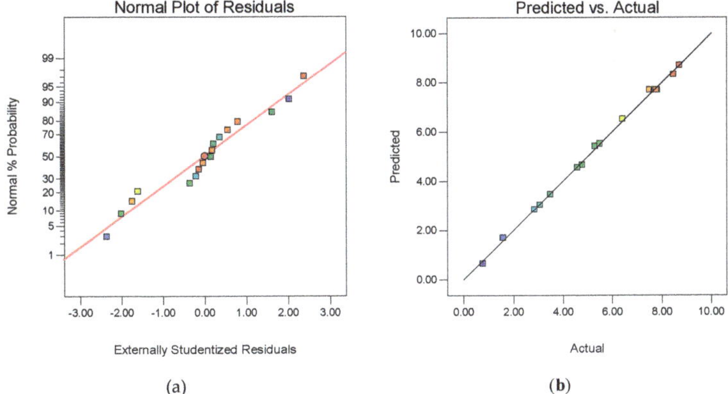

Figure 4. (**a**) Normal probability plot of residuals and (**b**) Predicted versus actual responses for material removal rate (MRR).

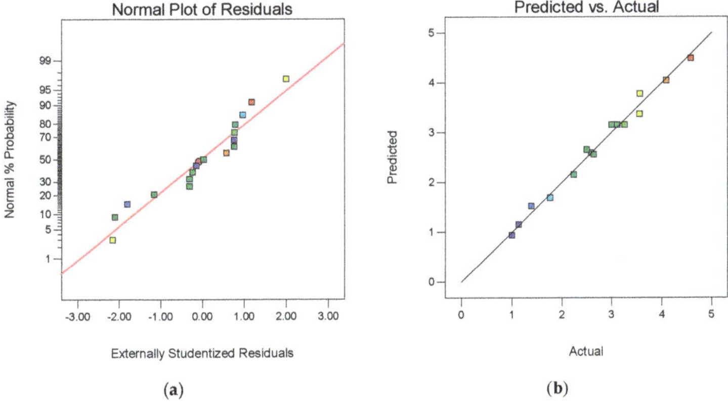

Figure 5. (**a**) Normal probability plot of residuals and (**b**) Predicted versus actual responses electrode wear rate (EWR).

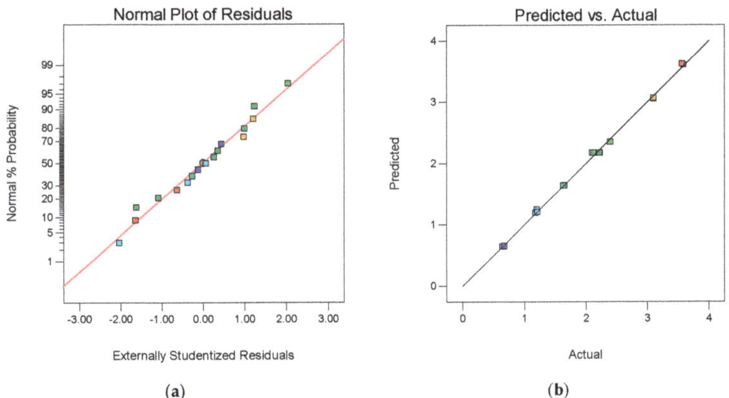

Figure 6. (**a**) Normal probability plot of residuals and (**b**) Predicted versus actual responses for surface roughness (SR).

To validate the established empirical models, six (6) verification experimental runs were designed by randomly selecting the values of process parameters within design space (the selected levels were different from the designed values used for model development). The results of validation experiments and corresponding percentage error have been provided in Table 5. The percentage error value was calculated by using Equation (6) [48–50].

$$\text{Percentage Error} = \left| \frac{\text{Actual value} - \text{Predicted value}}{\text{Predicted value}} \right| \times 100 \tag{6}$$

Table 5. Evaluation of predicted and actual performance measures.

Run No.	Process Parameters			Performance Measures						Percentage Error		
				Predicted			Actual					
	Pon	Current	Poff	MRR	EWR	SR	MRR	EWR	SR	MRR	EWR	SR
	μs	A	μs	mm³/min	mm³/min	μm	mm³/min	mm³/min	μm	%	%	%
1	300	12	80	5.75	2.30	1.66	5.96	2.24	1.61	3.65	2.6	3.01
2	500	12	80	7.00	3.04	1.96	7.18	2.93	1.92	2.57	3.61	2.04
3	300	12	120	5.18	2.26	1.8	5.38	2.24	1.88	3.80	0.89	4.4
4	500	14	120	8.10	3.92	2.51	8.18	3.87	2.57	0.98	1.28	2.39
5	300	14	120	5.73	2.55	2.27	5.56	2.63	2.31	2.9	3.14	1.76
6	500	14	80	8.45	3.56	2.77	8.57	3.64	2.8	1.42	2.25	1.08

The established mathematical models are found valid as the percentage error is under 5%. Hence, these established models are effectively applicable for the prediction of performance measures in future.

4.3. 3D Response Surface

The effects of process parameters on material removal rate (MRR), electrode wear rate (EWR) and surface roughness (SR) have been presented in 3D response surface plots (Figures 7–9).

4.3.1. Material Removal Rate (MRR)

The influence of both Pon and current on material removal rate (MRR) have been presented as a surface plot in Figure 7a. In the beginning, MRR increases as Pon increases up to a maximum value of 8.80 mm³/min and after that decreases. Contemporarily, a positive relationship exists between MRR and current, that is, MRR increases as current increases since maximum discharge energy enhances the material removal phenomena. Furthermore, Pon is the most influencing process parameter than current. Figure 7b describes the influence of Pon and Poff on MRR. Higher MRR is observed at the lower value of Poff and middle value of Pon. The correlation of MRR with Poff and current has been shown as 3D graph in Figure 7c. The graph confirms gradual increment in MRR along with increasing values of current. Conversely, Poff has an inverse effect on MRR. MRR increases as current and Pon increases because maximum discharge energy become available and deeper, and overlying craters are produced as a result of concentrated heat and localized melting. The trends are similar to those observed by Mandaloi et al. [7], Payal et al. [14], Sultan et al. [25] and Lin et al. [26].

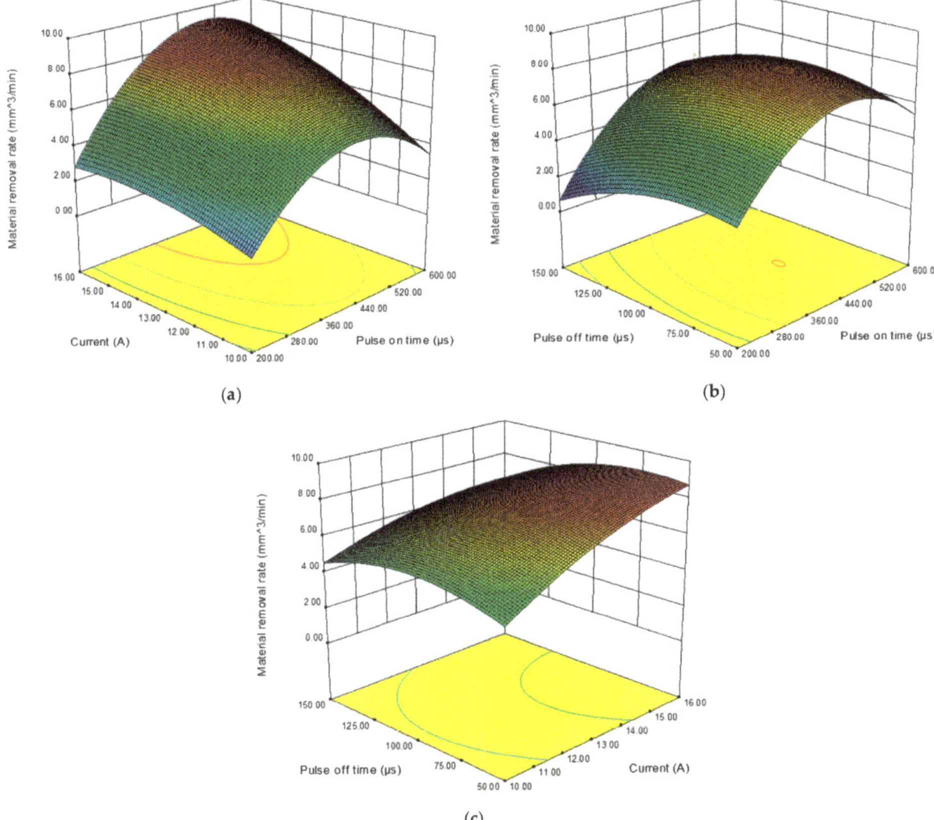

Figure 7. Response surface graphs showing the effects of (**a**) Pon and current, (**b**) Pon and Poff and (**c**) current and Poff on material removal rate.

4.3.2. Electrode Wear Rate (EWR)

Figure 8a presents the influence on electrode wear rate (EWR) when changing Pon and current. It is evident from the figure that Pon has more influence than current. Furthermore, EWR increases by increasing Pon and current. The highest value of EWR is observed at maximum values of Pon and current. The response of Poff and Pon on EWR have been presented in Figure 8b. The Figure clearly indicates that EWR is minimum at the lower value of Pon and high value of Poff and maximum at high level of Pon and low level of Poff. Moreover, EWR is relatively more affected by Pon than Poff. The surface plot of Poff and current (shown in Figure 8c) indicates that EWR is in direct relation with the current while Poff has an inverse effect on EWR. Maximum EWR resulted at higher values of current and Pon because more powerful discharging occurs with higher energy density that melts and removes more material from electrode. Similar effects have been observed by Sultan et al. [25] and Lin et al. [26].

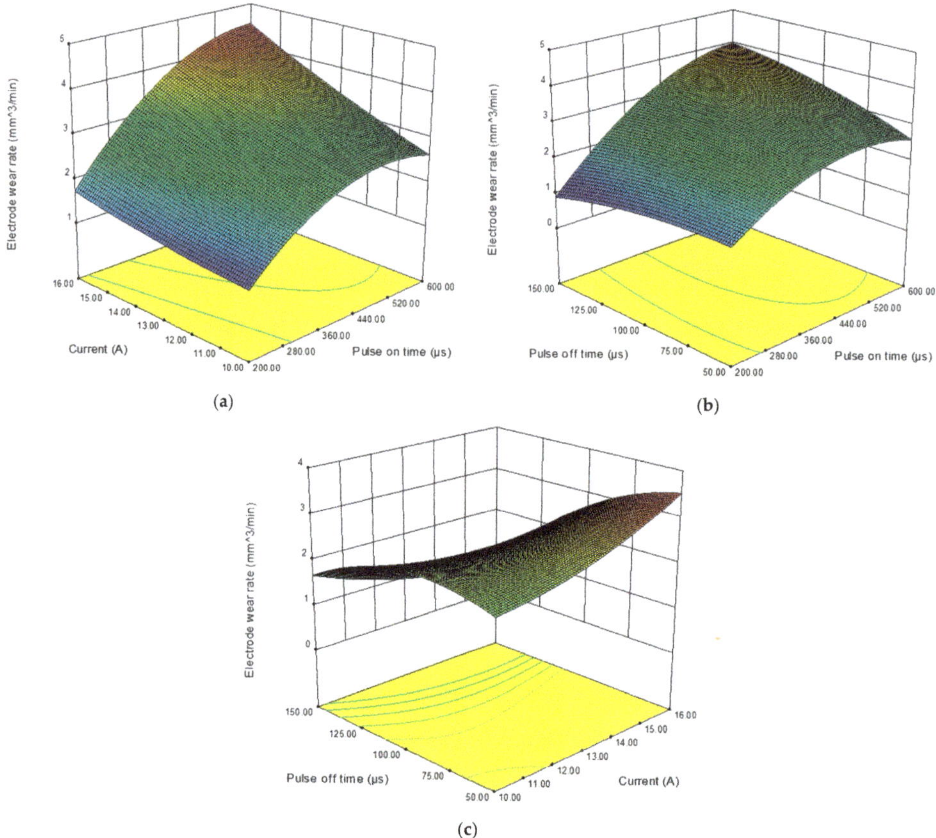

Figure 8. Response surface graphs showing the effects of (**a**) Pon and current, (**b**) Pon and Poff and (**c**) current and Poff on electrode wear rate.

4.3.3. Surface Roughness (SR)

The plot of surface roughness (SR) based on Pon and current indicates that SR increases as both current and Pon increase (Figure 9a). Furthermore, SR is significantly influenced by current than Pon. The 3D plot of Pon and Poff with SR is presented in Figure 9b. It is evident from the figure that minimum SR can be achieved at lower value of Pon and Poff. The effects of current and Poff is displayed in Figure 9c. The figure depicts that SR increases with the increase in current. Whereas, Poff has a negligible influence on SR. At higher levels of current, discharge energy becomes greater which enhances the erosion and melting of material and hence, surface roughness increases as stated by Sultan et al. [25] and Lin et al. [26].

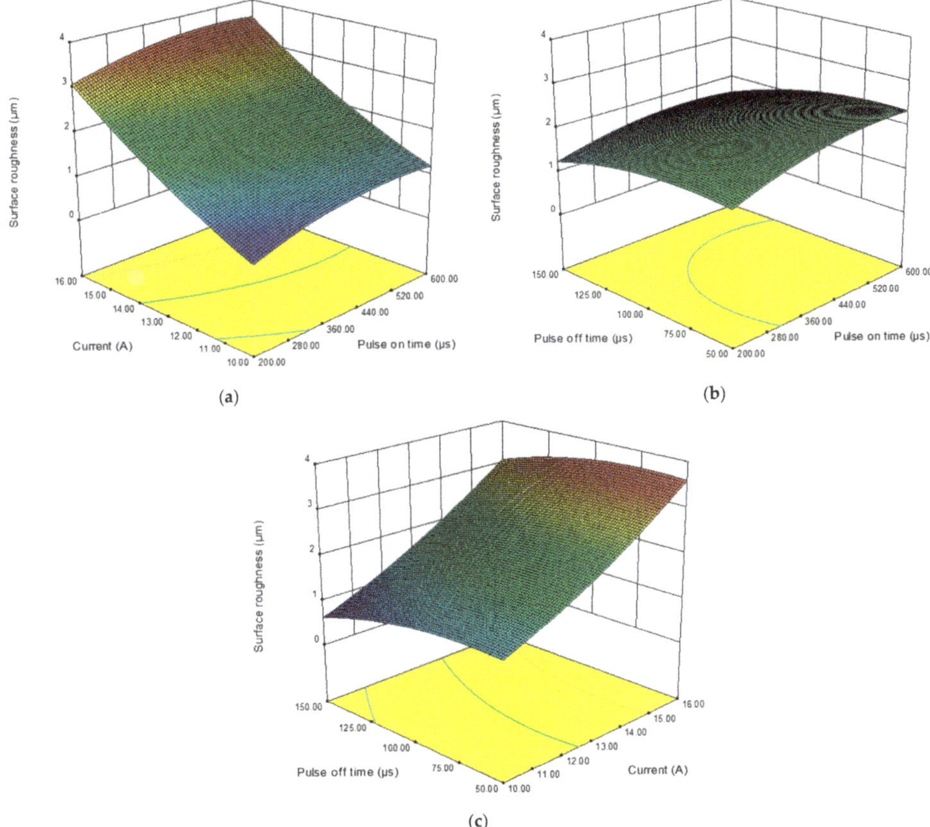

Figure 9. Response surface graphs showing the effects of (**a**) Pon and current, (**b**) Pon and Poff and (**c**) current and Poff on surface roughness.

5. Optimization Associated with Sustainability

Sustainable machining aims to achieve high production rate at minimum cost while maintaining the highest quality standards. Simultaneous optimization of these objective functions leads to minimum environmental damage and thereby assures sustainable production. The performance measures for the current research include MRR, EWR and SR. The sustainability function is the combination of three objective functions and is given by the relation 7.

$$\text{Sustainability} = \begin{cases} \text{Maximize MRR (Productivity)} \\ \text{Minimize EWR (Cost)} \\ \text{Minimize SR (Quality)} \end{cases} \tag{7}$$

On the basis of detailed analysis of empirical models and 3D response surfaces presented in previous sections, relationships of process parameters with respect to performance measures have been presented in Table 6. The table represents a comparison of two functions, "as-is" function versus "to-be" function. Here, "as-is" function characterizes achieved effects on performance measures (MRR, EWR and SR) while increasing level of the three process parameters (Pon, Current and Poff), for example, MRR, EWR and SR increase by increasing Pon and Current and vice versa. On the other hand, "to-be" function depicts the norm for desired sustainability. For instance, the objective function is to maximize

MRR, while minimizing EWR and SR, whereas, in reality, all performance measures (MRR, EWR and SR) increase with the increase in Pon. Similar results are achieved by increasing current. Increase in Poff, on the other hand, leads to sustainable EWR and SR with compromised MRR.

Table 6. As-is and To-be sustainability function.

Parameters	As-Is Function (Achieved Function)			To-Be Sustainability Function (Desired Sustainability Function)		
	MRR	EWR	SR	MRR	EWR	SR
A: Pon	↑	↑	↑	↑	↓	↓
B: Current	↑	↑	↑	↑	↓	↓
C: Poff	↓	↓	↓	↓	↑	↑

Notation: ↑ represents increasing function and ↓ represents decreasing function.

From the above discussion, it can be concluded that the simultaneous optimization of these performance measures cannot be attained directly. To overcome this problem, multi-objective optimization has been carried out considering desirability. The purpose of the desirability function is to combine the effects of multiple responses into a single desirability value using mathematical transformation. This multi-objective optimization-based desirability has been accomplished in two stages namely (i) desirability identification and (ii) formulation of combined desirability geometric mean (CDGM). During the desirability identification stage, each performance measure Yi is converted into a single desirable value di having range $0 \leq di \leq 1$, where 0 indicates the most undesirable value and 1 depicts the most desirable value. Once the desirability of individual performance measure has been obtained, they were combined into a single value using geometric mean. The desirability functions for maximizing MRR, minimizing EWR and SR and combined desirability geometric mean (CDGM) have been presented in Equations (8), (9) and (10) respectively [51–53].

$$\mathrm{di} = \begin{cases} 0, & Y_i \leq L_i \\ \left(\frac{H_i - Y_i}{H_i - L_i}\right)^w, & L_i < y_i < H_i \\ 1, & Y_i \leq H_i \end{cases} \tag{8}$$

$$\mathrm{di} = \begin{cases} 0, & Y_i \leq L_i \\ \left(\frac{H_i - Y_i}{H_i - L_i}\right)^w, & L_i < y_i < H_i \\ 1, & Y_i \geq H_i \end{cases} \tag{9}$$

$$\mathrm{DGM} = \left(\mathrm{d1} \times \mathrm{d1} \times \ldots \ldots \times \mathrm{d_n}^{W_n}\right)^{\frac{1}{n}}, \tag{10}$$

where H_i, L_i, w and n represent higher value, lower value, weight associated with a performance measure and number of performance measures respectively. The multi-objective optimization goals along with the conditions used for desirability approach have been provided in Table 7. All performance measures and process parameters are given equal weights (1) for both upper and lower limits and similarly equal importance value (3) for optimization. The process parameters values and achieved desirability are presented in Table 8. It is evident (from Table 8) that desirability values up to 75.5% have been achieved when all performance measures were given equal weights.

Table 7. Conditions for optimization.

Condition	Units	Goal	Limits		Weights		Importance
			Lower	Upper	Lower	Upper	
A:Pon	μs	In range	200	600	1	1	3
B:Current	μs	In range	10	16	1	1	3
C:Poff	A	In range	50	150	1	1	3
MRR	mm^3/min	Maximize	1.90	6.40	1	1	3
EWR	mm^3/min	Minimize	1.49	4.23	1	1	3
SR	μm	Minimize	1.47	5.11	1	1	3

Table 8. Achieved desirability.

No.	Pon	Current	Poff	MRR	EWR	SR	Desirability	Remarks
	μs	A	μs	mm^3/min	mm^3/min	μm		
1	220.21	13.17	50.00	4.47186	1.80364	2.01094	0.755	Selected
2	220.13	13.21	50.00	4.48082	1.80449	2.02074	0.755	
3	215.90	13.28	50.00	4.45033	1.7644	2.03568	0.755	

The corresponding values of process parameters and performance measures have been presented in Figure 10. The achievable ranges of performance measures are 1.9 mm^3/min to 6.4 mm^3/min for MRR, 1.5 mm^3/min to 4.2 mm^3/min for EWR and 1.47 μm to 5.10 μm for SR as shown in Figure 10. However, with maximum desirability of 75.5%, 4.47 mm^3/min of MRR, 1.8 mm^3/min of EWR and 2.01 μm SR can only be achieved. Practically on the shop floor, where machines exhibit different ranges, the process planners have the constraints of limited selection of process parameters values. For such critical situations, the contour plots (Figure 11a–c)) can be employed to use the available values with certain sustainability. For example, at 13A current and 400μs Pon, only 62.9% sustainability can be obtained (Figure 11a).

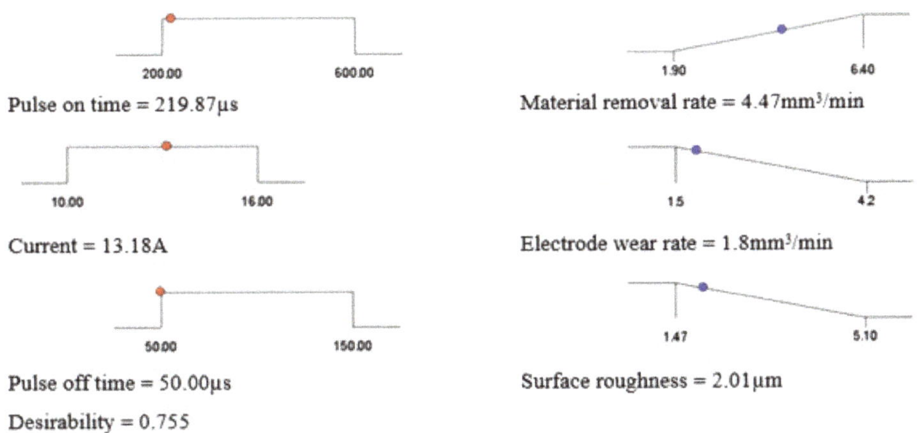

Figure 10. Achievable range of performance measures against specific process parameters range with desirability 0.755.

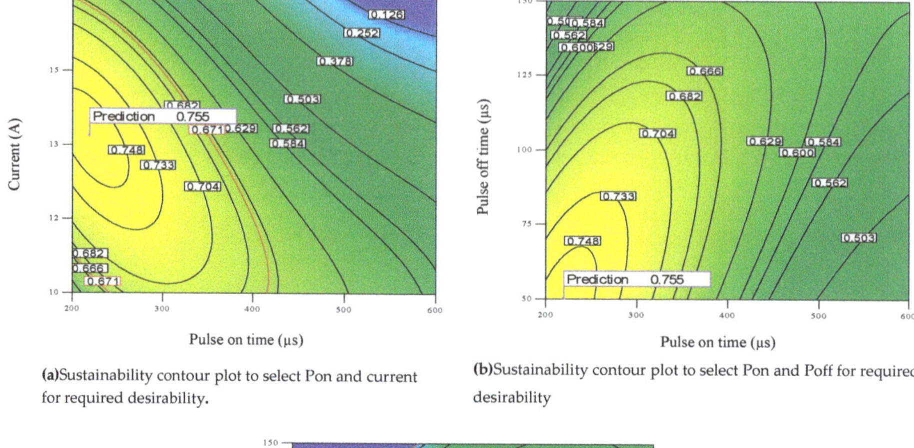

(a)Sustainability contour plot to select Pon and current for required desirability.

(b)Sustainability contour plot to select Pon and Poff for required desirability

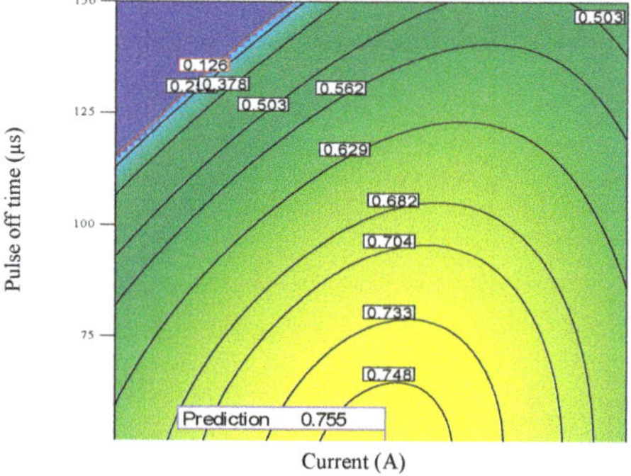

(c)Sustainability contour plot to select current and Poff for required desirability

Figure 11. Sustainability contour plot to select parameters for required desirability.

6. Microstructures Analysis

In order to have an explicit understanding of process parameters (Pon, current and Poff) on performance measures (MRR, EWR and SR), microstructures of machined parts have also been examined. Three samples were taken at low, middle and high levels of Pon and current while keeping Poff at the middle value of 100 μs. It is pertinent to mention that in this study, only Pon and current have been identified as the most significant process parameters as compared to Poff. The microstructures graphs of varying current and Pon have therefore been considered for detailed investigation. The microstructures graphs of samples are presented as Figure 12a–c). From Figure 12a, it is manifested that at lower levels of current and Pon (10 A and 200 μs) fewer numbers of craters, debris, globules, pits and voids are visible with minute level micro-cracks. Whereas an increase in the size of micro-cracks, debris, globules, pits craters and voids can be observed at middle levels of current (13 A) and Pon (400 μs), as presented in Figure 12b. Moreover, samples obtained at 16 A current (upper level) and Pon (600 μs) exhibited prominent micro-cracks, craters, debris, globules, pits and voids (Figure 12c). This clearly indicates that increase in current and Pon results in higher cracks, large globule size, pits and voids. This is due to increase vaporization at higher level of current and Pon.

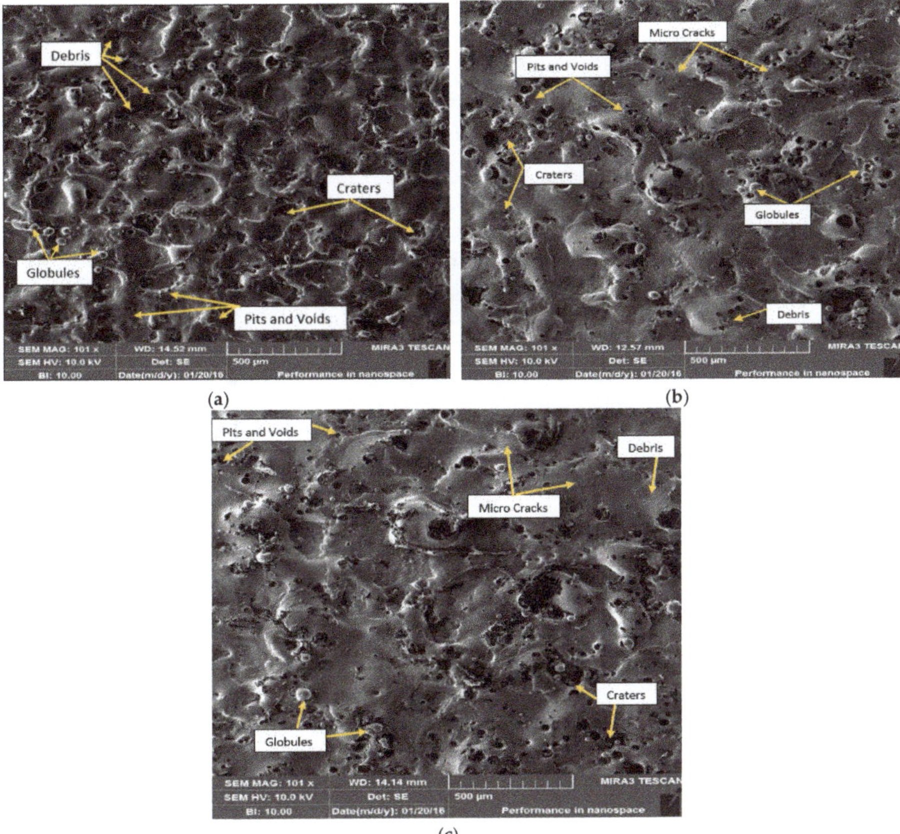

Figure 12. Scanning electron microscope (SEM) microstructures (**a**) at lower level of current (10 A) and Pon (200 μs), (**b**) at intermediate level of current (13 A) and Pon (400 μs), (**c**) at higher level of current (16 A) and Pon (600 μs).

7. Conclusions

The aim of this research was sustainable production by the enhancement of productivity and quality along with cost minimization during EDM of a low alloy tool steel (AISI L3). Initially, empirical models have been developed by analyzing the performance measures (MRR, SR and EWR) through response surface methodology. After that, multi-objective optimization, considering the sustainability, has been executed by instituting a compromise among productivity (MRR maximization), cost (EWR minimization) and quality (SR minimization). From the present investigation, the following interpretations can be concluded:

- Pon and current are the most significant process parameters influencing performance measures, MRR, EWR and SR, to a great extent.
- The higher values of MRR (productivity) can be achieved by keeping both Pon and current at their higher settings with Poff at its lower level. Conversely, lower values of SR and EWR (quality and cost) can be maintained at lower agreeable level of both Pon and current and upper level of Poff.
- By performing multi-objective optimization while incorporating the sustainability measures, maximum MRR of 4.47 mm^3/min, minimum EWR of 1.8 mm^3/min and SR of 2.01 μm is obtained

as compared to individual values obtained for maximum MRR (6.4 mm^3/min), minimum EWR (1.5 mm^3/min) and minimum SR (1.47 μm).

- The microstructure analysis highlighted that the increase in Pon and current results in prominent micro-cracks, craters, debris, globules, pits and voids due to increase in vaporization at the high level of Pon and current.
- The established sustainability contour plots can be employed successfully for feasible machine limits to attain a certain level of desirability.

Further research should evaluate the environmental aspect of sustainability using electric discharge machining. Besides, other performance measures like white/ grey recast layer and cost-based models should be investigated for improvement and enrichment of machining performance.

Author Contributions: Conceptualization, M.N., S.S. and W.A.; Data curation, M.N., S.S. and W.A.; Formal analysis, M.N. and S.S.; Funding acquisition, W.A. and K.S.; Investigation, M.N. and S.S.; Methodology, M.N.; Project administration, W.A. and E.S.; Resources, W.A.; Software, M.N., S.S. and W.A.; Supervision, W.A. and E.S.; Validation, W.A., E.S. and K.S.; Visualization, M.N., S.S., E.S. and K.S.; Writing–original draft, M.N.; Writing–review, editing, layout formatting M.N., S.S., E.S. and K.S. All authors have read and agreed to the published version of the manuscript.

Funding: The work is supported by the University of Engineering and Technology Taxila, Pakistan.

Acknowledgments: The authors would like to acknowledge the head and staff of Fracture and Mechanics Laboratory, University of Engineering and Technology Taxila, Pakistan for providing the experimental facility.

Conflicts of Interest: The authors declare no conflict of interest.

References

1. Torres, A.; Puertas, I.; Luis, C.J. Modelling of surface finish, electrode wear and material removal rate in electrical discharge machining of hard-to-machine alloys. *Precis. Eng.* **2015**, *40*, 33–45. [CrossRef]
2. Lin, Y.-C.; Cheng, C.-H.; Su, B.-L.; Hwang, L.-R. Machining characteristics and optimization of machining parameters of SKH 57 high-speed steel using electrical-discharge machining based on Taguchi method. *Mater. Manuf. Process.* **2006**, *21*, 922–929. [CrossRef]
3. Cverna, F.; Conti, P. *Worldwide Guide to Equivalent Irons and Steels*; ASM International: Cleveland, OH, USA, 2006.
4. Ho, K.; Newman, S. State of the art electrical discharge machining (EDM). *Int. J. Mach. Tools Manuf.* **2003**, *43*, 1287–1300. [CrossRef]
5. Gamage, J.R.; DeSilva, A.K.M. Assessment of research needs for sustainability of unconventional machining processes. *Procedia CIRP* **2015**, *26*, 385–390. [CrossRef]
6. Padhi, S.K.; Mahapatra, S.S.; Padhi, R.; Das, H.C. Performance analysis of a thick copper-electroplated FDM ABS plastic rapid tool EDM electrode. *Adv. Manuf.* **2018**, *6*, 442–456. [CrossRef]
7. Mandaloi, G.; Singh, S.; Kumar, P.; Pal, K. Effect on crystalline structure of AISI M2 steel using copper electrode through material removal rate, electrode wear rate and surface finish. *Measurement* **2015**, *61*, 305–319. [CrossRef]
8. Salonitis, K.; Stournaras, A.; Stavropoulos, P.; Chryssolouris, G. Thermal modeling of the material removal rate and surface roughness for die-sinking EDM. *Int. J. Adv. Manuf. Technol.* **2009**, *40*, 316–323. [CrossRef]
9. Mondal, R.; De, S.; Mohanty, S.K.; Gangopadhyay, S. Thermal energy distribution and optimization of process parameters during electrical discharge machining of AISI D2 steel. *Mater. Today Proc.* **2015**, *2*, 2064–2072. [CrossRef]
10. Guu, Y.H.; Hocheng, H.; Chou, C.Y.; Deng, C.S. Effect of electrical discharge machining on surface characteristics and machining damage of AISI D2 tool steel. *Mater. Sci. Eng. A* **2003**, *358*, 37–43. [CrossRef]
11. Dhobe, M.M.; Chopde, I.K.; Gogte, C.L. Optimization of wire electro discharge machining parameters for improving surface finish of cryo-treated tool steel using DOE. *Mater. Manuf. Process.* **2014**, *29*, 1381–1386. [CrossRef]
12. Gostimirovic, M.; Kovac, P.; Skoric, B.; Sekulic, M. Effect of electrical pulse parameters on the machining performance in EDM. *Indian J. Eng. Mater. Sci.* **2012**, *18*, 411–415.

13. Thomas, D.; Kumar, R.; Singh, G.K.; Sinha, P.; Mishra, S. Modelling of surface roughness in coated wire electric discharge machining through response surface methodology. *Mater. Today Proc.* **2015**, *2*, 3520–3526. [CrossRef]

14. Payal, H.S.; Choudhary, R.; Singh, S. Analysis of electro discharge machined surfaces of EN-31 tool steel. *J. Sci. Ind. Res.* **2008**, *67*, 1072–1077.

15. Singh, A.; Kanth Grover, N. Wear properties of cryogenic treated electrodes on machining of En-31. *Mater. Today Proc.* **2015**, *2*, 1406–1413. [CrossRef]

16. Baraskar, S.S.; Banwait, S.S.; Laroiya, S.C. Multiobjective optimization of electrical discharge machining process using a hybrid method. *Mater. Manuf. Process.* **2013**, *28*, 348–354. [CrossRef]

17. Kanlayasiri, K.; Jattakul, P. Simultaneous optimization of dimensional accuracy and surface roughness for finishing cut of wire-EDMed K460 tool steel. *Precis. Eng.* **2013**, *37*, 556–561. [CrossRef]

18. Tosun, N.; Cogun, C.; Tosun, G. A study on kerf and material removal rate in wire electrical discharge machining based on Taguchi method. *J. Mater. Process. Technol.* **2004**, *152*, 316–322. [CrossRef]

19. Pellicer, N.; Ciurana, J.; Delgado, J. Tool electrode geometry and process parameters influence on different feature geometry and surface quality in electrical discharge machining of AISI H13 steel. *J. Intell. Manuf.* **2011**, *22*, 575–584. [CrossRef]

20. Ferreira, J.C. A study of die helical thread cavity surface finish made by Cu-W electrodes with planetary EDM. *Int. J. Adv. Manuf. Technol.* **2007**, *34*, 1120–1132. [CrossRef]

21. Shabgard, M.R.; Seyedzavvar, M.; Oliaei, S.N. Influence of input parameters on characteristics of EDM process. *Strojniški vestnik-Journal Mech. Eng.* **2011**, *57*, 689–696. [CrossRef]

22. Amorim, F.L.; Weingaertner, W.L. The influence of generator actuation mode and process parameters on the performance of finish EDM of a tool steel. *J. Mater. Process. Technol.* **2005**, *166*, 411–416. [CrossRef]

23. Kiyak, M.; Çakır, O. Examination of machining parameters on surface roughness in EDM of tool steel. *J. Mater. Process. Technol.* **2007**, *191*, 141–144. [CrossRef]

24. Zarepour, H.; Tehrani, A.F.; Karimi, D.; Amini, S. Statistical analysis on electrode wear in EDM of tool steel DIN 1.2714 used in forging dies. *J. Mater. Process. Technol.* **2007**, *187–188*, 711–714. [CrossRef]

25. Sultan, T.; Kumar, A.; Gupta, R.D. Material removal rate, electrode wear rate, and surface roughness evaluation in die sinking EDM with hollow tool through response surface methodology. *Int. J. Manuf. Eng.* **2014**, *2014*, 1–16. [CrossRef]

26. Lin, J.L.; Lin, C.L. The use of grey-fuzzy logic for the optimization of the manufacturing process. *J. Mater. Process. Technol.* **2005**, *160*, 9–14. [CrossRef]

27. Bundel, B. Experimental investigation of electrode wear in die-sinking EDM on different pulse-on & off time (μs) in cylindrical copper electrode. *Int. J. Mod. Eng. Res.* **2015**, *5*, 49–54.

28. Dixit, A.C.; Kumar, A.; Singh, R.K.; Bajpai, R. An experimental study of material removal rate and electrode wear rate of high carbon-high chromium steel (AISI D3) in EDM process using copper tool electrode. *Int. J. Innov. Res. Adv. Eng.* **2015**, *2*, 257–262.

29. Annamalai, N.; Sivaramakrishnan, V.; Baskar, N. Response surface modeling of electric discharge machining process parameters for EN 24 low alloy steel. In Proceedings of the 5th International & 26th All India Manufacturing Technology, Design and Research Conference, Guwahati, India, 12–14 December 2014.

30. Mathew, N.; Kumar, D.; Beri, N.; Kumar, A. Study of material removal rate of different tool materials during EDM of H11 steel at reverse polarity. *Int. J. Adv. Eng. Technol.* **2014**, *5*, 25–30.

31. Vates, U.K.; Singh, N.K. Optimization of surface roughness process parameters of electrical discharge machining of EN-31 by response surface methodology. *Int. J. Eng. Res. Technol.* **2013**, *6*, 835–840.

32. Arunkumar, N.; Rawoof, H.S.A.; Vivek, R. Investigation on the effect of process parameters for machining Of EN31 (air hardened steel) by EDM. *Int. J. Eng. Res. Appl.* **2012**, *2*, 1111–1121.

33. Singh, S.; Maheshwari, S.; Pandey, P.C. Some investigations into the electric discharge machining of hardened tool steel using different electrode materials. *J. Mater. Process. Technol.* **2004**, *149*, 272–277. [CrossRef]

34. Purohit, R.; Rana, R.S.; Dwivedi, R.K.; Banoriya, D.; Singh, S.K. Optimization of electric discharge machining of M2 tool steel using grey relational analysis. *Mater. Today Proc.* **2015**, *2*, 3378–3387. [CrossRef]

35. Amorim, F.L.; Dalcin, V.A.; Soares, P.; Mendes, L.A. Surface modification of tool steel by electrical discharge machining with molybdenum powder mixed in dielectric fluid. *Int. J. Adv. Manuf. Technol.* **2017**, *91*, 341–350. [CrossRef]

36. Long, B.T.; Phan, N.H.; Cuong, N.; Jatti, V.S. Optimization of PMEDM process parameter for maximizing material removal rate by Taguchi's method. *Int. J. Adv. Manuf. Technol.* **2016**, *87*, 1929–1939. [CrossRef]

37. Molinetti, A.; Amorim, F.L.; Soares, P.C.; Czelusniak, T. Surface modification of AISI H13 tool steel with silicon or manganese powders mixed to the dielectric in electrical discharge machining process. *Int. J. Adv. Manuf. Technol.* **2016**, *83*, 1057–1068. [CrossRef]

38. Samanta, A.; Sekh, M.; Sarkar, S. Influence of different control strategies in wire electrical discharge machining of varying height job. *Int. J. Adv. Manuf. Technol.* **2019**, *100*, 1299–1309. [CrossRef]

39. Singh, V.; Bhandari, R.; Yadav, V.K. An experimental investigation on machining parameters of AISI D2 steel using WEDM. *Int. J. Adv. Manuf. Technol.* **2017**, *93*, 203–214. [CrossRef]

40. Gov, K. The effects of the dielectric liquid temperature on the hole geometries drilled by electro erosion. *Int. J. Adv. Manuf. Technol.* **2017**, *92*, 1255–1262. [CrossRef]

41. Lin, Y.-C.; Hung, J.-C.; Lee, H.-M.; Wang, A.-C.; Chen, J.-T. Machining characteristics of a hybrid process of EDM in gas combined with ultrasonic vibration. *Int. J. Adv. Manuf. Technol.* **2017**, *92*, 2801–2808. [CrossRef]

42. Dwivedi, A.P.; Choudhury, S.K. Effect of tool rotation on MRR, TWR, and surface integrity of AISI-D3 steel using the rotary EDM process. *Mater. Manuf. Process.* **2016**, *31*, 1844–1852. [CrossRef]

43. Tripathy, S.; Tripathy, D.K. Multi-response optimization of machining process parameters for powder mixed electro-discharge machining of H-11 die steel using grey relational analysis and topsis. *Mach. Sci. Technol.* **2017**, *21*, 362–384. [CrossRef]

44. Talla, G.; Gangopadhyay, S.; Biswas, C.K. Multi response optimization of powder mixed electric discharge machining of aluminum/alumina metal matrix composite using grey relation analysis. *Procedia Mater. Sci.* **2014**, *5*, 1633–1639. [CrossRef]

45. Shao, G.; Kibira, D.; Lyons, K. A virtual machining model for sustainability analysis. In Proceedings of the Volume 3: 30th Computers and Information in Engineering Conference, Parts A and B. (ASMEDC 2010), Montreal, QC, Canada, 15–18 August 2010; pp. 875–883.

46. Pradhan, M.K. Estimating the effect of process parameters on MRR, TWR and radial overcut of EDMed AISI D2 tool steel by RSM and GRA coupled with PCA. *Int. J. Adv. Manuf. Technol.* **2013**, *68*, 591–605. [CrossRef]

47. Niamat, M.; Sarfraz, S.; Aziz, H.; Jahanzaib, M.; Shehab, E.; Ahmad, W.; Hussain, S. Effect of different dielectrics on material removal rate, electrode wear rate and microstructures in EDM. *Procedia CIRP* **2017**, *60*, 2–7. [CrossRef]

48. Niamat, M.; Sarfraz, S.; Shehab, E.; Ismail, S.O.; Khalid, Q.S. Experimental characterization of electrical discharge machining of aluminum 6061 T6 alloy using Different Dielectrics. *Arab. J. Sci. Eng.* **2019**, *44*, 8043–8052. [CrossRef]

49. Sarfraz, S.; Jahanzaib, M.; Wasim, A.; Hussain, S.; Aziz, H. Investigating the effects of as-casted and in situ heat-treated squeeze casting of Al-3.5 % Cu alloy. *Int. J. Adv. Manuf. Technol.* **2017**, *89*, 3547–3561. [CrossRef]

50. Sarfraz, S.; Shehab, E.; Salonitis, K.; Suder, W. Experimental Investigation of Productivity, Specific Energy Consumption, and Hole Quality in Single-Pulse, Percussion, and Trepanning Drilling of IN 718 Superalloy. *Energies* **2019**, *12*, 4610. [CrossRef]

51. Davoodi, B.; Eskandari, B. Tool wear mechanisms and multi-response optimization of tool life and volume of material removed in turning of N-155 iron–nickel-base superalloy using RSM. *Measurement* **2015**, *68*, 286–294. [CrossRef]

52. Assarzadeh, S.; Ghoreishi, M. A dual response surface-desirability approach to process modeling and optimization of Al2O3 powder-mixed electrical discharge machining (PMEDM) parameters. *Int. J. Adv. Manuf. Technol.* **2013**, *64*, 1459–1477. [CrossRef]

53. Hanif, M.; Ahmad, W.; Hussain, S.; Jahanzaib, M.; Shah, A.H. Investigating the effects of electric discharge machining parameters on material removal rate and surface roughness on AISI D2 steel using RSM-GRA integrated approach. *Int. J. Adv. Manuf. Technol.* **2019**, *101*, 1255–1265. [CrossRef]

Article

Sustainability Assessment for Manufacturing Operations

Prateek Saxena [1], Panagiotis Stavropoulos [2], John Kechagias [3] and Konstantinos Salonitis [1,*]

[1] Manufacturing Department, Cranfield University, Bedfordshire MK43 0AL, UK; p.saxena@cranfield.ac.uk
[2] Laboratory for Manufacturing Systems & Automation, Department of Mechanical Engineering & Aeronautics, University of Patras, 26100 Patras, Greece; pstavr@lms.mech.upatras.gr
[3] Laboratory for Machine Tools and Manufacturing Processes, General Department, University of Thessaly, 41500 Gaiopolis, Greece; jkechag@teilar.gr
* Correspondence: k.salonitis@cranfield.ac.uk; Tel.: +44-1234-758347

Received: 31 March 2020; Accepted: 20 May 2020; Published: 29 May 2020

Abstract: Sustainability is becoming more and more important as a decision attribute in the manufacturing environment. However, quantitative metrics for all the aspects of the triple bottom line are difficult to assess. Within the present paper, the sustainability metrics are considered in tandem with other traditional manufacturing metrics such as time, flexibility, and quality and a novel framework is presented that integrates information and requirements from Computer-Aided Technologies (CAx) systems. A novel tool is outlined for considering a number of key performance indicators related to the triple bottom line when deciding the most appropriate process route. The implemented system allows the assessment of alternative process plans considering the market demands and available resources.

Keywords: sustainability; multi-criteria decision making; process planning

---

1. Introduction

Sustainability is becoming extremely relevant in all stages of the life cycle of a commodity. A variety of questions need to be addressed in the early stages of a product such as product design, manufacturing, use, disposal, and in its effect on the society. All these phases have specific sustainability insights and metrics that should be considered. There is a desire to maneuver towards a more sustainable design and means of development. However, these requirements need to be addressed very early in the design of new products. Approximately 80% of the cost of production is determined during the design process of the product [1], therefore, having a clear understanding of the production cost is important.

Sustainability is a broad concept which has been adopted to reflect the need for civilization to work within its own means and to use the resources and products in a manner that does not affect the quality and well-being of future generations [2]. It is characterized by the "triple bottom line", i.e., the need for sustaining and even reconciliating environmental, social, and economic demands. While environmental and economic expectations are conveniently exhaustive, the concept of social responsibility also encourages ethical action in relation to the structures alluded to above. The goal of sustainable social growth is to improve the rights and capabilities of individuals to lead lives that they have cause to admire, without undermining the capacity of future generations to fulfill their own human welfare needs. Such reconciliation cannot be accomplished without more effective solutions and technology which, in part, must be supported by manufacturing.

The manufacturing sector is already moving towards more sustainable practices and the research groups have proposed a range of approaches to enhance the performance of these sustainable practices. The sustainability helix [3] is indicative of a philosophy that is focused on comparing the use and reuse of raw material models to those of nature. In a number of recent studies [4,5] it has been noted that there is a need to consider sustainability as one of the key manufacturing attributes (Figure 1), essentially incorporating cost as one of the dimensions of sustainability. The other three typical qualities of production, i.e., quality, flexibility, and time are always important for the well-being of the industry.

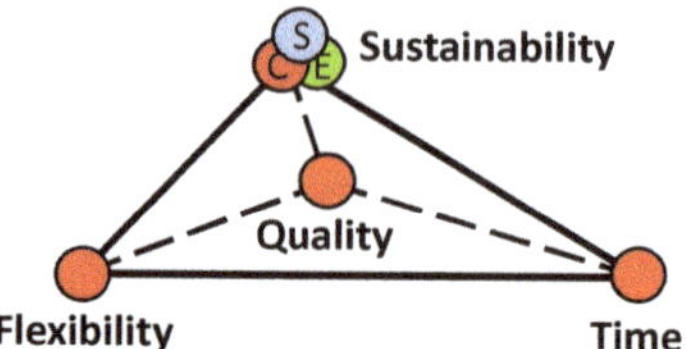

Figure 1. Decision-making attributes of manufacturing (social (S), cost/economical (C), and environmental (E) dimensions of sustainability).

A variety of factors, such as raw materials, supply chain requirements, production processes, use, distribution, and decommissioning influence the design of sustainable products. The social sustainability assessment brings importance to the overall viability of the product as it offers a detailed and focused review of the social footprint of the company. Nowadays, via global market boundaries and, consequently, globally distributed manufacturing cells, it is standard practice for businesses to follow a product life cycle approach aimed at preventing and shifting adverse impacts from one life cycle stage to another or from one social problem to another. Global sustainability metrics are all about well-being such as access to services, healthcare, social security, and the quality of the natural environment. Alternatively, sustainable manufacturing depends on the environmental friendliness of manufacturing plants and processes. This is measured in numerous ways, such as in terms of the energy use, water use, waste and emissions, the health and safety of its employees, etc.

The present paper discusses a framework for analyzing and assessing alternative process plans and strategies considering sustainability aspects as one of the key decision-making attributes. Such a framework would be used by a process engineer to draw up and review different process plans which could adapt to different scenarios of market needs, under the sustainability prism and performance. This paper is structured in a way to help "dive" into sustainable manufacturing from the higher level of product life cycle to product design and manufacturing system selection to boost the sustainability efficiency of an organization.

2. Life Cycle Considerations

As stated in the Introduction, product life cycle management (PLM) is essential for improving the performance of a product throughout its lifetime. However, the life of the product ranges from a few minutes in the case of aluminum cans, to more than 30 years in the case of airplanes, and even decades in the case of buildings and infrastructures. Today, one of the main principles focuses on prolonging the life of goods in order to reduce the need for new materials and the environmental effects of manufacturing. In comparison, reuse or even refurbishment of the parts before recycling and discarding will help minimize energy consumption and the effects on the environment.

The Ellen Macarthur Foundation [6] took this concept one step further by introducing the circular economy concept. A circular economy seeks to rebuild capital. The fundamental concept is to replace a linear manufacturing model with a re-use and recycle model which improved supply of goods and services. Enhanced use of bio-based products such as molded pulp products (MPP) instead of plastics can help to achieve a smooth transition from linear to the circular economy. MPP are obtained from the molding of paper pulp (similar to the injection molding of plastics) [7]. These products are an incredibly strong alternative to their toxic plastic counterparts, especially as a packaging material. The product life cycle of MPP is defined from the procurement of the raw materials (chipping of woods) to the end of life (disposal). A recent advancement in the category of MPP is the Green Fiber Bottle (GFB). The green product aims to replace the existing glass packaging for carbonated beverages with a sustainable paper-based packaging product [8,9]. From the energy perspective, efficient pulping methods could substantially reduce the cost of energy use in the preparation of paper pulp [10,11] and from the tooling perspective, additive manufacturing based tools for paper molding could further reduce the tooling costs and minimize the lead time for tool production [12].

Traditionally, the environmental impact of a product during its life is assessed through the life cycle assessment (LCA). The LCA is used to assess the various emissions that both the manufacturing and the use of the component generate over the entire life of the product. The results are presented in the form of cumulative indexes such as the "eco-indicator" factor. However, the execution of the complete LCA is time and computationally demanding, and outcomes are subject to uncertainties. Since 80% of the economic impact cost is calculated at the design stage when many decisions are still uncertain, the LCA can be used to classify which process is dominant, for example, in the case of a civil aircraft, almost 95% of the energy is consumed during the use of the machine, whereas for furniture production, most of the energy consumption occurs during its manufacturing.

It is clear that success in producing sustainable goods through sustainable production practices requires knowledge and effective management of the product life cycle. PLM packages are the basis for handling the entire product lifecycle from layout to final disposal. In fact, there are three stages within the life cycle of the product, the beginning of life (BoL), the middle of life (MoL) and the end of life (EoL) [13]. In conventional PLM systems, only BoL is considered. However, with a view to sustainability, the traditional PLM packages must be modified and the MoL and EoL phases must also be considered [14]. However, a variety of challenges preclude full incorporation [13–16], as pointed out in Figure 2.

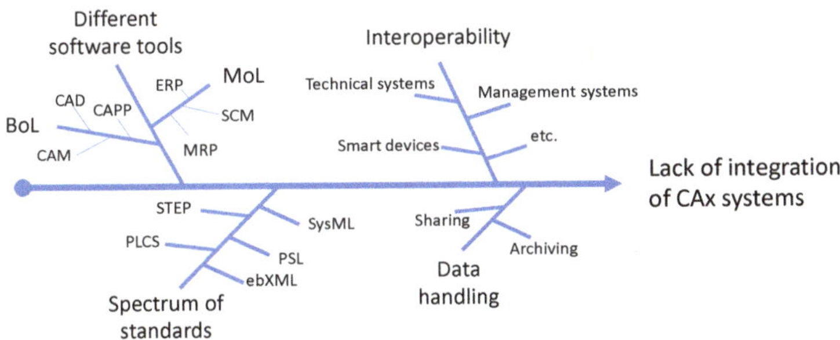

Figure 2. Challenges prohibiting the complete integration of Computer-Aided Technologies (CAx) systems.

3. Design for Sustainability

The starting point for a sustainable product is that it is designed having in mind the three main sustainability dimensions and the criteria imposed by these dimensions. Design for sustainability (D4S), also referred to as sustainable product design, is basically a set of guidelines and practices introduced for companies to apply when developing new products. In other words, from the early phase of a new product design, a firm addresses the environmental and social demands in addition to the obvious financial demands.

Allwood and Cullen [17] emphasized the influence of the weight of a part on its sustainability efficiency. A lightweight design is much more sustainable because of the more efficient raw material usage; scrap reduction and the case of moving parts require less energy during operation. However, when considering the light-weighting impact, the energy required during the manufacturing of the component needs to be considered. For example, if the energy required during the production of the primary material is considered, the use of lightweight material does not necessarily result in the most energy-efficient solution [18]. Therefore, light-weighting is not a solution for all sectors.

Since the beginning of the 1900s, the manufacturing world has been moving for standardization in order to take advantage of economies of scale related to tooling costs and the efficiency of continuous processes. Nevertheless, this standardization results in simplistic designs of components that are heavier than optimized ones. Allwood and Cullen [17] identified the following five technical criteria for the use of less raw materials that should be required for sustainable design:

- Support multiple loads with fewer structures where possible;
- Do not overspecify the loads (i.e., avoid over-engineering);
- Align loads with members to avoid bending if possible;
- If bending is unavoidable, optimize the cross-section along with the member;
- Choose the best material.

Alternatively, Ijomah et al. [19] introduced a set of recommendations for remanufacturing to further promote sustainable manufacturing. Ceschin and Gaziulusoy [20] discussed a quasi-chronological approach at four levels, namely product, product-service system, spatio-social system, and socio-technical system. Such a framework is useful to understand the evolution of sustainable manufacturing from being product-centric to system-level changes. In another work by Harik et al. [21], sustainability metrics for quantifying sustainability in manufacturing industries were presented, which were useful in integrating and analyzing sustainability in manufacturing operations.

4. Sustainable Manufacturing

Production and manufacturing systems have a great impact on all three dimensions of sustainability. Sustainable manufacturing ensures that the technologies and methods used to manufacture goods fulfill the criteria for the three principles of sustainability. Since there is no widely agreed concept of sustainable production, a recent study defines it as a process that leads to the following [22]:

1. Enhanced environmental friendliness;
2. Reduced prices;
3. Reduction in power consumption;
4. Reduced wastes;
5. Enhanced operational safety;
6. Improved health of workers.

Manufacturing facilities must meet the ever-increasing demand for consumer products as the quality of life continues to climb. Table 1 summarizes some examples of sustainable technologies.

Table 1. Sustainable technologies and their benefit over conventional technologies.

Sustainable Technology	Advantage over Conventional Technology	References
Fuel cells	Sustainable energy source	[23]
Lithium-ion batteries (TiO_2 anatase based)	High performance, low cost	[24]
Perovskite photovoltaics	Shortest Energy payback time (EPBT)/High power conversion efficiency	[25–27]
Paper-based packaging products	Reduced waste, recyclable, environmentally friendly	[8,28,29]
Through energy storage systems	Renewable electricity generation	[30]
Hybrid additive manufacturing	Low cost	[31]
Sustainable businesses	Sustainable value creation	[32]
Sustainable castings	Low energy consumption, reduced casting defects	[33,34]

Reducing energy use while increasing the use of renewable energies is important, since approximately one-third of the global energy demand and CO_2 emissions are linked to manufacturing activities. It needs recognition at the factory level and the use of clean energy-efficient actuators and components to their maximum degree, while also taking into account the whole supply chain, from raw materials to manufacturing of final components. Davé et al. [35] presented frameworks for the modeling of factory-level ecoefficiency and discussed the impact of the available data in the prediction of energy efficiency.

At the process stage, optimized technical changes are required to reduce energy and resource usage, hazardous wastes, workplace hazards, etc., to enhance product life by controlling process-induced surface integrity. For example, recycling of metal powder in metal additive manufacturing processes can optimize the resource consumption in three-dimensional (3D) printing technology [36,37].

Sustainable manufacturing can be accomplished by more energy-efficient technologies. A lot of work is underway, for example, to replace traditional methods that display high energy consumption with novel alternatives. For instance, for small batch sizes, the grind-hardening process is considered to be an alternative to the traditional heat treatment process [38].

5. Social Dimension

Traditionally, judgments in manufacturing environments are made based on a tradeoff between cost, speed, and quality. One of the well-known anecdotes is that manufacturing organizations strive for "faster, better, and cheaper" means of production. Cost is related to sustainability through the economic pillar. It has only been during the last decade that the environmental pillar has also become important.

Both the manufacturing of goods and the facilities needed for their development are typically produced through conventional processes, which, in certain situations, are inefficient and have negative impacts. From a production perspective, this topic has been illustrated in the literature, with a special focus on regulatory, health, and safety human concerns, rather than cultural and ethical decision-making criteria [39–41]. Bell and Morse [42] questioned the feasibility of measuring the social aspect of sustainability. Few findings have been addressed aimed at quantifying quality of life from both empirical and subjective viewpoints [43].

The Global Reporting Initiative's Sustainability Reporting Guidelines (the "G3 Guidelines") specify a formal structure for annual sustainability disclosure that is open to all forms of organizations [44]. The social performance indicators are divided into the following four main groups: labor practices and decent work, human rights, society, and product responsibility, as indicated in Figure 3.

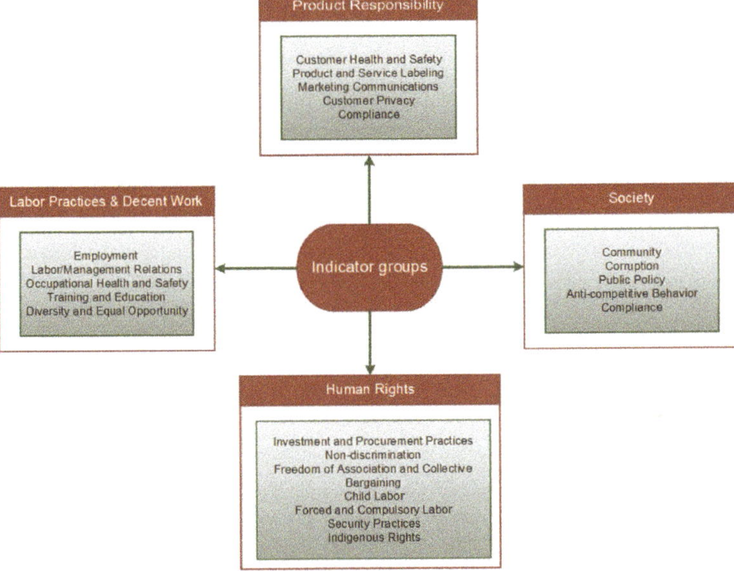

Figure 3. Social KPIs.

The efforts directed to integrate social aspects in sustainability measures can be classified into two key concepts as follows:

1. The movement towards social sustainability and social life cycle assessment supplements the environmental LCA [45–47] by determining social and political influences that lead to environmental problems. The core aspect of this approach is the study of the environmental resources relating environmental effects to social, cultural, and political platforms.

2. Investigation of the mechanisms of effect between organizational flows, such as inputs and outputs, and socially focused mid- and endpoints [48]. These strategies consider the effect of goods and services on individuals, in particular, the protection of good well-being, human dignity, and the satisfaction of basic needs. A two-tiered method is recommended, with a mandatory component governed by standardized local and country requirements and an optional component allowing the addition of parameters of special significance or importance.

For the purposes of the current study, which focused on the sustainability aspects within the process-planning phase, consolidation of the social KPIs was performed which was oriented on the manufacturing/processing level. In order to define the social KPIs and evaluate their weighting in the field of the manufacturing industry, a multi-stage study was carried out comprised of the steps depicted in Figure 4.

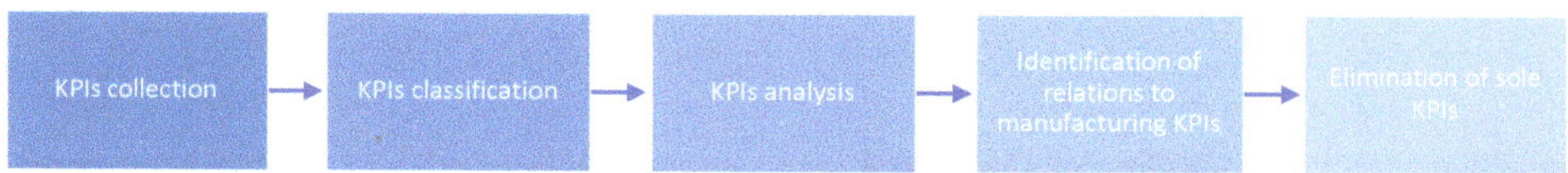

Figure 4. Selection of KPIs.

6. Framework for an Integrated CAx System towards Sustainability

The basic goal is to analyze a variety of potential process maps during the concept generation in order to facilitate effective decision-making. The system suggested, in its operating nature, would consist of a variety of modules, interfaces and information databases. The framework is composed of two major phases, that have different requirements in terms of data and information. During the first phase, the process plans are assessed qualitatively and roughly ranked. Figure 5 presents a block diagram of the proposed method.

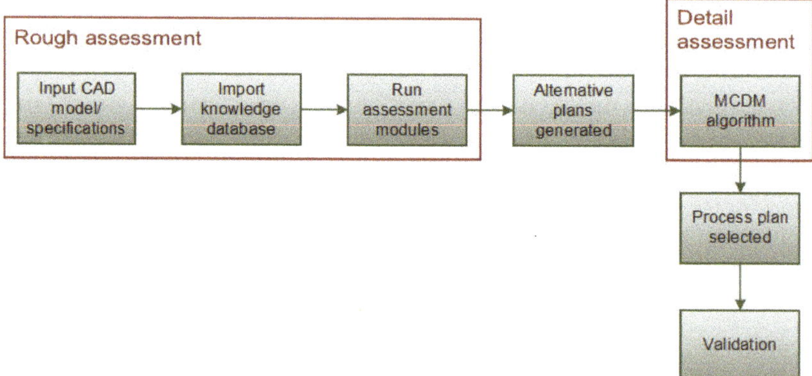

Figure 5. Block diagram of proposed method.

The general architecture of the system currently being applied is outlined in Figure 6. For the highest-ranking process plans, quantitative data are then collected for running a multi-criteria assessment. Figure 7 presents the four modules that are part of the first phase of the integrated CAx system. The modules are described in detail in the following paragraphs.

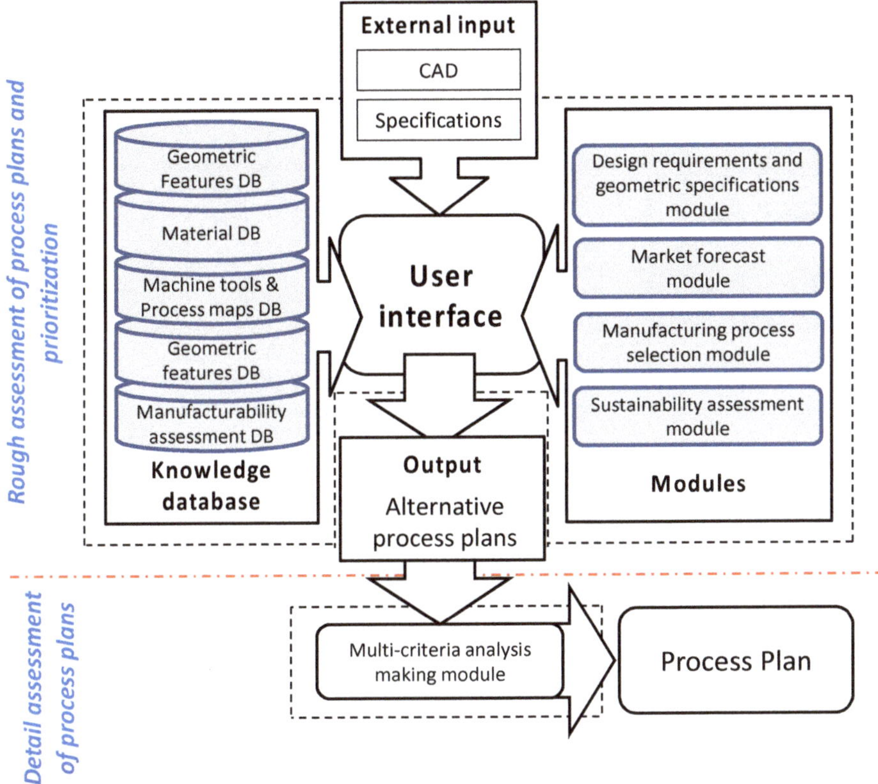

Figure 6. Architecture of the proposed integrated CAx system.

6.1. Design Requirements and Specifications Module

The first module focuses on capturing the design specifications and relevant data for assisting the decision-making process. The type of information captured includes data about the cost of the product and its production; manufacturing times such as cycle times, setup and changeover times; machine tools' utilization, uptime, and reliability; product quality; production volume; inventory levels; transportation requirements; etc. Such data can be captured through process mapping and value stream mapping. Furthermore, geometric feature data are captured from CAD files, including shape, feature name, width, and length.

6.2. Market Forecast Module

For the industrialization of the product and the ramping up of the production, information is required with regards to the market demand and the associated lead times. Depending on the foreseen market demand, different production routes could be adopted. In general, demand is stochastic; however, methods have been developed and presented for assessing and forecasting such demand profiles during the product's lifecycle. Within this module, such forecasting models are considered and incorporated in algorithms that detect potential alternate resources that could be used for the manufacturing of parts or the item itself, and compliance with the order specifications, including the quality specification for each lot of the order.

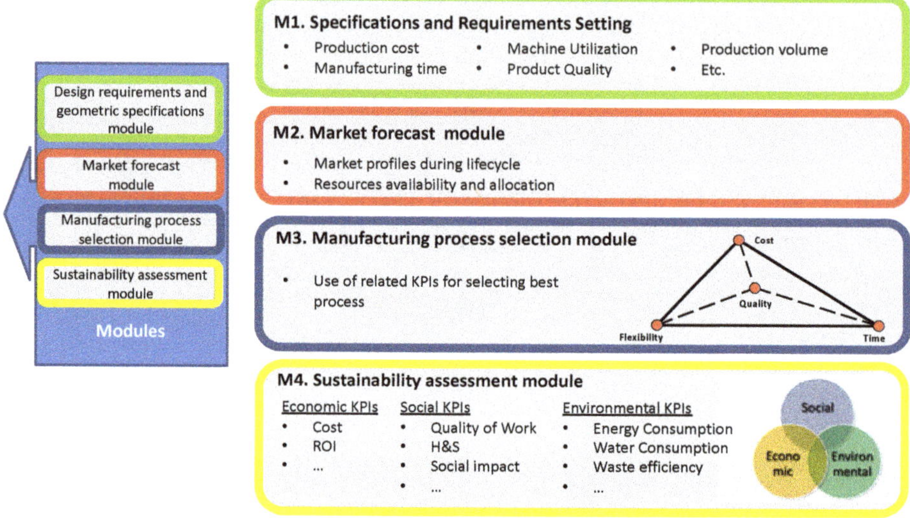

Figure 7. Module details.

6.3. Manufacturing Process Selection Module

The selection of the most appropriate production route requires a profound understanding of the capabilities of various manufacturing processes available for the production of the product. The selection of the most appropriate process requires, at least at the beginning, rough estimations of the key manufacturing attributes, i.e., cost, time, and quality. These are reflected in key performance indicators. For example, the "cost" attribute can be assessed capturing information about equipment and facility costs, depreciation costs, material and consumables cost, labor and overtime costs, overhead costs, etc. "Time" can be assessed through key performance indicators such as cycle time, setup and changeover time, throughput time, lead time, etc. Finally, the "quality" attribute is assessed through feature-related metrics such as surface roughness measurements, tolerances, and financial measures such as cost of quality, yield, etc.

6.4. Sustainability Assessment Module

The major KPIs that are used for assessing sustainability are linked to the triple bottom line aspects, namely:

- Social pillar related KPIs, as highlighted in Section 6 of the present paper, such as:

  o Work quality;
  o Health and safety;
  o Social impact.
  o ...

- Economic pillar related KPIs, such as:

  o Cost;
  o Return on investment;

- ○ CAPEx;
- ○ Use of renewable.
- ○ ...

- Environmental related KPIs, such as:

- ○ Energy consumption;
- ○ Waste efficiency;
- ○ Recycling efficiency;
- ○ Use of renewable energy sources.
- ○ ...

As highlighted by Salonitis and Stavropoulos [49], energy efficiency is a metric that can potentially be utilized as a sustainability index. It is associated with the triple bottom line aspects, as energy efficiency is related to the economic pillar (through the cost of consumed energy), the environmental pillar and the social pillar. However, Bunse et al. [50] highlighted that present concepts of energy efficiency could be inaccurate. Energy efficiency can be linked to both technologies that can potentially reduce energy consumption, as well as operations and procedures that can have the same impact.

A plethora of energy-related KPIs are available [35]. They can be categorized into energy consumption-focused metrics (such as energy usage per product, total on-site energy, total energy consumption, etc.), the environmental impact (such as carbon footprint, greenhouse emissions, energy profile, etc.), and the economics (such as energy cost). The focus can be in different levels, such as the manufacturing process level, machine tool or manufacturing plant, etc. However, deciding the level of analysis is important, as this also defines the data granularity, resolution, and how these will be used afterwards [51].

The social sustainability aspect can be evaluated on the basis of the H&S of workers, the overtime requirements, and the human toxicity potential. Such metrics can be estimated using life cycle analysis techniques.

6.5. Multi-Criteria Analysis Module

In the second phase of the framework, the quantitative assessment of the highest-ranked alternative process plans is conducted. It is clear that the selection of the optimal manufacturing plan is based on several attributes related to both sustainability and the other three key manufacturing decision attributes, i.e., time, quality, and flexibility. For this reason, a multi-criteria decision analysis method need to be applied. The Technique for Order of Preference by Similarity to Ideal Solution (TOPSIS) method was selected for this reason as it is considered to be a very powerful method when assessing conflicting attributes and can be used for dealing both with qualitative and quantitative attributes. For the present paper, the simplified method presented by Lozano-Minguez et al. [52] is employed (Figure 8). The TOPSIS method has been used in the past for assisting the decision making within the manufacturing environment, and a recent example is the work presented by Salonitis et al. [53].

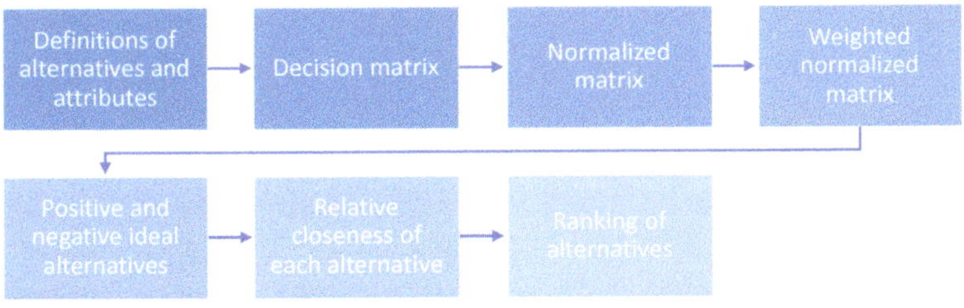

Figure 8. TOPSIS method flowchart.

In general, the following three steps are involved in the use of any decision-making method including numerical analysis of alternatives:

1. Identifying the appropriate criteria and alternatives;
2. Attaching empirical measurements to the relative significance of the criteria and the effect of the alternatives on all those criteria;
3. Processing the quantitative values to evaluate a ranking of each option.

As can be seen in Figure 8, a decision matrix is formulated, followed by normalization of every attribute of each option:

$$r_{ij} = \frac{x_{ij}}{\sqrt{\sum_{i=1}^{m} x_{ij}^2}} \tag{1}$$

where x_{ij} is the actual value of the i-th alternative in terms of the j-th criterion, and r_{ij} is the normalized value of the i-th alternative in terms of the j-th criterion. Next, criteria weights (w_j) have to be determined to indicate their relative importance and to calculate the weighted normalized values (v_{ij}) through:

$$v_{ij} = |w_j| * r_{ij} \tag{2}$$

The PIS (A+) and the NIS (A-), the ideal solutions are determined as:

$$A^+ = \left\{v_1^+, \ldots, v_j^+, \ldots, v_n^+\right\} = \left\{\left(\max_j v_{ij} \big| j = 1, \ldots, n\right) \big| i = 1, \ldots, m\right\} \tag{3}$$

$$A^- = \left\{v_1^-, \ldots, v_j^-, \ldots, v_n^-\right\} = \left\{\left(\min_j v_{ij} \big| j = 1, \ldots, n\right) \big| i = 1, \ldots, m\right\} \tag{4}$$

Then, the ranking of the alternatives is evaluated by calculating the relative distance of each solution from the PIS (S_i^+) and to the NIS (S_i^-), as:

$$S_i^+ = \sqrt{\sum_{j=1}^{n} \left(v_j^+ - v_{i,j}\right)^2} \tag{5}$$

$$S_i^- = \sqrt{\sum_{j=1}^{n} \left(v_j^- - v_{i,j}\right)^2} \tag{6}$$

The relative closeness of each solution to the ideal (C_i) is estimated as follows, and the most favorable is the one closest to 1:

$$C_i = \frac{S_i^-}{S_i^+ + S_i^-} \tag{7}$$

7. Validation

The proposed approach was developed in conjunction with Microsoft Access databases in Microsoft Excel (Figure 9). For the validation of the method, a case study of a simple metallic component was used. Four alternative routes of production were considered, as can be seen in Figure 10. In all four cases, the shafts were cut to length, and then their diameters were reduced by turning on a lathe. To achieve the necessary microstructure and hardness, they were heat treated. Four different solutions could be adopted, and thus the four different routes. The first two routes used conventional salt baths for the heat treatment, although in the first case the heat treatment was done in house, whereas, in the second case, the components were transferred to an external subcontractor. For both cases, the heat treatment took place in batches. The third and fourth options relied on non-conventional heat treatment processes such as grind hardening and hard turning. In both latter cases, the heat treatment was done on each component individually. Finally, after the heat treatment, grinding was required to achieve the dimensional accuracy and the surface roughness values required by the design specifications.

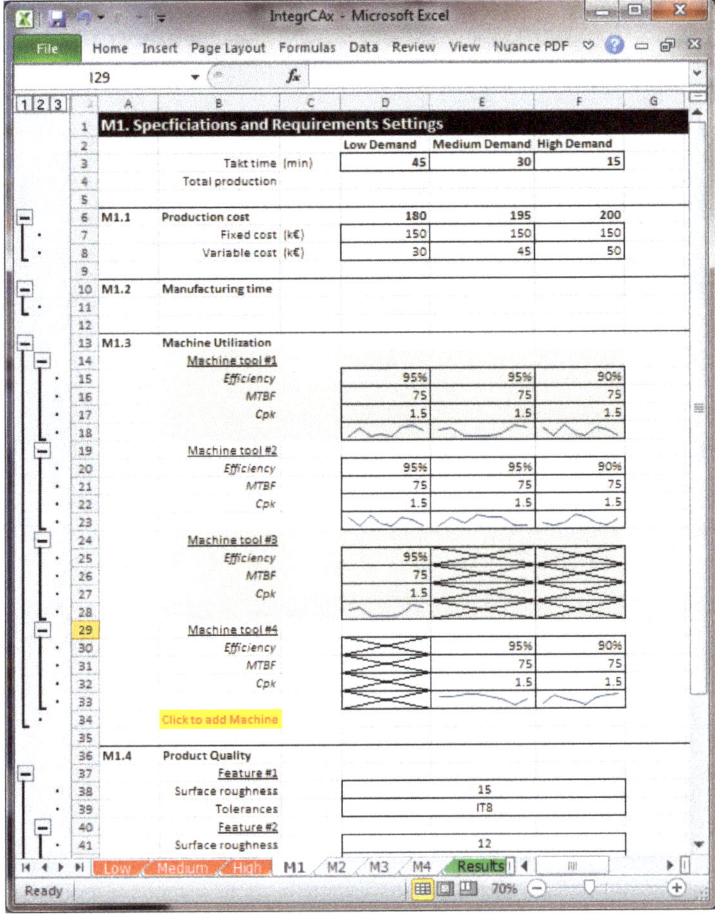

Figure 9. Sustainability assessment tool screenshot.

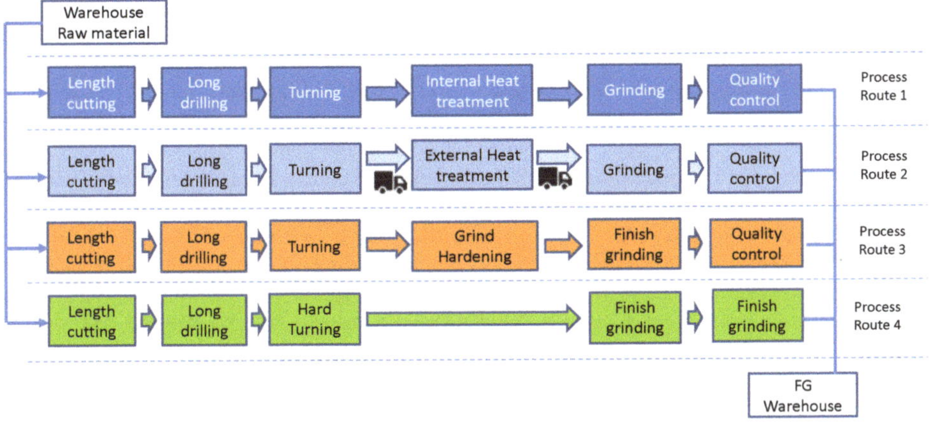

Figure 10. Validation scenario.

7.1. Rough Assessment of Process Routes (First Phase of the Framework)

Since the purpose of this analysis is to determine the best of the alternative process plans, the values for all chosen parameters must be obtained. Thirteen criteria were identified from the literature review (n = 13), which reflected the most significant aspects of the three pillars of sustainability and are shown in Table 2. Table 2 shows the case of low demand as an example. These tables were also produced for the other two demand scenarios. These criteria can be classified into negative or positive criteria, with negative criteria meaning that the lower their value the better the performance, whereas, for the positive ones, the higher their value the better the performance. Therefore, it is obvious that "energy efficiency", "raw material efficiency", "waste management", "H&S of workers", "work quality", "use of RES" and "cost efficiency" are positive criteria, whereas the rest are negative criteria. For the qualitative assessment of the various routes, the various criteria are assessed in three levels and compared to the rest of the alternative routes. The routes that are expected to have the highest performance with regards to the specific criterion are marked with "+++", the routes with the least with "+", and the routes in between with "++". As an example, on the one hand, for the "energy efficiency" criterion, Process Route 4 has the potential to have the highest energy efficiency, as the number of processes is minimized and there is no need. On the other hand, Process Routes 1 and 2 have the lowest expected performance with regards to energy efficiency, as they require heat treatment processes that are based in salt baths that require high energy input. This information can be collected from the production managers and is based on experience. Therefore, this needs to be considered carefully, as it most likely incorporates a degree of bias. However, this bias can be minimized, if the comparison and marking are done by more than one individual.

Table 2. Matrix for communicating alternatives (low demand).

	Process Route 1	Process Route 2	Process Route 3	Process Route 4
Energy efficiency	+	+	++	+++
Raw material efficiency	++	++	++	++
Waste management	+	+	++	++
CO_2 emissions	++	+++	++	+++
H&S of workers	+	+	++	++
Work quality	+	++	++	++
Overtime	++	+	+	+
Human toxicity potential	+++	+++	+	+
Human health	+	+	++	+++
Use of RES	++	++	++	++
Cost efficiency	+	+	++	++
ROI	+	+	++	++
Ecosystem quality	++	+	++	+++

As shown in Table 2, even for the simple component used for validation, no process route can be selected at first glance. Radar diagrams can be used to refine the solution and choose the correct process plan. Figure 11 presents such a diagram for one of the demand scenarios (low demand). For the development of the radar diagrams, practitioners with experience in the production (such as production managers and operators) are asked to rank the process routes on a scale from 0 to 6. In the radar diagram, the average score for each criterion is shown. All criteria are considered to be of equal importance (equal weights). Such diagrams allow the visual comparisons and visual evaluations of the different process routes.

The cumulative success for every process route can be quantified as the area within the region defined by the "radar" diagram for an initial rough evaluation. This can be easily calculated as per the following equation (generically presented for *n* criteria *CR*):

$$\text{Area} = \frac{1}{2}(CR_1 * CR_2 + CR_1 * CR_2 + \ldots + CR_{n-1} * CR_n + CR_n * CR_1) * \sin\left(\frac{360}{n}\right) \tag{8}$$

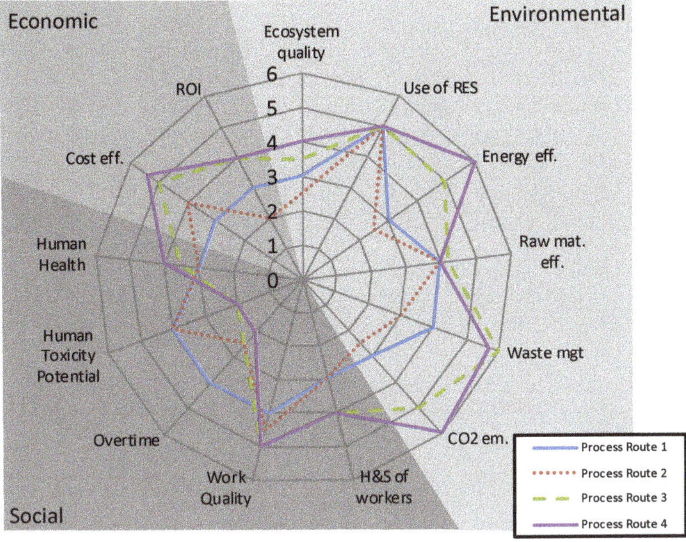

Figure 11. Radar diagram for comparing the four process routes for a low demand scenario.

In Figure 12, the index for each process route under three different market demand scenarios are presented.

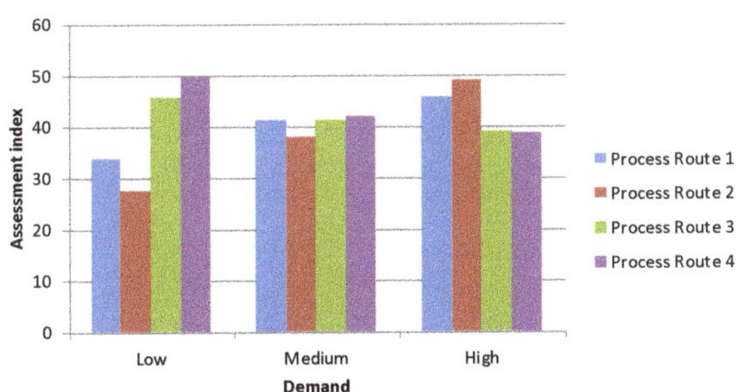

Figure 12. Overall sustainability assessment of the four process routes.

This initial rough assessment, however, treats all criteria with equal importance. By assigning weights, the relative importance of specific criteria can be highlighted. The weights can also help identify the process routes that perform better when interested in specific pillars of sustainability. Figure 13 compares the four different process routes when the focus is on the environmental, economic, or social pillars of sustainability. This is achieved by assigning a weight factor of one to the criteria of the respective pillar of interest, and 0.80 for the rest of the criteria.

Figure 13 highlights how the proposed method can be used to compare different manufacturing outputs and market demand scenarios under different criteria. For example, it is obvious that Process Route 2 is the preferred solution when the demand is high. That has to do with the fact that the production of shafts in large batches allows for economies of scale and makes use of an external partner for the heat treatment of the product. However, in low demands, such a process route is not favorable, as both the cost of the external subcontractor, as well as the environmental implications of transporting the small batches to the external location, reduces the overall assessment index.

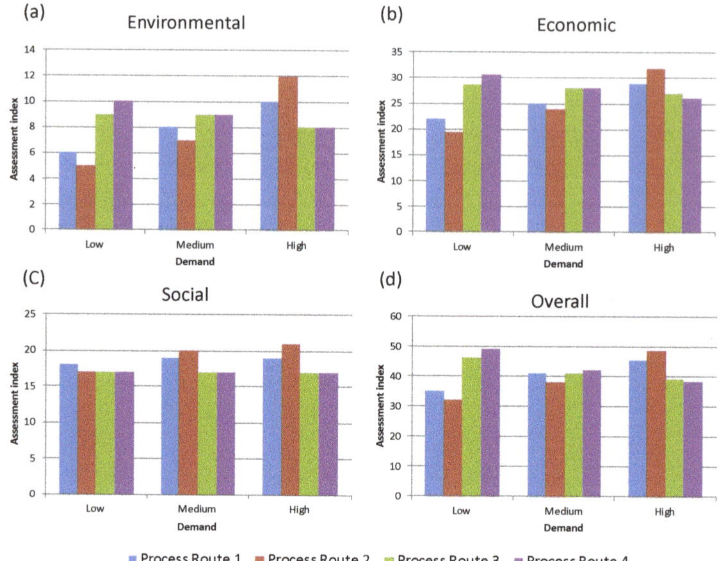

Figure 13. Ranking of process routes when the focus is on the (**a**) environmental, (**b**) economic, and (**c**) social pillar of sustainability for two different demand levels; (**d**) Presents the overall performance for equally weighted criteria.

An additional example of how such a dashboard can be used for decision making is, for example, the case of a low demand scenario. If the focus is on the economic or environmental performance, the best process route is the fourth one, whereas if the focus is on the social implications of the production, then the best processing route is the first one. Such information can help decision makers reach conclusions, revealing information that would be otherwise hidden if the decision was to be reached based on an aggregated index.

The validation scenario discussed in this study is applied to single component manufacturing and there is a one-to-one direct relationship between the batch size and energy consumption. The discussed approach is much more useful in a complex scenario where there is an inverse relationship, such as in additive manufacturing (AM) processes. In AM processes, batch size plays a vital role in deciding the cost of the part and energy costs. Therefore, the sustainability tool can be beneficial in deciding the suitability of AM processes over conventional for batch production at the industrial level, taking into account the factors of cost, time, quality, yield, energy consumption, and sustainability.

7.2. Detailed Assessment of Highest-Ranked Process Routes (Second Phase of the Framework)

The rough assessment of the various process routes determines their ranking. However, as mentioned, this rough ranking is based on qualitative comparison of the routes per criterion. For the more detailed and quantitative assessment, multi-criteria decision-making techniques such as the TOPSIS method described in Section 6.5 should be considered. Following on to the case presented, the process routes ranking changes as per the demand scenario (Figure 12). For low and high demand scenarios, the process route that outperforms the others is clear. However, for the case of medium demand, the differences in performance between the four options are slight and no confident decision can be reached. In such cases, more quantitative and detailed assessment is required. As per the framework outlined in Section 6, the next step is to perform TOPSIS specifically for the medium demand scenario. According to the process outlined in Figure 6, the decision matrix table is formulated. The variables' values are calculated (for example energy consumption can be calculated based on either energy audits or based on the physics of the process, CO_2 emissions can be calculated using life cycle analysis of all the stages involved) or assumed based on literature review findings (such as the case of human health and ecosystem quality). Table 3 presents the decision matrix and Table 4 presents the normalization of the variable values.

Table 3. Decision matrix for medium demand.

	Process Route 1	Process Route 2	Process Route 3	Process Route 4
Energy consumption per component	555	570	640	590
Raw material used per component	200	210	195	195
Waste per component	80	85	90	88
CO_2 emissions	850	1050	800	700
H&S of workers	3	3	3.5	4
Work quality	1	1	0.9	0.85
Overtime	7.5	4	15	16
Human toxicity potential	3	3	1	1
Human health impact	0.75	0.75	0.6	0.6
Use of RES	90	85	80	80
Cost	70	68	74	74.5
ROI	15	10	12	12
Impact on ecosystem quality	13	14	10	10

Table 4. Normalized matrix.

	Process Route 1	Process Route 2	Process Route 3	Process Route 4
Energy consumption per component	0.471	0.483	0.543	0.500
Raw material used per component	0.500	0.525	0.487	0.487
Waste per component	0.466	0.495	0.524	0.513
CO_2 emissions	0.494	0.611	0.465	0.407
H&S of workers	0.441	0.441	0.515	0.588
Work quality	0.532	0.532	0.479	0.452
Overtime	0.319	0.170	0.638	0.680
Human toxicity potential	0.671	0.671	0.224	0.224
Human health impact	0.552	0.552	0.442	0.442
Use of RES	0.537	0.507	0.477	0.477
Cost	0.488	0.474	0.516	0.520
ROI	0.606	0.404	0.485	0.485
Impact on ecosystem quality	0.547	0.589	0.421	0.421

The weight factors specifically affect the outcome of the process and are focused on the realistic technical skills of decision makers; thus, the more qualified the decision makers are, the more analytical the result. While most qualities can be represented in quantitative terms, this is a very daunting activity. For the needs of the present study, as in previous sections, the weighting factors are used in order to compare the routes when the focus is shifted to any one of the three pillars of sustainability. As a benchmark, the TOPSIS analysis results for the case of equally weighted criteria are shown in Figure 14. For this benchmarking case, the optimum process route is identified based on the calculation of the relative closeness of each alternative process route to the ideal solution. Process Routes 2 and 3 were found to provide the best option ($C_3 = C_4 = 0.51$), followed by Process Route 2 ($C_2 = 0.47$) and Process Route 1 ($C_1 = 0.44$).

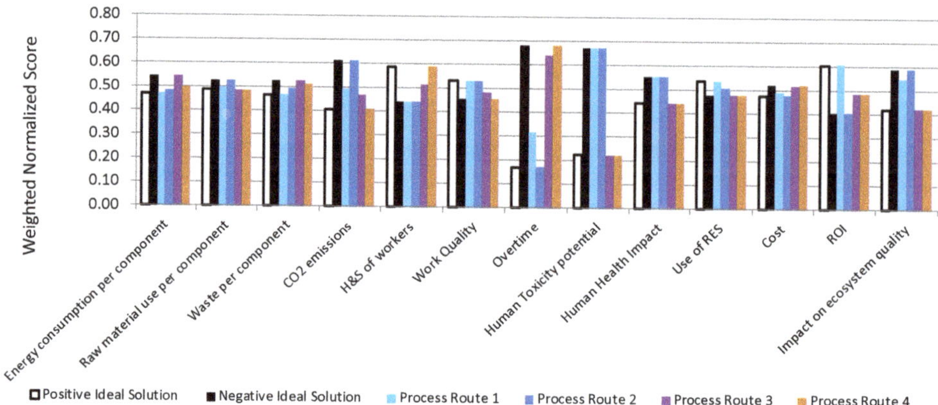

Figure 14. TOPSIS analysis results for equally weighted criteria for the medium demand scenario.

One of the key strengths of the TOPSIS method is the ease with which the importance is allocated to each variable. Therefore, if the focus was on the economic pillar of sustainability, by assigning a weighting factor for economic variables equal to one, and all the rest to 0.5, then, the preferred process route would be the first one. In Figure 15, the relative closeness is presented for the case of altering the importance of economic, environmental, and social respective variables. Process Routes 4 and 3 are preferred as processing routes, but if the focus is solely on the economic performance then Process Route 3 is preferable, and if the focus is on the environmental performance then Process Route 4 should be selected. Finally, if the focus is on the social performance, either Process Route 2 or 3 should be selected.

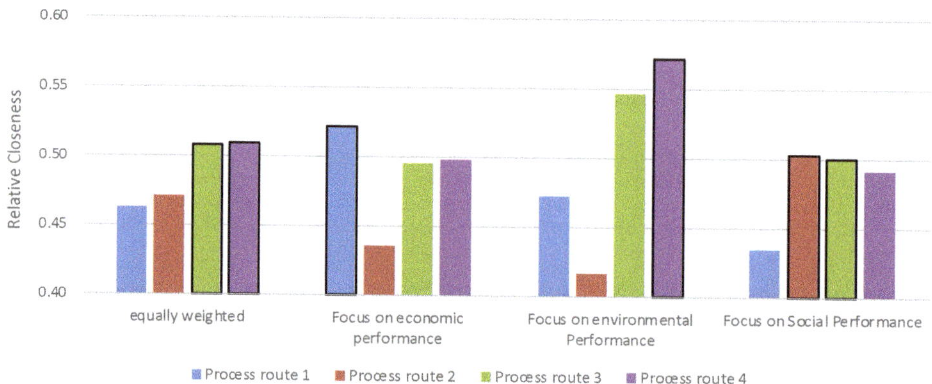

Figure 15. Relative closeness with changing weighting factors depending on where the focus of analysis is for the case of medium demand. Outlined in black, the process routes that are outperforming others for each scenario.

8. Conclusions

This paper presents and discusses a framework for assessing the impact of manufacturing operations' decisions on sustainable performance. It allows the ranking of different production routes, early in the design phase of a component, with regards to their performance in all three sustainability pillars. The methodology proposed helps in planning and scheduling manufacturing operations. The framework is composed of two phases, which include the rough ranking of the alternative process routes, in the first phase, and their detailed assessment and analysis, in the second phase. The tool was implemented in MS Excel coupled with MS Access databases. It was validated for the case of the manufacturing of hardened shafts using four different potential process routes. The validation highlighted that there was not an easy answer, and that it depended largely on the demand.

The potential for future innovations exists and a variety of topics have been defined, such as a material selection module to be integrated with the off the shelf software for selection of materials. The forecasting algorithms are implemented in MATLAB and interfaced with the market forecast module. Advance techniques such as artificial intelligence could further be explored and could be integrated for precise market forecast based on the historic dataset. The tool presented in this work can be used to couple sustainability and energy metrics which would be beneficial in identifying the optimal scenarios for energy-efficient manufacturing processes, particularly for energy-intensive processes (such as metal castings). Energy consumption costs could be reduced, while at the same time, identifying and incorporating clean manufacturing practices within manufacturing operations.

Author Contributions: Conceptualization, K.S. and P.S. (Panagiotis Stavropoulos); methodology, K.S.; software, K.S. and P.S. (Panagiotis Stavropoulos); validation, all; resources, K.S.; writing—original draft preparation, K.S. and P.S. (Panagiotis Stavropoulos); writing—review and editing, P.S. (Prateek Saxena) and J.K.; project administration, K.S. All authors have read and agreed to the published version of the manuscript.

Funding: This research received no external funding.

Conflicts of Interest: The authors declare no conflict of interest.

References

1. Grabowik, C.; Kalinowski, K.; Monica, Z. Integration of the CAD/CAPP/PPC systems. *J. Mater. Process. Technol.* **2005**, *164–165*, 1358–1368. [CrossRef]
2. Bruntland, G.H. World commission on environment and development. *Common Futur.* **1987**, *17*, 43–66.
3. Mullins, D. *Opportunities in Horticulture: A Proposal for Academic Posts in WMG*; WMG: Coventry, UK, 2004.
4. Mitchell, S.; O'Dowd, P.; Dimache, A. Manufacturing SMEs doing it for themselves: Developing, testing and piloting an online sustainability and eco-innovation toolkit for SMEs. *Int. J. Sustain. Eng.* **2019**, 1–12. [CrossRef]
5. Badurdeen, F.; Jawahir, I.S. Strategies for Value Creation through Sustainable Manufacturing. *Procedia Manuf.* **2017**, *8*, 20–27. [CrossRef]
6. Ellen Macarthur Foundation. Towards the Circular Economy. 2010. Available online: https://www.ellenmacarthurfoundation.org/publications (accessed on 21 March 2020).
7. Didone, M.; Saxena, P.; Brilhuis-Meijer, E.; Tosello, G.; Bissacco, G.; McAloone, T.C.; Cristina, D.; Pigosso, A.; Howard, T.J. Moulded Pulp Manufacturing: Overview and Prospects for the Process Technology. *Packag. Technol. Sci.* **2017**, *30*, 6. [CrossRef]
8. Meijer, E.B.; Saxena, P. The sustainable future of packaging: A biodegradable paper beer bottle. In Proceedings of the Sustain DTU 2015, Lyngby, Denmark, 17 December 2015; pp. 170–171.
9. Saxena, P.; Bissacco, G.; Stolfi, A.; de Chiffre, L. Characterizing Green Fiber Bottle prototypes using Computed Tomography. In Proceedings of the 7th Conference on Industrial Computed Tomography, Leuven, Belgium, 7–9 February 2017.
10. Pathak, S.; Ray, A.K.; Großmann, H.; Kleinert, R. High-energy electron irradiation of annual plants (bagasse) for an efficient production of chemi-mechanical pulp fibers. *Radiat. Phys. Chem.* **2015**, *117*, 59–63. [CrossRef]
11. Pathak, S.; Saxena, P.; Ray, A.K.; Großmann, H.; Kleinert, R. Irradiation based clean and energy efficient thermochemical conversion of biowaste into paper. *J. Clean. Prod.* **2019**, *233*, 893–902. [CrossRef]
12. Saxena, P.; Bissacco, G.; Meinert, K.Æ.; Danielak, A.H.; Ribó, M.M.; Pedersen, D.B. Soft tooling process chain for the manufacturing of micro-functional features on molds used for molding of paper bottles. *J. Manuf. Process.* **2020**, *54*, 129–137. [CrossRef]
13. Kiritsis, D.; Bufardi, A.; Xirouchakis, P. Research issues on product lifecycle management and information tracking using smart embedded systems. *Adv. Eng. Inform.* **2003**, *17*, 189–202. [CrossRef]
14. Ciceri, N.D.; Garetti, M.; Terzi, S. Product lifecycle management approach for sustainability. In Proceedings of the 19th CIRP Design Conference–Competitive Design, Bedford, UK, 30–31 March 2009.

15. FräMling, K.; HolmströM, J.; Loukkola, J.; Nyman, J.; Kaustell, A. Sustainable PLM through intelligent products. *Eng. Appl. Artif. Intell.* **2013**, *26*, 789–799. [CrossRef]

16. Kiritsis, D. Closed-loop PLM for intelligent products in the era of the Internet of things. *Comput. Des.* **2011**, *43*, 479–501. [CrossRef]

17. Allwood, J.M.; Cullen, J.M.; Carruth, M.A.; Cooper, D.R.; McBrien, M.; Milford, R.L.; Patel, A.C. *Sustainable Materials: With both Eyes Open*; UIT Cambridge Limited: Cambridge, UK, 2012.

18. Salonitis, K.; Jolly, M.; Pagone, E.; Papanikolaou, M. Life-Cycle and Energy Assessment of Automotive Component Manufacturing: The Dilemma between Aluminum and Cast Iron. *Energies* **2019**, *12*, 2557. [CrossRef]

19. Ijomah, W.L.; McMahon, C.A.; Hammond, G.P.; Newman, S.T. Development of design for remanufacturing guidelines to support sustainable manufacturing. *Robot. Comput. Integr. Manuf.* **2007**, *23*, 712–719. [CrossRef]

20. Ceschin, F.; Gaziulusoy, I. Evolution of design for sustainability: From product design to design for system innovations and transitions. *Des. Stud.* **2016**, *47*, 118–163. [CrossRef]

21. Harik, R.; Hachem, W.E.L.; Medini, K.; Bernard, A. Towards a holistic sustainability index for measuring sustainability of manufacturing companies. *Int. J. Prod. Res.* **2015**, *53*, 4117–4139. [CrossRef]

22. Jawahir, I.S.; Dillon, O.W., Jr. Sustainable manufacturing processes: New challenges for developing predictive models and optimization techniques. In Proceedings of the First International Conference on Sustainable Manufacturing, Montreal, QC, Canada, 18–19 October 2007; pp. 1–19.

23. Stambouli, A.B. Fuel cells: The expectations for an environmental-friendly and sustainable source of energy. *Renew. Sustain. Energy Rev.* **2011**, *15*, 4507–4520. [CrossRef]

24. Mancini, M.; Nobili, F.; Tossici, R.; Wohlfahrt-Mehrens, M.; Marassi, R. High performance, environmentally friendly and low cost anodes for lithium-ion battery based on TiO_2 anatase and water soluble binder carboxymethyl cellulose. *J. Power Sources* **2011**, *196*, 9665–9671. [CrossRef]

25. Gong, J.; Darling, S.B.; You, F. Perovskite photovoltaics: Life-cycle assessment of energy and environmental impacts. *Energy Environ. Sci.* **2015**, *8*, 1953–1968. [CrossRef]

26. Serrano-Lujan, L.; Espinosa, N.; Larsen-Olsen, T.T.; Abad, J.; Urbina, A.; Krebs, F.C. Tin- and Lead-Based Perovskite Solar Cells under Scrutiny: An Environmental Perspective. *Adv. Energy Mater.* **2015**, *5*, 1501119. [CrossRef]

27. Zandi, S.; Saxena, P.; Gorji, N.E. Numerical simulation of heat distribution in RGO-contacted perovskite solar cells using COMSOL. *Sol. Energy* **2020**, *197*, 105–110. [CrossRef]

28. Lokahita, B.; Aziz, M.; Yoshikawa, K.; Takahashi, F. Energy and resource recovery from Tetra Pak waste using hydrothermal treatment. *Appl. Energy* **2017**, *207*, 107–113. [CrossRef]

29. Mikkonen, K.S.; Tenkanen, M. Sustainable food-packaging materials based on future biorefinery products: Xylans and mannans. *Trends Food Sci. Technol.* **2012**, *28*, 90–102. [CrossRef]

30. Akinyele, D.O.; Rayudu, R.K. Review of energy storage technologies for sustainable power networks. *Sustain. Energy Technol. Assess.* **2014**, *8*, 74–91. [CrossRef]

31. Wang, Y.; Chen, X.; Konovalov, S.V. Additive Manufacturing Based on Welding Arc: A low-Cost Method. *J. Surf. Investig. X-ray Synchrotron Neutron Tech.* **2017**, *11*, 1317–1328. [CrossRef]

32. Shakeel, J.; Mardani, A.; Chofreh, A.G.; Goni, F.A.; Klemeš, J.J. Anatomy of sustainable business model innovation. *J. Clean. Prod.* **2020**, *261*, 121201. [CrossRef]

33. Zheng, J.; Zhou, X.; Yu, Y.; Wu, J.; Ling, W.; Ma, H. Low carbon, high efficiency and sustainable production of traditional manufacturing methods through process design strategy: Improvement process for sand casting defects. *J. Clean. Prod.* **2020**, *253*, 119917. [CrossRef]

34. Saxena, P.; Papanikolaou, M.; Pagone, E.; Salonitis, K.; Jolly, M.R. Digital manufacturing for Foundries 4.0. In *Light Metals 2020*; Springer: Cham, Switzerland, 2020; pp. 1019–1025.

35. Davé, A.; Ball, P.; Salonitis, K. Factory Eco-Efficiency Modelling: Data Granularity and Performance Indicators. *Procedia Manuf.* **2017**, *8*, 479–486. [CrossRef]

36. Gorji, N.E.; Saxena, P.; Corfield, M.R.; Clare, A.; Rueff, J.P.; Bogan, J.; Raghavendra, R. A new method for assessing the recyclability of powders within Powder Bed Fusion process. *Mater. Charact.* **2020**, *161*, 110167. [CrossRef]

37. Gorji, N.E.; O'Connor, R.; Mussatto, A.; Snelgrove, M.; González, P.G.M.; Brabazon, D. Recyclability of stainless steel (316 L) powder within the additive manufacturing process. *Materialia* **2019**, *8*, 100489. [CrossRef]

38. Salonitis, K.; Chryssolouris, G. Cooling in grind-hardening operations. *Int. J. Adv. Manuf. Technol.* **2007**, *33*, 285–297. [CrossRef]
39. Seuring, S. Integrated chain management and supply chain management comparative analysis and illustrative cases. *J. Clean. Prod.* **2004**, *12*, 1059–1071. [CrossRef]
40. Kleindorfer, P.R.; Singhal, K.; van Wassenhove, L.N. Sustainable operations management. *Prod. Oper. Manag.* **2005**, *14*, 482–492. [CrossRef]
41. Linton, J.D.; Klassen, R.; Jayaraman, V. Sustainable supply chains: An introduction. *J. Oper. Manag.* **2007**, *25*, 1075–1082. [CrossRef]
42. Bell, S.; Morse, S. *Sustainability Indicators: Measuring the Immeasurable?* Routledge: London, UK, 2012.
43. Stiglitz, J.E.; Sen, A.; Fitoussi, J.-P. *Report by the Commission on the Measurement of Economic Performance and Social Progress*; Citeseer: State College, PA, USA, 2009.
44. Global Reporting Initiative. Available online: https://www.globalreporting.org/ (accessed on 21 March 2020).
45. O'Brien, M.; Doig, A.; Clift, R. Social and environmental life cycle assessment (SELCA). *Int. J. Life Cycle Assess.* **1996**, *1*, 231–237. [CrossRef]
46. Hunkeler, D. Societal LCA methodology and case study (12 pp). *Int. J. Life Cycle Assess.* **2006**, *11*, 371–382. [CrossRef]
47. Schmidt, I.; Meurer, M.; Saling, P.; Kicherer, A.; Reuter, W.; Gensch, C.-O. Managing sustainability of products and processes with the socio-eco-efficiency analysis by BASF. *Greener Manag. Int.* **2004**, *45*, 79–94.
48. Dreyer, L.; Hauschild, M.; Schierbeck, J. A framework for social life cycle impact assessment. *Int. J. Life Cycle Assess.* **2006**, *11*, 88–97. [CrossRef]
49. Salonitis, K.; Stavropoulos, P. On the Integration of the CAx Systems towards Sustainable Production. *Procedia CIRP* **2013**, *9*, 115–120. [CrossRef]
50. Bunse, K.; Vodicka, M.; Schönsleben, P.; Brülhart, M.; Ernst, F.O. Integrating energy efficiency performance in production management–gap analysis between industrial needs and scientific literature. *J. Clean. Prod.* **2011**, *19*, 667–679. [CrossRef]
51. Davé, A.; Salonitis, K.; Ball, P.; Adams, M.; Morgan, D. Factory Eco-Efficiency Modelling: Framework Application and Analysis. *Procedia CIRP* **2016**, *40*, 214–219. [CrossRef]
52. Lozano-Minguez, E.; Kolios, A.J.; Brennan, F.P. Multi-criteria assessment of offshore wind turbine support structures. *Renew. Energy* **2011**, *36*, 2831–2837. [CrossRef]
53. Salonitis, K.; Vidon, B.; Chen, D. A decision support tool for the energy efficient selection of process plans. *Int. J. Mechatron. Manuf. Syst.* **2015**, *8*, 63–83. [CrossRef]

MDPI
St. Alban-Anlage 66
4052 Basel
Switzerland
Tel. +41 61 683 77 34
Fax +41 61 302 89 18
www.mdpi.com

Energies Editorial Office
E-mail: energies@mdpi.com
www.mdpi.com/journal/energies